FRACTALS IN MOLECULAR BIOPHYSICS

TOPICS IN PHYSICAL CHEMISTRY
A Series of Advanced Textbooks and Monographs

Series Editor, Donald G. Truhlar

F. Iachello and R.D. Levine, *Algebraic Theory of Molecules*
P. Bernath, *Spectra of Atoms and Molecules*
J. Cioslowski, *Electronic Structure Calculations on Fullerenes and Their Derivatives*
E.R. Bernstein, *Chemical Reactions in Clusters*
J. Simons and J. Nichols, *Quantum Mechanics in Chemistry*
G.A. Jeffrey, *An Introduction to Hydrogen Bonding*
S. Scheiner, *Hydrogen Bonding: A Quantum Chemical Perspective*
T.G. Dewey, *Fractals in Molecular Biophysics*

FRACTALS IN MOLECULAR BIOPHYSICS

T. Gregory Dewey

Department of Chemistry
University of Denver

Oxford New York • Oxford University Press 1997

Oxford University Press

Oxford New York
Athens Auckland Bangkok Bogota Bombay Buenos Aires
Calcutta Cape Town Dar es Salaam Delhi Florence Hong Kong
Istanbul Karachi Kuala Lumpur Madras Madrid Melbourne
Mexico City Nairobi Paris Singapore Taipei Tokyo Toronto Warsaw

and associated companies in
Berlin Ibadan

Published by Oxford University Press, Inc.
198 Madison Avenue, New York, New York 10016

Library of Congress Cataloging-in-Publication Data
Dewey, Thomas Gregory, 1952–
Fractals in molecular biophysics / by T. Gregory Dewey.
p. cm. – (Topics in physical chemistry)
Includes bibliographical references and index.
ISBN 0–19–508447–0
1. Molecular biology–Mathematical models. 2. Biophysics–Mathematical models. 3. Fractals. I. Title. II. Series. Topics in physical chemistry series.
QH506.D49 1997
574.8'8'0151474–dc20 96–46067

9 8 7 6 5 4 3 2 1

Printed in the United States of America
on acid-free paper

This book is dedicated to my wife Cindy,
and my three children
Thomas, John, and Frances.

Preface

Throughout its history, science has focused on a reductionist approach. A complicated phenomenon is studied by dissecting it into simpler components, ascertaining how these components work and then putting it all back together again. Most graduate students learn early in their careers to simplify. We are taught to control as many variables as possible, to change one variable at a time and that many simple experiments usually yield better results than one grand, but convoluted, experiment. Indeed, this reductionist approach is largely responsible for the great success of science. However, there comes a time when the limits of reductionism are reached. As one colleague complained, "The easy experiments have all been done."

We are now in a new era where the focus has shifted from the simple to the complex. This in part reflects the success in tackling simple problems. It also reflects a deeper sense that perhaps reductionism does not always work, that perhaps there are some problems that cannot by understood from their parts. This change in attitude has been spurred by the development of the appropriate mathematical tools to handle complexity. Just as the advent of calculus stimulated the development of Newtonian mechanics, the mathematics of complexity is promoting interest in problems that just a few years ago where deemed too messy for proper consideration. While "the calculus" provided a single unifying formalism for approaching a wealth of problems, the current era sees a plethora of mathematical techniques arising to comprehend complexity. There is now an arsenal of new approaches. These go by different names: chaos, fractals, nonlinear dynamics, and computational complexity are just a few of the subdisciplines that have arisen.

How do you measure complexity? In our current formative period, there is no single, easy answer to such a question. Most often, the mathematical techniques one uses are dictated by the problem at hand. Fractal geometry offers a particularly appealing, visual approach to understanding complex and disordered systems. Fractal geometry is, amongst other things, a mathematical technique for handling nonanalytic functions: the jagged, nondifferentiable curves that, for instance, occur in the path of a lightning bolt. This new mathematics provides a new tool, the fractal dimension, that in a sense is the equivalent of the derivative of calculus. This tool can be used to describe those untidy, jagged structures that occur in nature.

Although fractals have found their greatest applications in condensed matter physics, there have been a number of "little invasions" into other disciplines. One can see a smattering of applications in diverse fields ranging from astrophysics to organismal biology. This disordered advance of fractals into other fields has left its share of skeptics. It is the goal of this book to pull together diverse applications and to present a unified treatment of how fractals can be used in molecular biophysics. This book is intended for two audiences: the biophysical chemist who is unfamiliar with fractals, and the "practitioner" of fractals who is unfamiliar with biophysical problems. It is my hope that this will stimulate both groups to work in this infant field.

People not familiar with the mathematics of fractals often associate them with fantastic repetitive poster images or unusual scenes of landscapes. An obvious question then is: What do fractals have to do with molecular biophysics? There are a number of answers to this question. A theme that runs through this book is the close association of fractals and renormalization group theory. Renomalization group theory is intimately associated with phase behavior of polymers and aggregates (after all, much of biochemistry is polymer science). The renormalization group will appear in different guises throughout the book. Basic fractal concepts and their association with renormalization group theory are introduced in Chapter 1. This chapter is a sufficient introduction to the field to allow the book to be self-contained. However, Chapter 1 should be used by the interested laymen as an entrée into the fractal literature rather than as a comprehensive exposition.

In the following chapters (2 and 3) the association between polymer statistics and fractal concepts is made in a pedagogical fashion. This association is illustrated with examples describing the gross morphology of proteins as well as the loop structure of both proteins and nucleic acids. A number of other special topics are discussed as well. These two chapters might well have been written from a polymer theory point of view without reference to fractals. What fractals do in these examples is to provide a unifying formalism or umbrella for discussing complex structures. This formalism will be put to good use in the later chapters that discuss dynamic phenomena. Chapter 4 discusses the multifractal formalism, an extension of fractal concepts to probability distributions. It describes the origin of multifractal behavior and discusses its appearance in proteins. It provides a formal, mathematical association between the statistical mechanics of order–disorder transitions in biopolymers and a generalization of fractal concepts (multifractals).

In the following three chapters (5–7), the application of fractals to temporal phenomena is considered. Fractals provide a powerful connection between structure and dynamics, and this connection is used to advantage throughout the three chapters. It is seen that fractal descriptions naturally lend themselves to the complicated, nonexponential rate processes that are so pervasive in biological systems. Chapter 5 deals with the effects of dimensionality on diffusional processes. It considers chemical kinetics in restricted and fractal geometries. Applications to hydrogen isotope exchange in proteins and diffusion in biomembranes are considered. Chapter 6 discusses how and why protein dynamics might be fractal. It examines specific results for ion channel gating kinetics. Vibrational relaxation processes in proteins are considered in Chapter 7. This area is of great historical value in the development of concepts such as the fracton. We revisit some of these early applications with the hindsight accrued by the passage of time. In a treatment close in spirit to the three chapters on dynamics, we consider the fractal analysis of sequence data in proteins and nucleic acid in Chapter 8. This is an area of considerable recent interest, as well as controversy.

The final two chapters (9 and 10) are reserved for processes that are perhaps the best examples of fractal phenomena in biophysics. In Chapter 9, we deal with intrinsically fractal structures known as percolation clusters. It is seen that these arise in a number of biochemical settings: antibody receptor clustering, microdomains in biomembranes, and in the hydration of proteins. Chapter 10 presents a brief review of chaos in enzymatic systems. Chaos is a companion discipline to fractals and utilizes fractal descriptions to describe complex dynamics. The "strange attractors" characteristic of chaotic dynamics are fractals. Examples of two enzyme systems that exhibit chaos are discussed.

As in any new and rapidly developing field, it is extremely difficult to keep up with the literature. I apologize in advance for any oversight concerning references, old or new. While I did intend to survey much of the field, I also wished to present a cohesive and somewhat pedagogical treatment of fractals in molecular biophysics. I have on occasion omitted results that were inconclusive or were tangential to the themes of the book. I have had valuable discussion with a number of people and wish to thank P. Pfeifer, L. Liebovitch, M. Saxton, A.Y. Grosberg, F. Family, and M. Shlesinger for inputs of various kinds.

T. Gregory Dewey
Denver, 1996

Contents

FRACTALS IN MOLECULAR BIOPHYSICS

1

What Are Fractals?

River basins are fractal. So is the vasculation of the cornea. A lightning bolt shows a certain hierarchical jaggedness characteristic of fractals. Fractals are self-similar or scale-invariant objects. Regardless of the magnification with respect to a given variable, the structure remains statistically invariant. Often the fractal nature of such macroscopic phenomena, especially growth processes, is readily visualized. It is not so easy to visualize the fractality of a microscopic object such as a protein or a biomembrane. Temporal processes can also exhibit fractal time behavior. These too are not always easily visualized.

A fractal is said to have no characteristic scale, whether it be spatial or temporal. We are used to thinking in a "building block" fashion where smaller units are assembled into larger ones. At each level of construction, one has a characteristic length that appears in a natural way. An apartment building serves as a good example of an assemblage with several scales. Starting with bricks and mortar, one has, within an order of magnitude, a single size. Likewise, as a practical matter, most rooms in apartment buildings are more or less the same size. While the size of the building itself may vary considerably, usually these variations are not greater than an order of magnitude. The apartment building has at least three characteristic sizes: bricks, rooms, and the building itself. One can construct characteristic scales for many such familiar objects. It may, in fact, be difficult to think of something without characteristic sizes.

Consider a molecular liquid and the variety of characteristic lengths and volumes associated with it. A few examples are the size of the atoms, the bond lengths within the molecule, the specific volume of the molecule, and the nearest neighbor distance between molecules. If the liquid has long-range order, like water, then there are additional, more complicated spatial scales. For every spatial scale, there often is a temporal counterpart. Liquids have a complicated array of vibrational, collisional, and diffusive modes. Yet these modes would all appear to have their own characteristic time scale.

If the pressure and temperature are adjusted so the liquid approaches its critical point, it begins to show large-scale fluctuations. The fluid is caught between a liquid and gaseous state and fluctuates between the two states. The fluctuations at one level of magnification look essentially the same as those at another level. At the critical point, fluctuations occur at all length scales. This gives rise to a light-scattering phenomenon known as critical opalescence. The details of the various interactions in the liquid become unimportant and the fluctuations are governed by certain "universal" behavior. The "universality" is a result of the fluctuations between two phases, and the microscopic details of the system are washed out.

This phase transition phenomenon is in a broad sense a fractal phenomenon. It has been successfully treated theoretically using renormalization group theory (Wilson, 1979). Renormalization group theory appears again and again in physics and materials science. It reveals a plethora of phenomena that are scale invariant. One of the goals of renormalization group approaches is to establish the universality class of a set of phenomena. Once this is done, seemingly different phenomena can be related to each other and the role of dimensionality in the system is more clearly revealed.

A number of biochemical systems exhibit structures that are fractals. These include percolation clusters in biomembranes, antibody aggregates, and the "strange attractor" of chaotic enzyme reactions. In addition to these examples, we consider biophysical phenomena that are fractal in the sense of renormalization group theory. In the following chapters, scale invariance will appear in a range of settings that include the polymer statistics of proteins, the symbolic dynamics of DNA sequences, and the kinetics of the opening and closing of ion channels. It is the aim of this book to provide a unified treatment of such phenomena.

In this introductory chapter the basic concepts of fractals and multifractals are presented. This chapter is not meant to be a comprehensive discussion of the field but rather is intended to provide sufficient background for the following chapters. It will provide the interested reader with an entrée into the subject. Indeed, entire books are devoted to this material, and the interested reader is referred to the excellent introductory texts by Feder (1988), Vicsek (1992), and Falconer (1990). Of course, the very colorful *The Fractal Geometry of Nature* by Mandelbrot (1983) is also strongly recommended. The chapter begins with an introduction to basic ideas that will be used throughout the book. Chapter 2 provides a reinforcement of this introduction by showing a direct application to polymer configurational statistics, in general, and to the specific details of protein structure. An elementary discussion of fractals and set theory is also provided. Often, simple set theoretical arguments can be of great benefit when developing scaling laws. Such scaling arguments are employed in most of the following chapters. This section is followed by a discussion of self-affine fractals. Self-affine fractals are fractals that do not scale identically along each coordinate. These occur in growth and dynamical processes and are particularly important in discussing diffusion phenomena. Finally, a discussion of fractals and renormalization group theory is presented. Renormalization group methods are used throughout the book and, as discussed above, provide a physical underpinning for a number of examples. This chapter begins with a particularly simple example of such methods. Throughout this chapter, the motivation for using fractals as physical models is emphasized. This chapter focuses more on the underlying physics rather than on mathematical rigor.

1.1 Basic Concepts

Elementary geometry provides simple relationships between parameters of different dimensionalities. For simple forms, the perimeter, P, is related to the radius, R, as $P \sim R$. Similarly, surface area, S, scales as $S \sim R^2$ and volume, V, scales as $V \sim R^3$. Benoit Mandelbrot had the great insight to realize that complex forms, as often arise in nature, follow more general scaling laws. Often, relationships, such as $P \sim R^D$ with $1 < D < 2$ or $S \sim R^D$ with $2 < D < 3$, will arise. In these instances, D is a fractional exponent or fractal dimension. The object described by this "fractional geometry" is a fractal. The fractal dimension can be used as a quantitative descriptor of the morphology of a structure and, occasionally, can provide insight into how the structure is formed. To gain some intuition regarding the nature of D, it is helpful to examine simple, fractal structures.

One such example of an object with fractional scaling laws is the Sierpenski carpet shown in Figure 1.1. This is a "deterministic fractal" because it is created from an iterative process that completely dictates the final structure. Starting with a triangle of area A and perimeter P, a new, large triangle is constructed by the assembly of three smaller triangles. The perimeter of this first iterate, P_1, is given by $P_1 = 2P$. The area, A_1 is given by: $A_1 = 3A$. As seen in Figure 1.1, the second iterate is constructed in a fashion identical to the first iterate, but now the first iterate serves as the unit of construction. In the second "go around," one now has $P_2 = 4P = 2^2 P$ and $A_2 = 9A = 3^2 A$. It can be seen that for the nth iteration, $P_n = 2^n P$ and $A_n = 3^n A$. If a fractal dimension, D, is defined as $A \sim P^D$ then one has $D = \ln A_n / \ln P_n$. Giving the initial triangle in the Sierpenski carpet a perimeter and area equal to unity, the fractal dimension, D, is simply $\ln 3 / \ln 2 \sim 1.585$. Thus, the simple algorithm presented in Figure 1.1 results in a structure with fractional scaling. An easy way of thinking about a dimension of 1.585 is that it takes up more space than a line ($D = 1$) and less space than a plane ($D = 2$). Since this structure is constructed on the plane, it is said to be "embedded" in a Euclidean dimension of 2.

It is interesting to note what happens in the above calculation when the central triangle is included in the construction. In this case, the construction fills the plane and should have a dimension of 2. This is also seen in the iterative accounting of perimeter and area. The first iterate, P_1, will still have a perimeter of $P_1 = 2P$. As in the Sierpenski carpet, higher order iterations obey: $P_n = 2^n P$. The area, on the other hand, behaves differently. The first iterate now gives: $A_1 = 4A$ and higher order constructs follow: $A = 4^n A$. Seeking a relationship of the form $A \sim P^D$ gives

FIGURE 1.1 Construction of the deterministic fractal known as the triangular Sierpinski carpet. The fractal is created by an iterative process of scaling and replacement of units from the previous iteration. Each step is known as a prefractal and the fractal is formed after an infinite number of iterations. The triangular carpet in the bottom of the figure has concentric circles superimposed on it and allows the mass fractal dimension to be calculated.

$D = \ln A_n / \ln P_n = \ln 4/\ln 2 = 2$. Thus, the iterative scheme gives the expected result that the triangles cover the plane and, therefore, have a fractal dimension of 2.

While the Sierpinski carpet is a very specific type of fractal, it does show properties common to all fractals. It is constructed by two operations: scaling and replacement. The initial triangle (top of Figure 1.1) is scaled to the size of the triangle in the second iteration. The central triangle is then replaced with an open region. This second structure is then scaled to the size of the third triangle, and again all the central triangles are removed. The structures formed by these iterations are called prefractals. Technically, a fractal is the infinite iteration of this process. Figures 1.2 and 1.3 show two other popular, deterministic fractals. They are the triadic Koch curve and the middle-third Cantor set, respectively. Again, in these cases the initial structure is scaled (reduced in this instance) and line segments in the original structure are replaced by the scaled structure.

All these examples are deterministic fractals. "Random" fractals can also be generated by preserving the statistical properties or randomness at all magnifications. In this case, as the magnification is changed the structure is not identical but does have a similar jaggedness. The jaggedness reflects the randomness of the process used to form the fractal and the fractal dimension can be used as a measure of that jaggedness. Lightning or dielectric breakdown is an example of such a random fractal, with statistical similarity on many different scales. An example of such a fractal is illustrated in Figure 1.4. Fractals form "naturally" in a number of different settings. Some physiological structures are almost deterministic fractals. The mammalian lung is an example of one of these (Shlesinger and West, 1991). Often, a growth phenomenon, whether it is biological or chemical, leads to a fractal

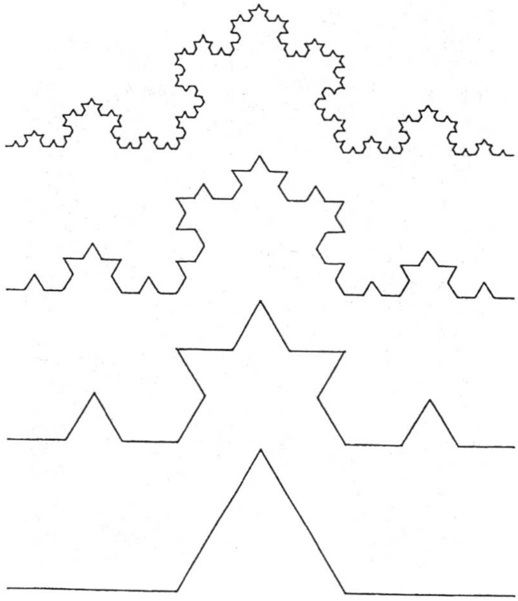

FIGURE 1.2 Construction of the triadic Koch curve. Each line segment is replaced by a scaled version of the previous iteration.

structure. A common growth process that forms a fractal is diffusion-limited aggregation. A diffusion-limited aggregate is shown in Figure 1.5. These structures have been observed in electrodeposition experiments and it has been possible to simulate them on the computer. How and why these fractals form under otherwise homogeneous conditions is one of the unsolved problems of fractal physics.

Although these physical systems may appear self-similar over a range of magnifications, ultimately there will be lower and upper length cutoffs at which fractal behavior cannot exist. Physical structures will not be infinitely repetitive. At first appearance this would seem to limit greatly the ability of fractals to model a physical phenomenon. In actuality this is not a serious constraint. The reason for this

FIGURE 1.3 Construction of the middle-third Cantor set. A unit line segment has the middle third interval removed. Each subsequent iteration removes the middle third of the remaining line segments.

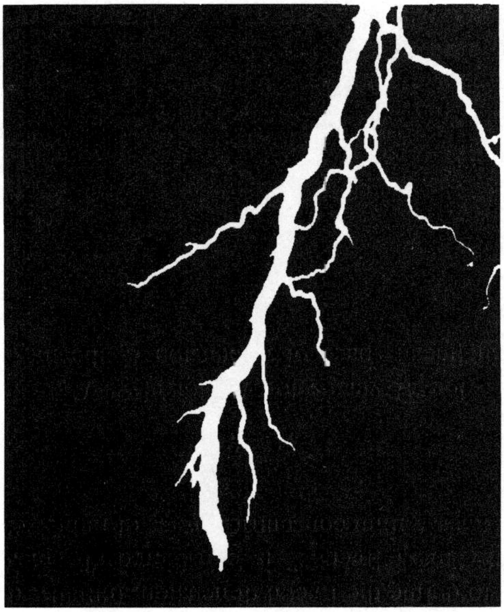

FIGURE 1.4 Image of a lightning bolt. An example of a random fractal.

FIGURE 1.5 Image of a diffusion-limited aggregate. Structure is formed by aggregation of diffusing particles. Diffusing particles perform a random walk, and when they hit the aggregate they stick. This simple physical process creates a fractal, and has been observed both experimentally and in computer simulations.

is that the properties of prefractals (the finite iteration) rapidly approach those of the true fractal. The examples shown in Figures 1.1–1.3 are all prefractals. Technically, it is only when the iteration goes to infinity that a fractal is formed. Although the limit of infinity rarely appears in physical applications, this turns out not to be a serious constraint. Properties of the finite iteration so closely capture the true fractal that the distinction between these two structures is often unimportant. Typically, fractal models are used to predict scaling laws over several orders of magnitude of an experimental parameter. The models will break down when the "length" scale is as small as the fundamental unit or as large as the entire structure. Thus, there should be well-defined limits for any fractal model. If enough physical insight exists on a given problem, then the "crossover" points between regimes can be predicted. This can greatly assist any effort to assess whether a limited scaling regime is due to fractal behavior.

Another important consequence of the scale-invariant property of fractals is that the functions describing them can never be smooth (witness the Koch curve, Figure 1.2). A common first approach to a problem in mathematical physics is to expand a function to first order and solve the linearized problem. Because fractals are scale invariant, one cannot find a scale on which they are smooth enough to linearize. Any expansion will not generate a smooth, local region, but rather will generate an identical structure, a consequence of self-similarity. Fractals intrinsically are nonanalytic functions, i.e., they have undefined derivatives. This makes them particularly useful in describing irregular shapes found in a range of natural phenomena. Before the advent of fractal geometry, such functions were considered to be mathematically "pathological."

Because of the nonanalytic behavior of fractals, the common tools of differential calculus are inapplicable. In their place is the concept of the fractal dimension. The irregular shape of a fractal is characterized by the fractal dimension(s) associated with it. There are a number of different ways of defining this parameter and, depending on the type of fractal one is dealing with, these definitions are not always equivalent. Indeed, even the definition of a fractal itself has often been rather loose. K. J. Falconer makes the analogy that a mathematician defining fractals is akin to a biologist defining life. While no simple or general definition has come forth, workers in the field still have a good sense of the attributes of the respective phenomenon. Thus, the nebulous status of the definition of a fractal should not be misleading. In most applications a very specific class of fractals is considered, which can be rigorously defined. This lack of a general definition is also a reflection of the openness of the ongoing research in this area.

In the present treatment, a pragmatic approach is taken and simple operational definitions are used to define the fractal dimension. Initially, the fractal dimension, D_M, associated with "mass fractals" is considered. To measure a mass fractal dimension, the total mass within concentric circles of radius, R, is determined. This mass scales as:

$$M \propto R^{D_M} \tag{1.1}$$

Thus, the slope of a log M versus log R plot gives D_M. Note that for uniform distributions in a Euclidean space of dimension E, the mass scales as $M \sim \rho R^E$

and $D = E$. For the Sierpinski carpet, in which the darkened triangles have uniform mass, the mass dimension is readily determined (see Figure 1.1). Placing the origin at one of the vertices of the carpet, concentric circles can be drawn which include an integer number of triangles. For the carpet shown in Figure 1.1, the total mass is taken as unity and the edge has unit length. For $R = 1/4$, one triangle is enclosed and $M = 1/9$. For $R = 1/2$, three triangles are enclosed, giving $M = 1/3$. For an infinite carpet the mass will scale as $M = (1/3)^n$ and $R = (1/2)^n$. Using Eq. 1.1, the fractal dimension is now given as $D_M = \ln 3/\ln 2$, the same number determined from area–perimeter relationships. Note that the density, $\rho(R)$, of a mass fractal embedded in a Euclidean space of dimension E is a function of the radius. It scales as:

$$\rho(R) \sim \frac{M(R)}{V} = \frac{R^{D_M}}{R^E} = R^{D_M - \mathrm{E}} \tag{1.2}$$

Density distribution as in Eq. 1.2 occur for a variety of aggregation and growth phenomena and have been studied extensively both experimentally and through computer simulations.

There are situations where one has an image or set of points and does not know the mass of the objects. For instance, an electron micrograph of the clustering of proteins in a biological membrane gives information on point distributions but not on the mass content. A similar situation arises when observing the distributions of stars in a galaxy. In such cases, it is more convenient to define a different fractal dimension known as the correlation dimension. This correlation dimension is closely related to the mass dimension. To determine the correlation dimension, the number of pairs of points, $C(r)$, that lie within a radius, r, of each other is counted (Grassberger and Procaccia, 1983). This quantity scales as:

$$C(r) \sim r^{D_C} \tag{1.3}$$

and the correlation dimension is determined from the slope of a log–log plot of $C(r)$ versus r. This definition has found popular application in determining the fractal dimensions of the "strange attractors" that characterize a chaotic system. Chaos in enzymatic systems is discussed in Chapter 10.

The fractal dimension may be defined in other ways. One such definition that is computationally very useful is the "box-counting" dimension. Consider a set of points (as in Figures 1.1–1.3) and find the minimum number, N, of boxes of edge length δ needed to cover all the points in the set. The covering of a smooth curve is illustrated in Figure 1.6. (From a more technical point of view, hypercubes are used to encompass the structure. For a structure embedded in a Euclidean space of dimension E, the hypercube will have E orthogonal edges.) Now, the box size is changed to determine how N depends on the length, δ. For a fractal structure this dependence will be described by an inverse power law such that:

$$N(\delta) \propto \delta^{-D_\mathrm{B}} \tag{1.4}$$

where D_B is known as the "box-counting" dimension. The slope of a plot of log N versus log δ provides this fractal dimension.

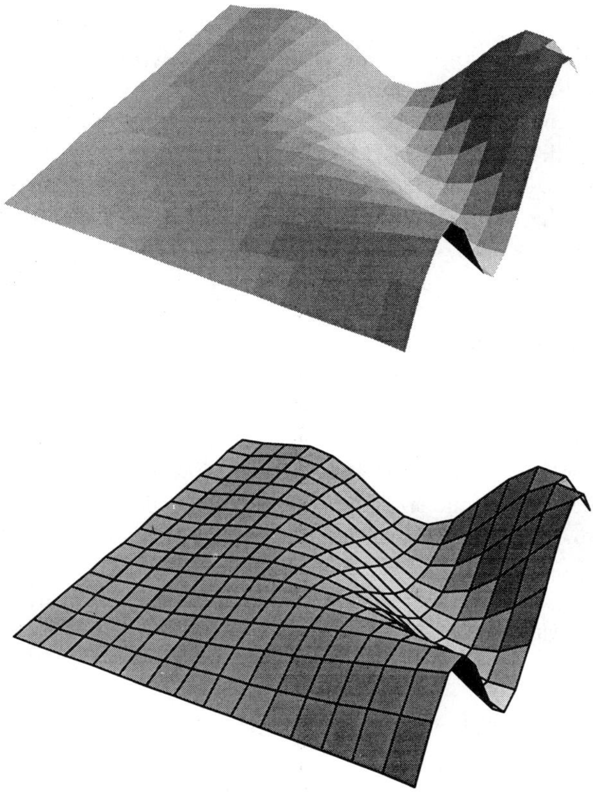

FIGURE 1.6 Example of the covering of a smooth curve with "boxes."

At first sight, this box-counting algorithm appears quite different from the one for mass fractals. However, for self-similar, deterministic fractals both procedures give the same dimension. As a simple example of the determination of a fractal dimension by box counting, the triadic Koch curve shown in Figure 1.2 is treated. Let the end-to-end length of the curve be of unit length. With a box of edge length equal to 1/3, the number of boxes required to cover the curve is 4. Note that we require the boxes to be centered on the line segments. Likewise, 16 boxes of length 1/9 are needed to cover the curve. It is seen that for boxes of length $(1/3)^n$, $N = 4^n$. From Eq. 1.4 the fractal dimension for the triadic Koch curve is given by $D = \ln 4/\ln 3 \sim 1.2628$. The curve "fills" more space than a line, $D = 1$, but less space than a plane, $D = 2$. Consider now the Cantor set shown in Figure 1.3, again let the total length be unity. Two boxes of diameter 1/3 can cover the set. Four boxes of diameter 1/9 also cover the set. It is seen that 2^n boxes of length $(1/3)^n$ are required to cover the set. Using Eq. 1.4 again, the fractal dimension is $D = \ln 2/\ln 3 \sim 0.6309$. The Cantor set or "dust" has a dimension less than 1 because it has fewer points than a line.

For self-similar fractals (as in Figures 1.1–1.3) the box-counting dimension is identical to the more rigorous description of a fractal using the Hausdorff–Bensicovitch dimension (see Mandelbrot, 1983). With the relationship given in Eq. 1.4, a number of other scaling laws can be established, and these allow for "experimental" determinations of the fractal dimension. For example, the apparent length of a curve, $L(\delta)$, will depend on the size of the ruler. This length is simply given by the ruler size, δ, multiplied by the number of steps, $N(\delta)$. From our definition in Eq. 1.4, $N(\delta) = \delta^{-D}$, one has:

$$L(\delta) = N(\delta)\delta \propto \delta^{1-D} \tag{1.5}$$

Because the length of the curve is dependent on the size of the ruler, it is referred to as an apparent length. For smooth curves, as the ruler size approaches zero the length converges to a finite limit. For fractals, no such convergence occurs, rather it diverges as the inverse power law of Eq. 1.4. This is the famous conundrum of the "coastline" problem (Mandelbrot, 1967), which shows that the length of a coastline or boundary depends on the size of the measuring device. By measuring the length of a curve with different step sizes, the fractal dimension is readily determined from a log–log plot.

Similarly, a surface area, $S(\delta)$, can be measured by covering the region of interest with squares the edges of which have length δ. This gives:

$$S(\delta) = N(\delta)\delta^2 \propto \delta^{2-D} \tag{1.6}$$

Likewise a volume, $V(\delta)$, is measured with cubes of edge length δ, giving:

$$V(\delta) = N(\delta)\delta^3 \propto \delta^{3-D} \tag{1.7}$$

The relationships shown in Eqs 1.5–1.7 establish the noninteger power law behavior of fractal structures. This is a direct consequence of self-similarity (see Pietronero, 1988). When measuring, say the volume, of a fractal, if the length scale, δ, is changed by a factor λ, i.e. $\delta \to \lambda\delta$, then one finds that:

$$V(\lambda\delta) = (\lambda\delta)^{3-D} = \lambda^{3-D}V(\delta) \tag{1.8}$$

Fractals generally obey functional equations of the form:

$$f(x') = f(\lambda x) = \lambda^{\alpha} f(x) \tag{1.9}$$

where α is an exponent that usually will not be an integer. Scaling relationships as in Eq. 1.9 are common throughout thermodynamics. The most elementary example is the scaling of state functions, such as the free energy, with extensive variables of the system. In many cases, $\alpha = 1$ and Euler's theorem is used to establish the differential form of the state function.

A more interesting situation occurs for systems near their critical point. Renormalization group theory leads to the following general expression for the free energy, F, of the system:

$$F(x) = \lambda^{-\alpha}F(\lambda x) + G(x) \tag{1.10}$$

where G is a regular function of a scaling field, x, of the system. Inhomogeneous functional equations of the form of Eq. 1.10 arise in a number of applications. These include the theory of phase transitions (in both magnetic and nonmagnetic systems), as well as the description of polymers near the theta point. As will be discussed in the section on renormalization group approaches, the solution to the homogeneous equation (equations of the form of Eq. 1.9) is similar to the inhomogeneous equation (Eq. 1.10). The homogeneous solution is readily obtained by differentiating Eq. 1.10 with respect to λ and using the chain rule to give:

$$x \frac{dF(x)}{dx} = \alpha\lambda^{\alpha-1}F(x) \tag{1.11}$$

Integrating Eq. 1.11 and setting $\lambda = 1$ now gives the power law:

$$F(x) = x^{\alpha}e^{C} \tag{1.12}$$

where C is the constant of integration. In most physical applications, $C = 0$.

From relationships such as Eq. 1.12, it is seen that self-similarity also implies noninteger power law behavior. Power laws do not have a characteristic scale. This might not be obvious, but can be readily seen by comparing a power law with an exponential function that does have a "characteristic length." Typically, one does not have a physical quantity, say l, raised to a fractional power unless it is a dimensionless quantity. Therefore, a scaling relation as in Eq. 1.12 should really contain a term such as: $(l/L)^{\alpha}$ where L is a constant with the same physical units as l. So why isn't L the characteristic length of the system? The lack of length in power law scaling can be seen by comparison with exponential behavior, $\exp\{-(l/L)\}$. Consider the behavior on the interval, $(l/L) = 1$–10. In the exponential case the ratio of the final value to the initial value is $\exp\{-9\}$, while the ratio for the power law is 10^{α}. Now, consider the next decade, 10–100. The exponential behavior shows the ratio, $\exp\{-90\}$, while the power law is unchanged, 10^{α}. The exponential is greatly diminished over the second decade. On the other hand, the power law behaves identically over the second decade as over the first. It is said to have no characteristic length.

Equations 1.9–1.12 show that the self-similar property of fractals implies nonanalytic behavior. There is a wealth of natural phenomena that can be described by power laws or by exponentials of power laws. This has provided strong motivation for the adoption of fractal models. However, the converse of this proposition, i.e., power laws imply fractals, is not necessarily true. Consider the common example of a density distribution that varies as $1/r$, where r is the radial distance from the origin. In this case the inverse power law does not imply a self-similarity. If the center of the distribution is moved from the origin to another point, the distribution for all other points changes accordingly. A self-similar fractal, on the other hand, is independent of the origin. Self-similarity means that the same structure will be generated regardless of location or magnification. Thus, our density distribution clearly does not behave as a fractal and yet it has a power law associated with it. The translational invariance property of self-similar fractals can be a serious restriction when considering real physical models. This is especially true for finite systems in which edge effects are important. For a system to be considered fractal, power law behavior is not sufficient; rather, some degree of self-similarity must be established. A rigorous demonstration of fractality involves demonstrating scale invariance in operations of dilation, translation, and rotation.

Returning to Eq. 1.12, it can be shown that this is still not the most general solution to the homogeneous equation. Often solutions to functional equations have a "unit periodic" term associated with them. These solutions are of the form:

$$F(x) = x^{\alpha} \cos(\phi \ln x) \tag{1.13}$$

where $\phi = 2\pi/\ln \lambda$. This solution shows logarithmic oscillations in x.

If the parameter λ is not uniquely determined by the problem, it can take any value. In these cases, there is no physical reason for choosing a given ϕ and Eq. 1.12 is taken as the solution. However, in a range of problems λ is restricted. For instance, in the box covering that was used for the Koch curve and the Cantor set (see Figures 1.2 and 1.3), the scaling behavior was simplified by using boxes of diameter $(1/3)^n$. These specific values fit with the self-similar structure of the specific fractals. Thus, in these instances, $\lambda = (1/3)$. Logarithmic oscillations will occur in the mass covered as the radius is smoothly varied over the fractal.

Equation 1.13 can also be written as:

$$F(x) = \mathrm{Re}(x^{\alpha + i\phi}) = \mathrm{Re}(x^{z}) \tag{1.14}$$

where z is a complex parameter. This shows that the fractal dimension need not be restricted to real numbers, but can be complex as well. Even Eq. 1.14 is not a completely general solution to the problem because the higher harmonics must also be considered. The general solution is:

$$F(x) = \mathrm{Re}\left(\sum_{m} A(m) x^{\alpha + im\phi}\right) \tag{1.15}$$

where $A(m)$ are the Fourier components that define the detailed shape of the oscillations.

As seen in Eqs 1.14 and 1.15, fractal models can give rise to logarithmic oscillations in addition to power laws. Have such oscillations ever been observed? Numerical simulations of certain random walk problems display this behavior. Oscillations have also been demonstrated experimentally for measurements of the magnetoresistance of fabricated, submicron Sierpinski gaskets that are made of aluminum (Doucot et al., 1986). Perhaps the most striking example of logarithmic oscillations appears in the structure of the human lung. The human lung is a fractal structure that has 23 generations of branches. (This is an unusually high number of branching generations. Most deciduous trees have only 7 or 8 generations!) Although the lung is asymmetric, it does show fractal behaviour. If the diameter of the bronchial tube is plotted versus branch (or generation) number, logarithmic oscillations are seen, as shown in Figure 1.7. Shlesinger and West (1991) have analyzed this with a model that predicts a fractal dimension that is a complex number as in Eq. 1.14. A number of protein dynamical phenomena show "wiggles" in log–log plots, and these have also been analyzed with a generalized noise model that yields complex fractal dimensions (Dewey and Bann, 1992).

In this section, we have discussed how to determine the fractal dimension of objects. This is the basic tool of fractal geometry. However, the mathematics of

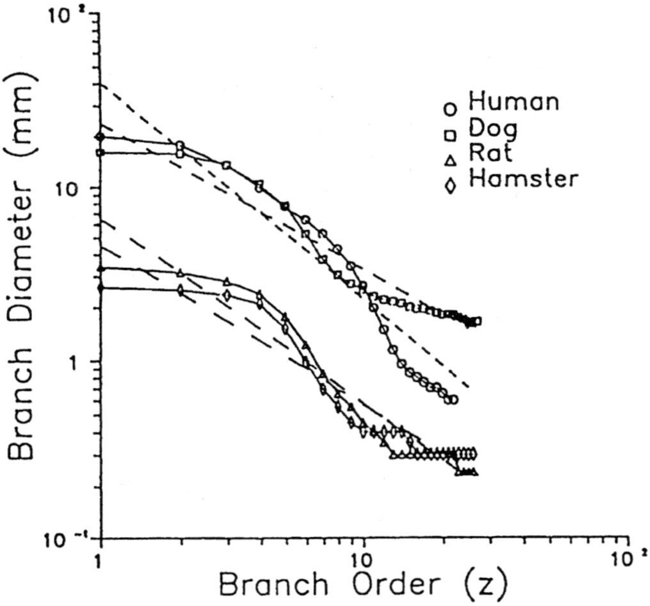

FIGURE 1.7 Power law behavior and logarithmic oscillations in the mammalian lungs. The log of the average bronchial diameter versus log of the branch order (or generation number). Overall dependence is an inverse power law (––); an oscillatory component (–) is superimposed on this. (Figure from Shlesingen and West, 1991.)

fractals is richer than just a set of algorithms for assessing scale invariance. For instance, the theory of fractal sets provides mathematical relationships that would be extremely difficult to derive by more conventional approaches. These relationships add to the predictive power of a given model and are often easily implemented. This is the subject of the next section.

1.2 Fractals and Set Theory

To demonstrate the utility of set theory in the study of fractals, two operations are considered: the product of fractal sets and the intersection of fractal sets. In both instances, new fractal structures are formed and, often, the fractal dimension of the new structure is readily determined if the dimensions of the initial fractals are known. As a word of caution, theorems involving fractal sets are often given as inequalities but frequently the limiting equality is used. It is sometimes possible to find a counterexample that is not consistent with the equality. The interested reader is referred to Falconer's book for a rigorous approach to these problems (Falconer, 1990). As will be seen in Chapters 2 and 5, simple set theoretical arguments can be used to advantage when considering surface properties of proteins.

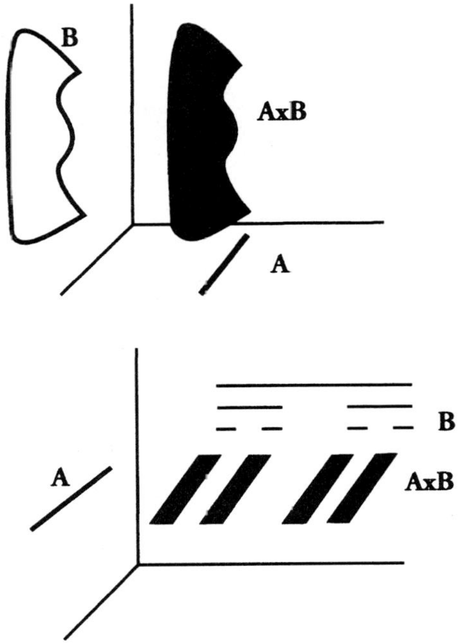

FIGURE 1.8 Example of the product of two sets. (Top) The product of a unit interval, **A**, with a curve, **B**, to form an area, **A** × **B**. It is often easier to determine the fractal dimension of the curve **B** rather than the area, **A** × **B**. (Bottom) The product of the unit interval, **A**, with the triadic Cantor set, **B**.

First, the product of two fractal sets is treated. The product of two sets, **A** and **B**, is defined as the set of points in which the first coordinate is any point in **A** and the second coordinate is any point in **B**. The dimensionality of the product is given by:

$$\dim(\mathbf{A} \times \mathbf{B}) \leq \dim \mathbf{A} + \dim \mathbf{B} \tag{1.16}$$

For cases in which the curves are "smooth," the equality holds in Eq. 1.16. A simple example is a line segment along the x-axis (set **A**) and a second line segment (set **B**) along the y-axis. The product, **A** × **B**, is a rectangle in the xy plane. Each line segment has dimensions of 1 (dim **A** = dim **B** = 1) and the product gives a rectangle of dimension 2 (dim (A × B) = 2). Figure 1.8 (bottom) shows the product of a line segment and the triadic Cantor set. The product of a line segment and the Cantor set gives a set of dimension, $1 + \ln 2/\ln 3$. The Cantor set example shows that the product of a fractal set with a nonfractal set is a fractal.

At first glance, the product theorem may not appear to be particularly useful. Certainly, it can generate a number of new and unusual fractals and it provides a relationship for determining the resulting fractal dimension. The utility of the product theorem goes beyond this. Consider the situation in Figure 1.8 (top) where a product (set **C**) of a line segment (set **A**) and a curve (set **B**) is shown. To determine the fractal dimension of the area (set **C**), the covering algorithm (Eq. 1.3) for a surface could be used. Computationally, it is simpler and more efficient to cover the curve (set **B**) and use Eq. 1.16. From the product theorem it is seen that dim (**C**) = 1 + dim (**B**). Thus, to determine the fractal dimension of the area in Figure 1.8 (top), one merely has to measure the dimension of the curve. This was done in a study on the surface dimension of proteins. Pfeifer et al. (1985) measured the boundary dimension of cross-sectional areas of the enzyme, lysozyme. It was 1.17. Using the product theorem, the surface dimension is given as 2.17. The product theorem provides a useful relationship between structures of different fractal dimensionalities. Surface dimensions of proteins are discussed more fully in Chapter 2.

Fractal sets may also be generated by the intersection of two sets. The intersection, set **C**, of two sets, **A** and **B**, is the set of points that are common to both **A** and **B**. The dimensionality of the intersection is always less than that of the sets. For instance, two surfaces intersect to form a curve. Usually, the intersection of two two-dimensional sets yields a single one-dimensional set. The exception is when the two surfaces miss each other completely, giving a zero-dimensional set. Also, one can envision a situation where the surfaces approach each other and intersect at a single point. The relative position of the surfaces are crucial to the intersection, and this is the reason why most set theoretical relationships are inequalities rather than equalities. The law of "dimensional coadditivity" states that the dimension of set **C**, the intersection of **A** and **B**, is given by:

$$\dim \mathbf{C} \geq \dim \mathbf{A} + \dim \mathbf{B} - E \tag{1.17}$$

where E is the Euclidean dimension in which the sets are embedded. Again, the equality in Eq. 1.17 does not generally hold, yet there are a large number of cases in which it does.

As an example of a physical application of the law of dimensional coadditivity, the interactions of polymers can be considered. As will be discussed more completely in Chapter 2, scaling relationships from polymer physics can be used to find the fractal dimension of a polymer configuration. The value of the fractal dimension depends on whether the polymer is in a good, ideal, or poor solvent. Excluded volume polymers are polymers in a good solvent (above the "theta point") and have a fractal dimension of 5/3. Ideal polymers occur at the theta point with a dimension of 2. Polymer collapse occurs below the theta point and is characterized by a fractal dimension of 3. As will be demonstrated in Chapter 2, globular proteins behave as collapsed polymers with fractal dimensions close to 3.

Now consider the geometry of complexes formed by the interpenetration of polymers. For protein–protein interpenetration (dim **A** = dim **B** = 3), the region of overlap (set **C**) will have a dimension of 3, provided the interactions do not alter the original dimensionality of the structure. For the interpenetration of a protein with an excluded volume polymer or an ideal polymer, the overlapping region has the same dimensionality as the polymer (5/3 or 2, respectively). Another interesting construction is the intersection of a plane, perhaps representing a biological membrane, with a polymer. For collapsed polymers this intersection has a

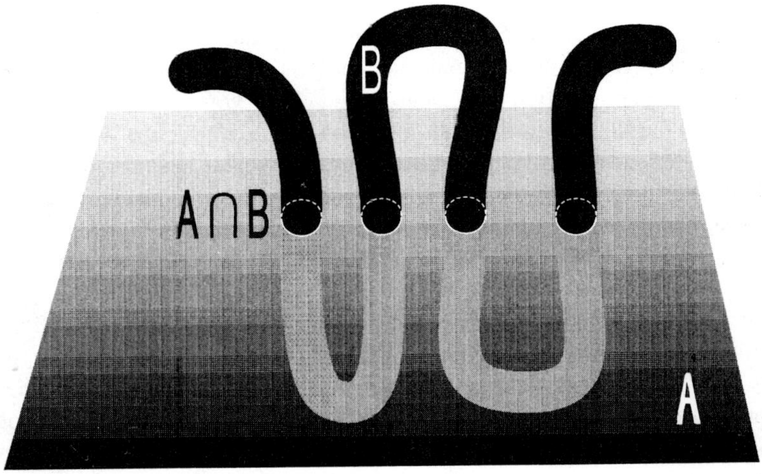

FIGURE 1.9 Example of the intersection of two sets. A set of points corresponding to a polymer intersects the plane of a membrane. The dimension of the intersection set (**A** × **B**) is determined from Eq. 1.17.

dimension of 2. The intersection of a polymer with a membrane is illustrated in Figure 1.9. This result suggests that the membrane surface excluded by a membrane-bound protein will not be rugged. There will be few protein "fjords" in which membrane lipids are trapped. A wide range of situations can be treated with the intersection theorem. This theorem gives results that would be laborious to obtain by conventional statistical mechanical theories. In these instances, there is a real utility to the fractal approach.

A useful concept from set theory is that of the codimensional space. If one has a fractal of dimension D, there is a space excluded by the fractal, its complement, that is also a fractal. This is the codimensional space. The fractal and the codimensional space have no overlap or intersection. Returning to Eq. 1.17, when the union of these two sets is the null set, $\mathbf{C} = 0$. If set **A** has fractal dimension D, then the codimensional space, set **B**, will have fractal dimension of $E - D$. The two-dimensional diffusion-limited aggregate shown in Figure 1.5 has a fractal dimension of 1.71. The fjords surrounding this aggregate make up the codimensional space. This space has a dimension of $2 - 1.71 = 0.29$. Note that this space is less then 1 because of the disconnected regions that comprise it. In Chapter 5, codimensional space will become important when considering hydrogen exchange kinetics in proteins. In this situation, the exchangeable protons are at the fractal protein surface, and the hydroxy ions that they react with is are in the surrounding aqueous environment of the codimensional space. As discussed in Chapter 2, the fractal surface of a protein is approximately 2.2. The surrounding aqueous region occupies a complementary space of dimension 0.8.

1.3 Self-Affine Fractals

In the "box-counting" algorithm, fractal sets are covered with boxes or "hyper-cubes." What would happen if a different "covering" were used – spheres for instance? Spherical coverings yield a fractal dimension known as the capacity dimension. For self-similar fractals, the capacity dimension and the box-counting dimension are the same. There is no dependence on the geometry or the nature of the covering object. For another class of fractals, known as self-affine fractals, this is not true. These fractals give different dimensions depending on the specific geometry of the covering. Self-affine fractals depend on more than one variable and show different scaling behavior with different variables. Despite this difficulty, it is still possible to determine the different fractal dimensions associated with each variable.

Subsets within a self-similar fractal can be related to each other via a similarity transformation. In such a transformation a point, $\mathbf{x} = (x_1, x_2, \ldots, x_{\mathrm{E}})$, is transformed into a second point, $\mathbf{x}'$, by $\mathbf{x}' = (rx_1, rx_2, \ldots, rx_E)$, where E is the Euclidean dimension that supports the structure and r is the scaling ratio. In these instances, r is identical for all dimensions. For a self-affine fractal, subsets are related via an affine transformation. In this case, the scaling ratio can be different for each spatial variable. Thus, the point $\mathbf{x} = (x_1, x_2, \ldots, x_E)$ is transformed into $\mathbf{x}'$ by $\mathbf{x}' = (r_1x_1, r_2x_2, \ldots, r_Ex_E)$, where the r_i terms are not all equal. This results in an asymmetric stretching of the figure. The consequence of this affine transformation is that the fractal scales differently in each direction. Following Barabási and Stanley (1995), the construction of a self-affine fractal is demonstrated in Figure 1.10. If one magnifies a segment of the fractal, one now does not recover the original structure. This is because the figure must be stretched more in one direction, a result of anisotropic scaling. In general these fractals are more difficult to characterize, and algorithms for determining their fractal dimensions require more computational effort. These have been discussed extensively by Schepers et al. (1992).

Self-affine fractals have attracted recent attention because they occur frequently . surface growth phenomena (Barabási and Stanley, 1995). For instance, in surface eposition experiments, the surface becomes uniformly covered in the plane while ractal growth occurs in the direction perpendicular to the surface. Such growth phenomena are intrinsically self-affine because all the growth occurs in one direction. Progress has been made in establishing the universality of a number of different growth processes. Such processes include fluid flow in porous media, molecular beam epitaxy, etching, and spreading of flame fronts. Many of these processes are well described by two critical exponents. The scaling laws for these exponents can often be associated with a nonlinear growth equation know as the KPZ equation.

Another well-studied example of a self-affine fractal is the trajectory of a particle undergoing Brownian (or fractional Brownian) motion. In this instance, there is a different scaling between the spatial and temporal variables. The number of boxes, N, required to cover the trajectory will be proportional to the time, $t \sim N$. The Einstein relationship provides a well-known relationship between the root mean squared displacement, $<x>$, and the time, $\langle x \rangle \propto t^{1/2}$. Thus, the box covering along a

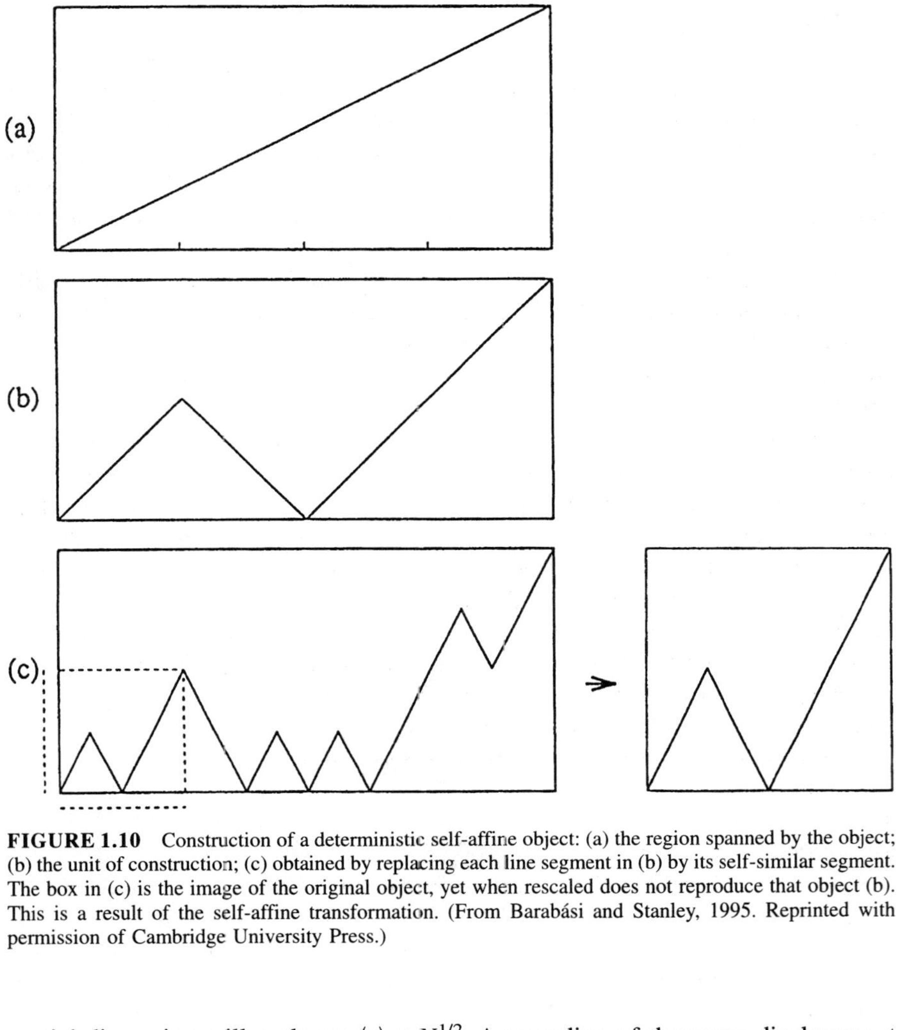

FIGURE 1.10 Construction of a deterministic self-affine object: (a) the region spanned by the object; (b) the unit of construction; (c) obtained by replacing each line segment in (b) by its self-similar segment. The box in (c) is the image of the original object, yet when rescaled does not reproduce that object (b). This is a result of the self-affine transformation. (From Barabási and Stanley, 1995. Reprinted with permission of Cambridge University Press.)

spatial dimension will scale as: $\langle x \rangle \propto N^{1/2}$. A recording of the mean displacement versus time will then be a self-affine fractal with the scaling exponent (as determined by box counting), being 1/2 along the spatial axes and 1 along the temporal axes. More complicated diffusional processes can occur and can be described as fractional Brownian motion. In these instances, the root mean squared displacement shows a more general scaling of the form: $\langle x \rangle \propto t^{\alpha}$, where α can now be a number between 0 and 1. The fractal dimension of the trajectory is now given by: $D = 2 - \alpha$. These more complicated diffusional processes are considered in Chapter 5 where chemical kinetics in restricted environments is discussed. They are also discussed in Chapter 7 when dealing with vibrational relaxation in polymers and the concept of the "fraction." As will be seen in Chapter 8, sequence information from DNA can be decoded into a "DNA walk." DNA walks are trajectories that are created using the base sequence as a set of instructions for the walker. DNA walks show self-affine trajectories and follow a scaling law consistent with fractional Brownian motion. These have been related to correlations within the sequence. Finally, Chapter 9 deals with anomalous diffusion in biomembranes, which is related to diffusion on fractal structures known as percolation clusters.

1.4 Multifractals and the Partition Function

By the mid-1980s, fractal concepts had fueled a wealth of applications in the physical sciences. Yet by this time there was an increasing awareness of their limitations. Fractal descriptions of physical models use sets of points, while it is often more convenient to work with probability or density distributions. Although parameters such as the box-counting dimension provide useful information for characterizing complex forms, they clearly are limiting. For instance, it is possible to have two structures with identical fractal dimension but which appear very different to the eye. Often they will have a different texture. A number of pioneering studies recognized that a "generalized fractal dimension" could be established (for introductory material see Feder, 1988; Te'l, 1988; Stanley and Meakin, 1988). In this new approach, probability distributions or densities are treated and a spectrum of fractal dimensions can be determined using the moments of these distributions. The box-counting dimension is a special case and represents only one of many fractal dimensions. Because of this range of dimensions, these constructs were termed "multifractals."

To show how multifractals are calculated, we return to the box-counting algorithm, as illustrated in Figure 1.11. In determining the box-counting fractal dimension, each box is given equal weight and the sum, $N(\delta)$, is calculated. Yet, clearly, each box does not contain the same number of points. Thus, information concerning the distribution of points in the boxes is being lost. To regain this information, the density of points, μ_i, in each box is determined. A "weighted" number of boxes, $Z(1)$, can be calculated as a function of box size, δ:

$$Z(1) = \sum_i \mu_i = \delta^{-\tau(1)} \tag{1.18}$$

where, by analogy with Eq. 1.4, $\tau(1)$ is a generalized fractal dimension. Of course, there is no need to restrict oneself to the first moment of the distribution. More generally the qth moment can be considered:

$$Z(q) = \sum_i \mu_i^q = \delta^{-\tau(q)} \tag{1.19}$$

Notice that when $q = 0$, the sum in Eq. 1.19 is merely the number of boxes needed to cover the curve, and the original, box-counting fractal dimension (Eq. 1.2) is recovered, i.e., $\tau(0) = D_B$. Equation 1.19 now gives a fractal dimension for each moment of the distribution; hence the term multifractal. Some applications allow q to be both a positive and negative integer. In these cases, the positive moments are dominated by boxes of high probability densities, while the negative moments are dominated by those boxes with low probability densities. Interestingly, for scale-invariant fractals such as the deterministic ones considered in Figures 1.1–1.3, the

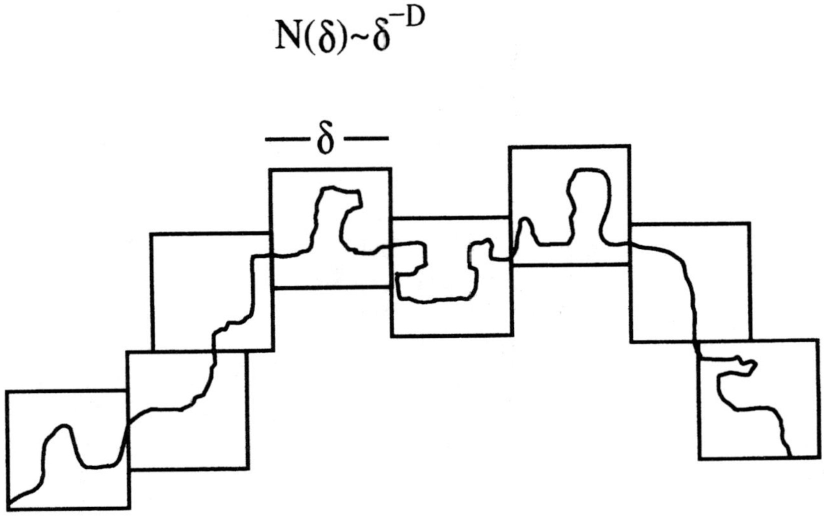

FIGURE 1.11 Illustration of the box-covering algorithm for determining the generalized fractal dimension. The normal fractal dimension is determined by counting the minimal number of boxes of length δ required to cover the curve. The dimension is given by the scaling law in Eq. 1.17. A more complete description of the curve requires that the density in each box be considered. The multifractal spectrum results from the moments of the distribution as a function of δ. The multifractal spectrum is calculated using Eq. 1.18.

probability distribution does not change from box to box, so the moments of the distribution do not provide more information. In these cases, a single fractal dimension is obtained rather than a spectrum.

The quantity, $Z(q)$, has a strong analogy to the partition function of statistical mechanics, and this can be exploited to reveal the nature of this sum. A number of descriptions of this analogy have been given (see Stanley and Meakin, 1988; Te'l, 1988), and there are some notational differences throughout the literature. Here we follow the notation of Falconer (1990). $Z(q)$ may also be written as:

$$Z(q) = \sum_{p} n(p)p^{q} = \sum_{p} e^{-F(p)} \tag{1.20}$$

where $F(p) = \ln(n(p)) + q \ln(p)$ and the quantity in the rightmost equality is essentially a partition function. The sum is over all probabilities, and $n(p)$ is the number of boxes with probability p. In many statistical applications, the sum is dominated by the most probable distribution, $p = p^*$, so that:

$$Z(q) = n(p^*)p^{*q} \tag{1.21}$$

The physics (and morphology) of the problem enters from the dependence of $n(p^*)$ and p^* on the size of the system. Assuming scaling behaviour of the form:

$$n(p^*) = \delta^{-f} \tag{1.22}$$

$$p^* = \delta^{\alpha} \tag{1.23}$$

it is demonstrated from Eqs 1.19 and 1.21 that:

$$\tau(q) = f - q\alpha \tag{1.24}$$

From the partition function form of Eq. 1.20, $\tau(q)$ is associated with a "generalized free energy," with q and f being the "generalized" temperature and energy, respectively. With this analogy, it is natural to establish the "generalized entropy" as:

$$\alpha(q) = \frac{-\,d\tau\,(q)}{dq} \tag{1.25}$$

This formalism allows one to speak of the entropy of a curve. Bear in mind that this is an analogy based on mathematical form and does not represent a true, thermodynamic entropy. A more detailed discussion of Eqs 1.24 and 1.25 using Legendre transforms and undetermined multipliers can be found elsewhere (Mandelbrot, 1988). The plots of $f(\alpha, q)$ versus $\alpha(q)$ are similar to the "extrathermodynamic" relationships seen in some physical systems where free energy and entropy are related.

The parameters $f(\alpha)$ and α are central to the interpretation of a multifractal analysis. They are a pairwise representation of the fractal dimension, f, associated with a singularity, α, of the fractal structure. Typically, a multifractal spectrum (f vs α) shows a single maximum with well-defined intercepts. The normal box-counting dimension is a single point on this curve (Figure 1.12). The very existence of this

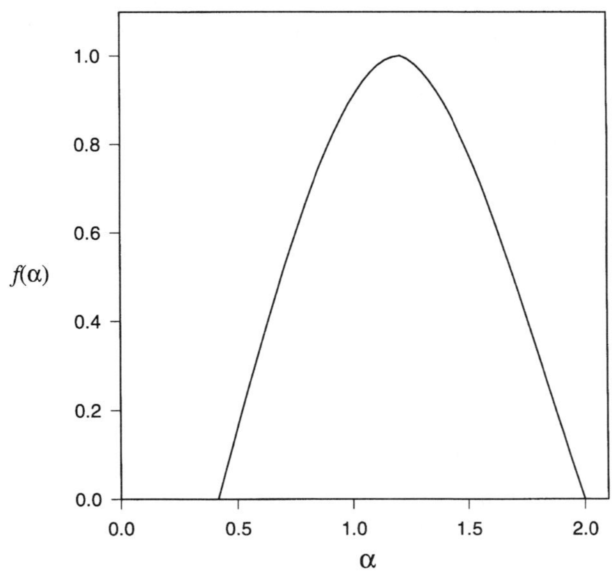

FIGURE 1.12 Example of a multifractal spectrum. Multifractal spectrum (f vs α) for a binomial multiplicative process with p = 0.25.

spectrum indicates that the observed phenomenon has an underlying hierarchical nature. Random processes or structures give very narrow multifractal spectra, while hierarchical assemblies give broad spectra. A range of complex physical phenomena has been analyzed using the multifractal approach. Studies on diffusion-limited aggregation and on turbulence have received particular attention. In these instances, hierarchical models are used to generate multifractal spectra similar to those obtained from experimental data.

One of the simplest hierarchical models is the random multiplicative process, and it is instructive to explore this model. Consider a one-dimensional distribution over a unit length. The length is divided in half and a probability $\mu_0 = p$ is associated with the first half, while a probability $\mu_1 = (1-p)$ is associated with the second half. This process is now repeated with each interval giving new interval lengths of 1/4, i.e., $\delta = (1/2)^2$). The probability for each box is now: $\mu_0\mu_0$, $\mu_0\mu_1$, $\mu_1\mu_0$, $\mu_1\mu_1$. The second and third boxes have degenerate probabilities, and further repetition of the process will generate larger numbers of boxes with the same probability. This number of degenerate boxes, $n(p)$, is given simply by the binomial coefficient. Thus Eq. 1.19 becomes:

$$Z(q) = \sum_k \binom{N}{k} (\mu_0^k \mu_1^{N-k})^q = (\mu_0^q + \mu_1^q)^N \tag{1.26}$$

where N is the total number of boxes into which the unit length was divided, and $\delta = 1/2^N$. Using Eq. 1.19 and eliminating N, the generalized fractal dimension is obtained:

$$\tau(q) = \ln(\mu_0^q + \mu_1^q)/\ln 2 = \ln[p^q + (1-p)^q]/\ln 2 \tag{1.27}$$

The "generalized entropy" is found using Eq. 1.25 and is:

$$\alpha(q) = \frac{-p^q \ln p + (1-p)^q \ln(1-p)}{(p^q + (1-p)^q)\ln 2} \tag{1.28}$$

and $f(q)$ is given by Eq. 1.24. Figure 1.12 shows a plot of f versus α for a multiplicative binomial process in which $p = 0.25$. This plot is highly characteristic of a multifractal spectrum.

Simple random multiplicative processes may at first appear to be little more than mathematical devices that show multifractal behavior. However, such models have been used to analyze such diverse processes as the conductivity of percolating clusters (see Stanley and Meakin, 1988) and thermal convection in fluids (Jensen et al., 1985). In Chapter 4, a multifractal analysis of protein structure and of helix – coil transitions in biopolymers is developed. These simple models of biopolymers can also be represented by the random multiplicative process discussed here. For polymer problems, multifractals can be used to provide a more complete classification of subpopulations within an ensemble of biopolymers. This classification reveals the complexity of such subpopulations. More general aspects of multifractals and protein structure are also discussed in Chapter 4. Interestingly, it is seen that naturally occurring structures give broad spectra, and "misfolded" proteins give narrow spectra. The multifractal spectrum can be used as a diagnostic tool for assessing the quality of protein folding algorithms. Chapter 8 develops a correspondence between multifractals and random walk parameters. These are used to discuss encoded walks from protein and DNA sequence data.

1.5 Fractals and the Renormalization Group

As discussed in the introduction to this chapter large-scale fluctuations occur in a system at its critical point. Because these fluctuations occur over all length scales of the system, the system no longer has a characteristic scale parameter. A consequence of this loss of scale, or scale-invariance, is that the various correlation functions characterizing the system show nonanalytic behavior. Typically, these functions follow fractional power laws. These fractional exponents show "universal" behavior in that they apply to a diverse range of systems. They depend on the dimensionality of the system rather than on its microscopic details. Thus, the fluctuations of the system dominate the thermodynamic behavior, and the specificity of molecular interactions is not as important.

The renormalization group (RG) theory was originally developed by Wilson to describe critical phenomena. Since its inception, applications have expanded rapidly to other areas of theoretical physics (for excellent introductions to RG theory see: Ma, 1976; Creswick et al., 1992). As in fractal geometry, the concept of scale invariance plays a major role in the RG approach. RG transformations involve two steps: decimation or coarse graining, and scaling. In the decimation process the degrees of freedom of the system that vary over short scales are averaged. Since the system is dominated by fluctuations, these short-range interactions are unimportant. After the length of the shortest interactions has been removed by the decimation process, the system must be rescaled to regain its original length scale. In this step all the lengths are redefined so that the coarse-grained unit corresponds to the original unit. Figure 1.13 illustrates a real-space renormalization process used in treating the configurational statistics of a polymer. For an excluded volume polymer, the physical structure on a global scale appears similar to that on a local scale. The polymer can be considered to be made up of a "string of pearls," yet if one looks inside each pearl one sees a polymer structure much like the overall structure. The renormalization procedure captures this self-similarity by a process of coarse-graining and rescaling. This process is iterated until a given parameter, often the free energy, of the system no longer changes. This is called the "fixed point" and represents the scale-invariant limit.

To put these concepts into more concrete form, a simple and popular example (see Maris and Kadanoff, 1978; Chandler, 1987) involving the renormalization of the one-dimensional Ising model is considered. This development is primarily heuristic, as the system does not show a phase transition. The main point of the example is that it will produce an equation that is an analog to functional equations describing fractals (Eq. 1.8). This analogy will be discussed in more detail below. It serves our purposes well, as it introduces both Ising models and the decimation procedure. Ising models of biopolymers will be discussed in more detail in Chapters

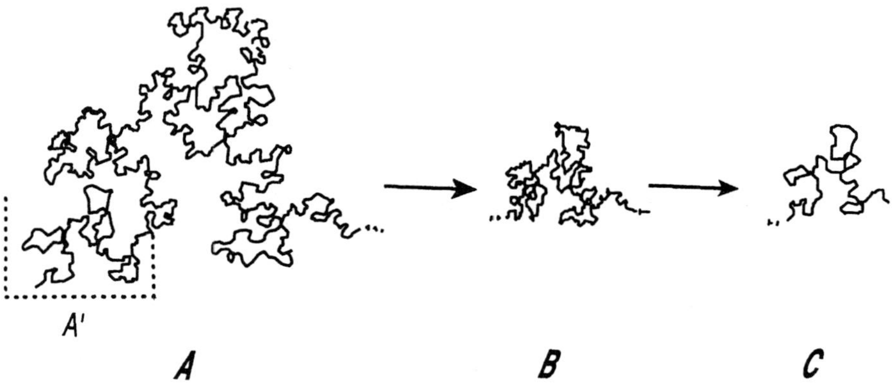

FIGURE 1.13 Real-space renormalization description of a polymer. The two operations in the renormalization process, scaling followed by a coarse graining step, are shown. (Diagram from Freed, 1987. Reprinted with permission of John Wiley & Sons.)

3 and 4. Applications of renormalization group theory arise throughout this book. In Chapter 3, a small cell real space renormalization (a decimation procedure) is used to calculate the statistics of loop formation in biopolymers. In Chapter 7, a decimation approach is used to calculate the vibrational density of states for lattice models applicable to alpha helices and double helices. In Chapter 9, we again turn to renormalization methods when considering percolation models of biomembranes.

To proceed with the one-dimensional Ising model, consider a linear array of spins in the absence of a magnetic field. Each spin can assume a value of $\sigma = \pm 1$ and only interacts with its nearest neighbor. The partition function, Z, is given by:

$$Z = \sum_{\text{spins}} e^{J(\sigma_1\sigma_2 + \sigma_2\sigma_3 + \sigma_3\sigma_4 + \dots)/kT} \tag{1.29}$$

where J is the energy contribution from parallel spins, T is temperature, and k is the Boltzmann constant. The summation is over all spin states. It is convenient to define the coupling constant, K, as $K = J/kT$ and to group even indices in the following fashion:

$$Z = \sum_{\text{spins}} e^{K(\sigma_1\sigma_2 + \sigma_2\sigma_3)}\, e^{K(\sigma_3\sigma_4 + \sigma_4\sigma_5)} \dots \tag{1.30}$$

The decimation step of the RG transformation is achieved by evaluating the summation for every other lattice point. Evaluating all the even-numbered lattice positions ($\sigma_{\text{even}} = \pm 1$), now gives the decimated partition function as:

$$Z = \sum_{\substack{\text{odd}\\ \text{spins}}} \left[e^{K(\sigma_1 + \sigma_3)} + e^{-K(\sigma_1 + \sigma_3)}\right] \left[e^{K(\sigma_3 + \sigma_5)} + e^{-K(\sigma_3 + \sigma_5)}\right] \dots \tag{1.31}$$

The next step of the transformation is to assert that the new decimated partition function (Eq. 1.31) must have the same form as the original function (Eq. 1.29). This is the scaling step and requires that:

$$e^{K(\sigma_1+\sigma_3)} + e^{-K(\sigma_1+\sigma_3)} = f(K)\, e^{K'\sigma_1\sigma_3} \tag{1.32}$$

Analogous expressions hold for all the lattice positions. The goal now is to determine the form of $f(K)$ and K' that "renormalizes" the decimated lattice. This is done by considering all possible values (±1) for σ_1 and σ_3 in Eq. 1.32. This gives two independent equations:

$$K' = (1/2)\cosh(2K) \tag{1.33}$$

$$f(K) = 2\cosh^{1/2}(2K) \tag{1.34}$$

The energy of the system will be proportional to N, the number of spins, and is given by:

$$N\zeta = \ln Z(N,K) \tag{1.35}$$

where ζ is the unit energy and is independent of N. Using Eqs 1.31 and 1.32, it is seen that:

$$Z(N,K) = f(K)^{N/2} \sum e^{K'\sigma_1\sigma_3}\, e^{K'\sigma_3\sigma_5} \ldots \tag{1.36}$$

$$= f(K)^{N/2} Z(N/2,K') \tag{1.37}$$

Substituting Eq. 1.37 into Eq. 1.35 gives:

$$\zeta(K) = (1/2)\zeta(K') + (1/2)\ln f(K) \tag{1.38}$$

which is the inhomogeneous analog of the fractal scaling equation (Eq. 1.19).

The pair of equations (Eqs 1.33 and 1.34) provide the basic iterative scheme for the RG analysis. If a limiting value for the partition function is known, Eq. 1.33 allows the recursive calculation of a new K, i.e., K'. The new energy, $\zeta(K')$, is given by Eq. 1.38. In systems that show a phase transition, this recursion converges to a "fixed point," and the free energy near the critical point can be determined. In the above example, the fixed point is zero and is not of great physical interest. An inverse iteration yields a second fixed point at infinity, which again is of no physical interest. This is a result of the one-dimensional Ising model not supporting a true phase transition.

The key point for our discussion is that the renormalization approach gives functional equations of the form:

$$\zeta(K) = L^{-E}\zeta(\lambda K) + \gamma(K) \tag{1.39}$$

where L is the decimation length, E is the Euclidean dimension, and γ is an eigenvalue of the functional equation.

Functional equations having the form of Eq. 1.39 appear not only in the RG approach to phase transitions but also in a number of different models of random processes. These have been discussed by Shlesinger and Hughes (1981), and our present discussion follows their approach. Iteration of Eq. 1.39, n times, gives:

$$\zeta(K) = \lim_{n\to\infty}\left[L^{-nE}\zeta(\lambda^n K) + \sum_{k=0}^{n-1} L^{-kE}\gamma(\lambda^k K)\right] \tag{1.40}$$

For most physical situations, the boundary conditions are such that the first term on the right-hand side of Eq. 1.40 will go to zero and the summation is responsible for the nonanalytic behaviour of the solution. This singular part is given by:

$$\zeta_{\text{sing}}(K) = A(K)\, K^{E/y} \tag{1.41}$$

Substitution into Eq. 1.39 gives $y = \ln \lambda / \ln L$ and:

$$A(K) = A(\lambda K) = \Sigma\, A_n\, e^{2\pi i n \ln K/\ln \lambda} \tag{1.42}$$

Thus the renormalization group equations give both the fractional power law behavior and the logarithmic oscillations described earlier for deterministic fractals (Eq. 1.15). For any given physical problem the existence of logarithmic oscillations will depend on whether λ represents a physically realistic eigenvalue. Nevertheless, it is seen that the physical description of phase transitions is centered on the concept of scale invariance. The mathematical consequence of this is fractional power laws. Power laws are the mathematical manifestations of the loss of characteristic scale.

1.6 Summary

In this chapter we have introduced the basic tools to be used throughout this book. We have seen how simple geometric constructs that are self-similar can yield fractional scaling laws. These constructs can be characterized by the fractal dimension. Computational methods for determining the fractal dimension were discussed, with emphasis on the mass dimension and the box-counting dimension. Simple set theoretical relationships are also useful in determining fractal dimensions and can prove very convenient computationally. Fractals are invariant to dilation, translation, and rotation. Yet there are limitations to this definition. For growth and diffusion problems, it is useful to introduce the concept of self-affine fractals. These fractals show an anisotropy in their scaling properties. The fractal concept must also be generalized when dealing with probability distributions. This generalization has led to the concept of multifractals, and an overview of that formalism has been presented. Finally, the strong ties between fractal geometry and renormalization group theory have been discussed, and a simple example of a renormalization calculation presented.

References

Barabási, A. L., and Stanley, H. E. 1995. *Fractal Concepts in Surface Growth.* Cambridge University Press, Cambridge.

Chandler, D. 1987. *Introduction to Modern Statistical Mechanics.* Oxford University Press, New York, p. 139.

Creswick, R. J., Farach, H. A., and Poole, Jr., C. P. 1992. *Introduction to Renormalization Group Methods in Physics*. Wiley, New York.

Dewey, T. G., and Bann, J. G. 1992. Protein Dynamics and 1/*f* Noise. *Biophys. J.* **63**, 594–598.

Doucot, B., Wang, W., Chaussy, J., Pannetier, B., Rammal, R., Vareille, A., and Henry, D. 1986. First Observation of the Universal Periodic Corrections to Scaling: Magnetoresistance of Normal-Metal Self-Similar Networks. *Phys. Rev. Lett.* **57**, 1235–1238.

Falconer, K. 1990. *Fractal Geometry: Mathematical Foundations and Applications*. Wiley, Chichester.

Feder, J. 1988. *Fractals*. Plenum Press, New York.

Freed, K. F. 1987. *Renormalization Group Theory of Macromolecules*. Wiley, New York.

Grassberger, P., and Procaccia, I. 1983. Measuring the Strangeness of Strange Attractors. *Physica* **9D**, 189–208.

Jensen, M. H., Kadanoff, L. P., Libchaber, A., Procaccia, I., and Stavans, J. 1985. Global Universality at the Onset of Chaos: Results of a Forced Rayleigh–Bénard. *Phys. Rev. Lett.* **55**, 2798–2801.

Ma, S. K. 1976. *Modern Theory of Critical Phenomena*. W. A. Benjamin, Reading, MA.

Mandelbrot, B. B. 1967. How Long is the Coast of Britain? Statistical Self-similarity and Fractional Dimension. *Science* **155**, 636–638.

Mandelbrot, B. B. 1983. *The Fractal Geometry of Nature*. W. H. Freeman, New York.

Mandelbrot, B. B. 1988. An Introduction to Multifractal Distribution Functions. In *Random Fluctuations and Pattern Growth*. ed. H. E. Stanley and N. Ostrowsky, Kluwer, Dordrecht, p. 333.

Maris, H. J., and Kadanoff, L. P. 1978. Teaching the Renormalization Group. *Am. J. Phys.* **46**, 652–657.

Pfeifer, P., Welz, U., and Wipperman, H. 1985. Fractal Surface Dimension of Proteins: Lysozyme. *Chem. Phys. Lett.* **113**, 535–540.

Pietronero, L. 1988. Fractals in Physics: Introductory Concepts. In *Order and Chaos in Nonlinear Physical Systems*, ed. S. Lundqvist, N. H. March and M. P. Tosi. Plenum Press, New York, pp. 277–295.

Schepers, H. E., van Beek, J. H. G. M., and Bassingthwaigthe, J. B. 1992. Four Methods to Estimate the Fractal Dimension from Self-Affine Signals. *IEEE Eng. Med. Biol.*, **11**, 57–71.

Shlesinger, M. F., and Hughes, B. D. 1981. Analogs of Renormalization Group Transformations in Random Processes. *Physica* **109A**, 597–608.

Shlesinger, M. F., and West, B. J. 1991. Complex Fractal Dimension of the Bronchial Tree. *Phys. Rev. Lett.* **67**, 2106–2108.

Stanley, H. E., and Meakin, P. 1988. Multifractal phenomena in physics and chemistry. *Nature* **335**, 405–409.

Te'l, T. 1988. Fractals, Multifractals, and Thermodynamics. *Z. Naturforsch, Teil a* **43**, 1154–1174.

Vicsek, T. 1992. *Fractal Growth Phenomena*, 2nd edn. World Scientific, Singapore.

Wilson, K. G. 1979. Problems of Physics with Many Scales of Length. *Sci Am.* **241** (2), 158–279.

2

Fractal Aspects of Protein Structure

Proteins are individually sculpted molecules the three-dimensional structure of which is dictated by their amino acid sequence. They are heteropolymers with a variable composition of 20 different amino acids. The amino acid sequence is referred to as the primary structure. For most proteins, the information contained in the primary structure is sufficient to dictate the three-dimensional folding of the protein. The peptide linkages that join the amino acid residues are planar and have two characteristic dihedral angles associated with them. These bond rotations are referred to as the Φ and Ψ angles. For steric reasons, the bond angles of proteins fall in localized regions of $\Phi\Psi$ space. These regions can be used to classify the secondary structural units. Commonly discussed secondary structural units are the alpha helix, beta sheet, beta turn, and random coil. The specification of the folded configuration of the chain is called the tertiary structure. The varied composition and nature of amino acid side-groups result in a range of possible interactions within the polymer. Because of the compact structure of most proteins, amino acid residues that are at distant points in the sequence can be in close proximity in the folded state. These distant residues can be linked by disulfide bridges, salt bridges (electrostatic), or hydrogen bonds. These long-range interactions help dictate the final three-dimensional structure of the molecule.

Given the strong dependence on sequence, it is not obvious that the statistical laws of polymer physics are applicable. Scaling relations in polymer physics are usually derived for large, homopolymers. By polymer physics standards, proteins are very small, consisting of 50–300 amino acids, and are also of varied composition. Given that scaling relationships may not hold for proteins, how could they be expected to be fractal? In the present chapter we explore the sense in which proteins are fractals. Traditionally, protein chemists have looked for relationships linking primary, secondary, and tertiary structural properties. Fractal behavior is not to be found within these classifications. For instance, alpha helices are not made up of smaller helices. As will be seen, proteins have an intrinsic self-similarity with regard to the compactness and packing of their structures. Although this is a simple form of fractal behavior, it has important consequences for the gross morphology of a protein and for the statistical thermodynamics of protein folding.

As an introduction, a general description of polymers as mass fractals is presented. Section 2.1 provides a classical derivation of scaling laws for polymers, and shows the relationship between these laws and the definition of the fractal dimension. From these general polymer considerations, we move to fractal descriptions of protein backbones (Section 2.2) and of protein surfaces (Section 2.3). These straightforward applications of fractal geometry are readily visualized and have considerable heuristic value. The scaling of end-to-end distances in a protein is seen to be identical on a global and local scale, and this provides the justification of treating proteins as fractals. The algorithms by which this scaling is obtained are discussed in Section 2.2. The fractal dimension of protein backbones is shown to be very similar for all proteins studied to date. Proteins are a very special form of collapsed polymer known as "crumpled globules." Crumpled globules have the biological advantage of not forming knots. Methods for determining the fractal surface dimension of a protein from high resolution X-ray structures are discussed in Section 2.3. Again, similar results are obtained over a variety of proteins and computational algorithms. This suggests that universal behavior is also observed with respect to this parameter. Collectively, these results indicate that the gross morphology of proteins is indeed governed by statistical laws. The experimental determination of fractal dimensions using low-angle scattering techniques is discussed in Section 2.4. This is a general technique that has seen limited application to proteins, but provides a concrete example of a physical measurement that can yield a fractal dimension. Finally, in the more speculative Section 2.5, scaling laws for membrane proteins are derived by considering the polymer confinement problem. Implications of these results for transmembrane signaling are discussed.

The scaling laws determined in this chapter will be of great utility when considering structural issues in later chapters. In Chapter 3, the configurational statistics of loop formation is considered. A number of different loop topologies exist and these will have characteristic scaling laws. These scaling laws are dependent on the mass fractal dimension of the polymer. Considering the complexity of protein structure, it is not surprising that protein dynamics are also quite complicated. Such issues will be taken up in Chapter 5 and 7. In Chapter 5, the surface fractal dimension of protein is used to develop a chemical kinetic model of hydrogen isotope exchange. The model relates structure and kinetics via a fractal dimension, known as a spectral dimension. Chapter 7 considers a similar connection for vibrational relaxation in proteins.

2.1 Polymers as Mass Fractals

2.1.1 The Radius of Gyration and the Fractal Dimension

To introduce the application of fractal concepts to proteins, a simple discussion of polymer scaling laws is useful. As will be seen, these laws are readily recast in the fractal formalism and provide guideposts for the range of mass fractal dimensions that can be obtained for polymers. This, of course, does not necessarily prove that polymers are fractal. For a polymer to be a fractal, it must be self-similar over an extended scaling regime. Additionally, it must be translationally invariant. The success of the renormalization group in describing an excluded volume polymer suggests that, statistically, such a polymer has similar configurational structures at different levels of scaling. The assumption of self-similarity is inherent in the successful renormalization group theory. For translational invariance to hold, the statistical properties must be independent of the specific position within the polymer. The theoretical manifestation of this is the superposition properties of probability distribution functions. These properties provide a useful tool in polymer physics (deGennes, 1979).

Central to the discussion of polymer statistics is the scaling of the radius of gyration, R_g, of the polymer chain with number of units in the chain, N. The radius of gyration is defined as the root mean squared distance of the constituent masses of a polymer from the center of mass. This dependence is given by:

$$R_g \sim N^{\nu} \tag{2.1}$$

where ν is a fractional exponent. The scaling exponent, ν, takes on different values depending on the solvent conditions. For a "good" solvent, the polymeric units have a preferential interaction with the solvent rather than with each other. As a result, the polymer is very extended and is referred to as an "excluded volume polymer." The value of ν is 3/5. In an ideal or "theta" solvent, the polymer interacts with the solvent with identical strength as it interacts with itself. Consequently, the polymer is more compact and $\nu = 1/2$. Polymer collapse occurs in the presence of a "poor" solvent. In this case, polymer–polymer interactions are more favorable than polymer–solvent interactions. The polymer forms a collapsed or globule state with $\nu = 1/3$.

To describe polymers in terms of fractals, it is convenient to consider the definition of the mass fractal dimension. The operational definition of a mass fractal dimension, D, is considered first. As discussed in Chapter 1, to define such a dimension, one draws concentric spheres around the center of an object. The mass contained within each sphere is then determined. The mass fractal dimension is defined by the scaling relationship:

$$M \sim R^{D} \tag{2.2}$$

where M is mass and R is radius. Typically, D would be obtained from the slope of a plot of ln R versus ln M.

The contour of an excluded volume polymer backbone describes a mass fractal. Instead of determining the mass within our hypothetical spheres, the behavior of the radius of gyration, R_g, is considered. The mass of a polymer, M, will be proportional to the number of units, N, in the chain. The dependence of the radius of gyration, R_g, on chain length is given by:

$$N \sim M \sim R_g^D \tag{2.3}$$

From Eqs 2.1 and 2.3, it is seen that:

$$\nu = 1/D \tag{2.4}$$

Thus, the excluded volume polymer will have a fractal dimension of 5/3, and an ideal polymer has a dimension of 2. A collapsed polymer ($\nu = 1/3$) will have a fractal dimension of 3, i.e., it is space filling. It might at first be surprising to have a one-dimensional structure such as a linear polymer chain assume a fractal dimension of 3. Yet the familiar example of a ball of yarn readily shows how this is possible. In a sense, a collapsed polymer has a "trivial" fractal dimension because it is a homogeneous, space-filling object. Within the interior of the polymer globule, it is scale invariant and translationally invariant in the same sense as found in simple geometric objects such as a line segment or a plane.

2.1.2 Scaling of Excluded Volume, Ideal, and Collapsed Polymers

The scaling laws for polymers can be derived in a simple (and fortuitous) manner. These were originally presented by Flory for the ideal and excluded volume polymer (see Flory, 1953) and later extended by deGennes (1975, 1978) to collapsed polymers. This derivation employs a minimization of the configurational free energy with respect to a chain deformation parameter, α. The free energy is determined from two terms: a local or deformation term, and a nonlocal term involving interactions between polymer units at distant locations along the chain (see Chan and Dill, 1991).

Flory determined the form of the deformation term from simple scaling considerations. A full derivation is not presented here, but the outline of the argument is given. A polymer in a theta solvent has a chain configuration that is essentially a self-avoiding random walk. That is, the chain can be viewed as the trajectory of a series of fixed steps that can occur in random directions. These random steps are only restricted by the requirements that they cannot overlap or occupy the same space. For bonds of fixed length, it can be shown that the probability distribution function describing the end-to-end distance is a Gaussian (Flory, 1953). The probability distribution of an end-to-end vector from the chain origin to the location of the ith unit (x_i, y_i, and z_i) is given by:

$$P(x_i, y_i, z_i) = \left(\frac{\sqrt{3/2\pi}}{R_0}\right)^3 \exp\{-(3/2)\,(x_i^2 + y_i^2 + z_i^2)\,R_0^{-2}\} \tag{2.5}$$

where

$$R_0 = N^{1/2}a \tag{2.6}$$

and the parameter a is the bond length and N is the number of units in the polymer.

We now let this ideal chain be deformed isotropically as a result of nonideal solvent interactions. Upon isotropic deformation, the probability distribution function, P_{def}, is given by:

$$P_{\text{def}}(x_i, y_i, z_i) = P\,(x_i/\alpha, y_i/\alpha, z_i/\alpha) \tag{2.7}$$

where the dependence of the deformation factor, α, on N remains to be determined from an energy minimization procedure. The deformation factor is defined as:

$$\alpha = \frac{R}{R_0} \tag{2.8}$$

with R and R_0 being the radius of gyration for the deformed and the ideal polymer, respectively. In the ideal case, the free energy of deformation is determined solely by an entropic contribution and has close parallels with the compression of an ideal gas. For both cases, this is given by:

$$\Delta S = k_B n \ln\left(\frac{V}{V_0}\right) \tag{2.9}$$

where k_B is Boltzmann's constant, n is the number of polymers with a specific end-to-end vector, V is the volume after deformation (or compression), and V_0 is the ideal (or initial) volume. Using the probability functions in Eqs 2.5 and 2.7, and calculating the work of deformation as in Eq. 2.9, the deformation free energy, F_{def}, is shown to be:

$$\frac{F_{\text{def}}}{kT} = 3\left(\frac{\alpha^2}{2} - \ln\alpha\right) \tag{2.10}$$

The nonlocal energy term, F_{nl}, is determined by the interaction of polymer units brought into close proximity by the folding of the chain. It will depend upon the density of these units, and is best represented as a series expansion in powers of the density. This term is given by:

$$\frac{F_{\text{nl}}}{kT} = \left(\frac{N}{2}\right)[\rho W_1(T) + \rho^2 W_2(T) + \ldots] \tag{2.11}$$

where the density is:

$$\rho = \frac{\kappa N}{R^3} = \kappa N^{-1/2} a^{-3} \alpha^{-3} \tag{2.12}$$

with κ being a numerical constant, and the right-hand equality in Eq. 2.12 is obtained using Eqs 2.6 and 2.8. The parameter W_i is the ith virial coefficient related to the interaction of subunits. Usually, only the first two coefficients are considered, and these will account for two- and three-body interactions, respectively. The two-body interaction term, sometimes called the excluded volume coefficient, has the following temperature dependence:

$$W_1(T) = \frac{v\,(T - \Theta)}{\Theta} \tag{2.13}$$

with Θ being the theta temperature and v is the volume associated with the interaction. This volume will be roughly the size of a monomeric unit. For ideal polymers, $T = \Theta$ and $W_1 = 0$. For excluded volume polymers, $T > \Theta$ and the W_1 term dominates; in this case, W_2 is usually ignored. In the treatment of collapsed polymers, the W_2 term becomes very important; in this case, $T < \Theta$ and W_1 is negative. This attractive term by itself would result in collapse of the polymer to unrealistic densities. Consequently, the positive W_2 term must be included to offset this effect and obtain a realistic density distribution.

The total free energy to be minimized is now given by:

$$\frac{F}{kT} = 3\left(\frac{\alpha^2}{2} - \ln \alpha\right) + \frac{N}{2}\,[\rho W_1(T) + \rho^2 W_2(T)] \tag{2.14}$$

Setting the first derivative with respect to α equal to zero, one obtains:

$$\frac{\alpha^5 - \alpha^3 - k^2 W_2}{(a^2\alpha)^3} = kN^{1/2} W_1 a^{-3} \tag{2.15}$$

Three cases are now considered: $W_1 = 0$, an ideal polymer; $W_1 >> 0$, an excluded volume polymer; and $W_1, << 0$, the collapsed polymer. In the first case, where $W_1 = 0$ and W_2 is small, one has $\alpha = 1$. From Eqs 2.6, 2.8, and 2.15, it is seen that $R \sim N^{1/2}$, giving $\nu = 1/2$ and a fractal dimension of 2. When $W_1 >> 1$, the α^5 term on the left-hand side of Eq. 2.15 will dominate and $\alpha \sim N^{1/10}$. This gives $R \sim N^{3/5}$. Thus, $\nu = 3/5$ and the fractal dimension is 5/3. For the case when $W_1 << 1$, the dominant term of the left-hand side of Eq. 2.15 is the α^{-3} term. The value of ν is 1/3, and the fractal dimension is 3. Thus the entire range of fractal dimensions for polymers can be derived from Flory's simple energy minimization scheme.

In the above derivation of the fractal dimensions of polymers, scaling laws were obtained from a very traditional energy minimization scheme. No intrinsic fractal properties, apart from the definition of a mass fractal, were evoked in the derivation. An obvious question is: Are polymers legitimately fractals or have we just renamed the scaling exponents so that they appear as fractal dimensions? The heart of the answer lies in renormalization group (RG) theory and the scale-invariant properties of polymers. As mentioned in the previous chapter, the RG approach represents a succession of scaling and coarse-graining operations that, in an appropriate application, converge to a fixed point. In a scale-invariant system this can be used to extract the dominant behavior from the microscopic details. "Coarse-graining" has been a dominant concept in polymer physics since the early approaches of Kuhn (see Freed, 1987). As discussed in Chapter 1, a collection of subunits along the chain can often be treated as a single "ideal" unit (creating a single pearl in a string of pearls). The details of the interactions within a unit are not important in determining the overall configurational properties of the polymer. Quite often the actual size of the unit is unimportant. The RG approach, especially in the real-space treatment, extends these concepts by demonstrating the scale invariance intrinsic in the system. In the excluded volume case, when scaled and coarse-grained, a large polymer provides an adequate description of a segment within the polymer. Thus, the renormalization procedure is not only a mathematical device, but also a physical representation of the polymer, and highlights the scale invariance of the structure. As will be seen in the next section, the self-similarity of proteins is very different from that of an excluded volume polymer. The experimental basis of this self-similarity is also discussed.

2.2 Fractal Dimension of the Protein Backbone

2.2.1 Measuring the Fractal Dimension

The previous discussion of fractal dimensions of polymers was based on statistical arguments for homopolymers. Proteins, on the other hand, are heteropolymers and each one has a specific primary sequence. There is a wealth of X-ray crystallographic data on globular proteins that provide detailed structural information. From this database, it is seen that each protein has a specific conformational structure that is dictated by its respective amino acid sequence. The great structural specificity of proteins is in apparent contrast to most synthetic homopolymers for which the laws of polymer physics were derived. Nevertheless, there are common structural motifs that recur in proteins, irrespective of the primary sequence (Chothia, 1984). Also, it has long been recognized that globular proteins have certain universal features with regard to surface areas, volumes, and packing densities (for a review see Richards, 1977).

The question remains whether proteins can be described by statistical laws. Figure 2.1 shows a log–log plot of the radius of gyration, R_g, versus number of residues, N, for 43 different proteins (data from Abad-Zapatero and Lin, 1990). The linearity of this plot demonstrates scaling behavior. The value of ν is determined directly from the slope. A value of 0.35 ± 0.03 is found, giving a fractal dimension of 2.85. The data in Figure 2.1 are slightly scattered, and this may reflect the variable amino acid compositions. Within experimental error this plot gives a fractal dimension of 3 and demonstrates that proteins are collapsed polymers.

In previous work (Isogai and Itoh, 1984; Wang et al., 1990), the fractal dimensions of individual proteins were determined from a direct, geometric approach involving an analysis of X-ray structures. This early application of fractal geometry turns out to be incorrect, and it is instructive to see how it can be corrected (Dewey, 1993a, 1995). The data from the more extensive study (Wang et al., 1990) are summarized in Table 2.1. In these works, the length, L, of the backbone was measured by a stepwise connection of straight lines between the C^{α} atoms of the protein backbone (Figure 2.2). This length was measured for different intervals of m residues. For $m = 3$, for instance, a straight line was used to connect the C^{α} atoms

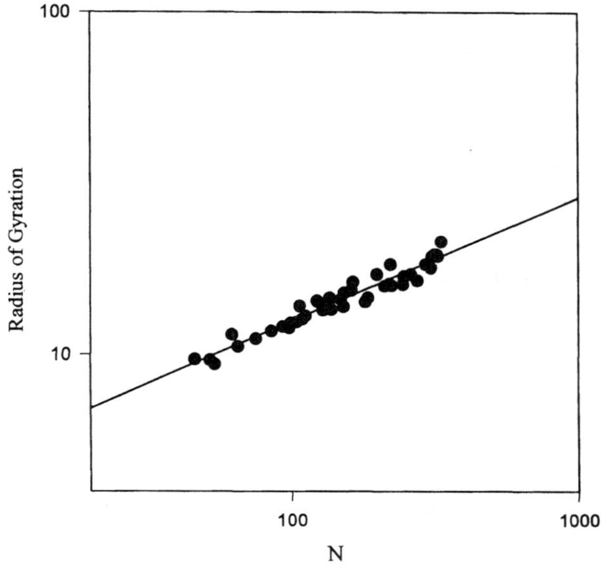

FIGURE 2.1 Log–log plot of the radius of gyration, *R*g, versus number of residues (amino acids) for 45 different proteins. A fractal dimension of the protein backbone of 2.86 is determined from the slope of this plot. (Data from Abad-Zapatero and Lin, 1990.)

TABLE 2.1 Fractal dimensions of protein backbones*

	Number of Structures	ν	D
Class of Protein[1]			
Alpha helices	19	0.39 ± 0.03	2.6
Beta sheets	25	0.31 ± 0.07	3.3
Alpha/beta mix	22	0.31 ± 0.04	3.3
Small disulfide-rich	9	0.39 ± 0.04	2.6
Small metal-rich	15	0.39 ± 0.04	2.6
All categories	90	0.35 ± 0.05	2.9
Theoretical models[2]			
Excluded volume polymer ($T > \Theta$)		0.6	1.67
Ideal polymer ($T = \Theta$)		0.5	2.0
Collapsed polymer ($T < \Theta$)		0.33	3.0

* Only data for the "global" dimension are used. This corresponds to D_2 of Isogai and Itoh, (1984). See text for details.

[1] See Isogai and Itoh (1984) for classification scheme and results for individual protein structures.

[2] T, temperature; Θ, theta temperature of the polymer.

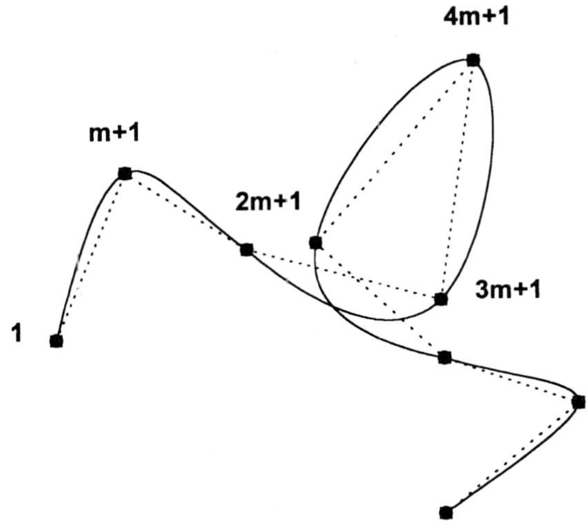

FIGURE 2.2 Schematic representation of an algorithm to determine the fractal dimension of a protein backbone. Line segments are drawn between the C^{α} atoms in the backbone that are spaced m amino acids apart. Total length, L, is computed as a function of the step size m. Fractal dimensions are determined from the slope of log L versus log m, as in Figure 2.3. (Figure modified from Isogai and Itoh, 1984.)

on every third residue. The length of the backbone determined in this fashion depends on the interval size, m. The fractal dimension of the backbone was obtained from the slope of a log L versus log m plot. For 90 protein structures, this log–log plot was bilinear with a slope of –0.38 (for $m < 10$) and –0.65 (for $m > 10$) (see Figure 2.3 for a specific example). The bilinear plot is attributed to a "short" scaling regime (small m) that crosses over into a "long" regime (large m). Note that even the "long" regime is not a global property, as it consists of lengths much shorter than the entire protein.

As discussed in Chapter 1 (Eq. 1.4), the fractal dimension of a curve may be defined by measuring the length, L, with rulers of fixed length, ε. The length is a function of ε and is given by:

$$L(\varepsilon) = N(\varepsilon)\,\varepsilon \tag{2.16}$$

where $N(\varepsilon)$ is the number of steps of length ε that are needed to cover the curve. In general, $N(\varepsilon) \sim \varepsilon^{-D}$ and $L \sim \varepsilon^{1-D}$, where D is the fractal dimension of the curve. This scaling law was used to determine fractal dimensions of 1.38 (short) and 1.65 (long) from the experimental data. The global dimension corresponds to the large "ruler" size, and the suggestion was made that the gross features of a protein are those of an excluded-volume polymer ($D = 5/3$). This, of course, runs counter to the experience of crystallographers who observe very compact structures. This approach is inappropriate, since line segments drawn between C^{α} atoms on a protein are not fixed-length increments and do not scale with m in a linear fashion (Dewey,

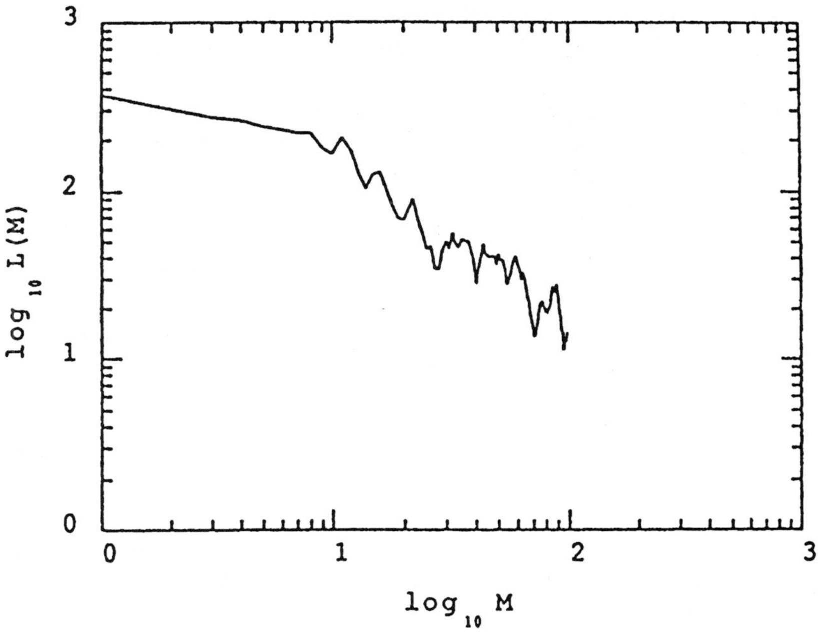

FIGURE 2.3 Representative data for determining the fractal dimension of a protein. Plot of log L versus *log m*, where L is the total length of the protein and m is the step size (as in Figure 2.2). The plot shows two linear regions, which have been described as local and global. Analysis of the slope of this plot yields the fractal dimension and is presented in the text. Numerical results are summarized in Table 2.1. (From Wang et al., 1990. Reprinted by permission of Wang Cun Xin.)

1993a). In actuality, the algorithm uses a variable-length ruler rather than a fixed-length one.

For sufficiently long line segments, a scaling Ansatz similar to Eq. 2.2 is made. The relationship between the root mean squared length of the segment, δ, and the number of residues, m, is taken as $\delta = am^{\nu'}$, where ν' is now a "local" exponent and reflects the scaling of length within the protein. This is in contrast to ν, which gives the global scaling as determined by the radius of gyration. When measuring protein backbones by the method described above, the number of line segments, $N(\delta)$, is given simply by N_P/m where N_P is the number of amino acid residues in the protein. On average, the length of the protein, $L(\delta)$, is given by:

$$L(\delta) = N\,(\delta)\,\delta = \left(\frac{N_{\mathrm{P}}}{m}\right) am^{\nu'} \tag{2.17}$$

The log L versus log m plot will now have a slope of $\nu' - 1$ or $1/D - 1$. The previous analyses that predicted an exponent of $1 - D$ do not properly determine the fractal dimension for the protein backbone.

In principle, both the "short" and "long" scaling regimes in Figure 2.3 can be interpreted. It is difficult to analyze the "short" region ($m < 10$) in a meaningful fashion because the units are not rigid enough to act as a fixed-length ruler (or "ε ruler" in our notation). Also, these units are not long enough that the statistical dependence $\delta = am^{\nu'}$ will hold. Thus, they will not be good "δ rulers" either. No attempt is made to interpret this "local" scaling regime. As m increases, the scaling Ansatz will be applicable, and this "global" regime can be interpreted using Eq. 2.17.

The average value of ν' for 90 protein structures is 0.35, giving a fractal dimension of 2.86. This is in excellent agreement with the value of 2.85 from the log–log plot shown in Figure 2.1. In Table 2.1, the data are grouped into specific structural classes of proteins. Interestingly, it appears that the fractal dimension of the backbone is independent of the secondary structural content of the protein. This suggests that all classes of secondary structure formed in a protein are compatible with collapsed conditions. This observation has been noted before in the context of the energetics of protein folding (Chan and Dill, 1989).

2.2.2 Significance of the Fractal Dimension

At this point, a protein crystallographer might well ask: What has the fractal approach told us that we do not already know? Viewing proteins as collapsed polymers is completely consistent with long-standing ideas of protein structure. Crystallographers have observed that proteins are extremely compact structures with virtually no void volume. The density within a protein is almost identical to the density of an amino acid crystal. Proteins generally have a dense, hydrophobic core the units of which prefer to interact with each other rather than with the aqueous solution. This behaviour is consistent with the thermodynamics of polymer collapse. The effects of denaturants on proteins are also consistent with this view. Most denaturants are solutes, such as urea or alcohols, that have some affinity for the protein subunits. These solutes create interactions between the polymer and the solution that are favorable. The solution is now no longer a poor solvent and a transition occurs out of the collapsed state.

It might appear that the length analysis and the radius of gyration data provide the same information regarding the fractal dimension of a protein, especially since they agree so closely. However, this is not true. The scaling of the radius of gyration with number of subunits in the entire polymer (Eq. 2.3) shows the behavior of the end-to-end distance of the polymer. The length analysis provides information on the scaling of two points separated by m units within the polymer. This distinction and the dependence on length for a classical collapsed polymer and for a protein is shown in Figure 2.4. A classical polymer will locally be "quasi-ideal" (deGennes, 1979). This means that it will be a Gaussian coil with $\nu' = 1/2$. These local Gaussian coils will then pack globally into a collapsed polymer. This is sometimes referred to as the string of pearls or bead model. The beads are made up of globules of polymer that are ideal. When polymer collapse occurs these globules collapse and interpenetrate. Despite this coalescence, the statistics within the globule still give $\nu' = 1/2$. As m increases there will no longer be a local scaling, eventually scaling as the entire polymer of length N. This results in the flat region in Figure 2.4. In this region, snipping a small end of the polymer does not dramatically change R_g. The

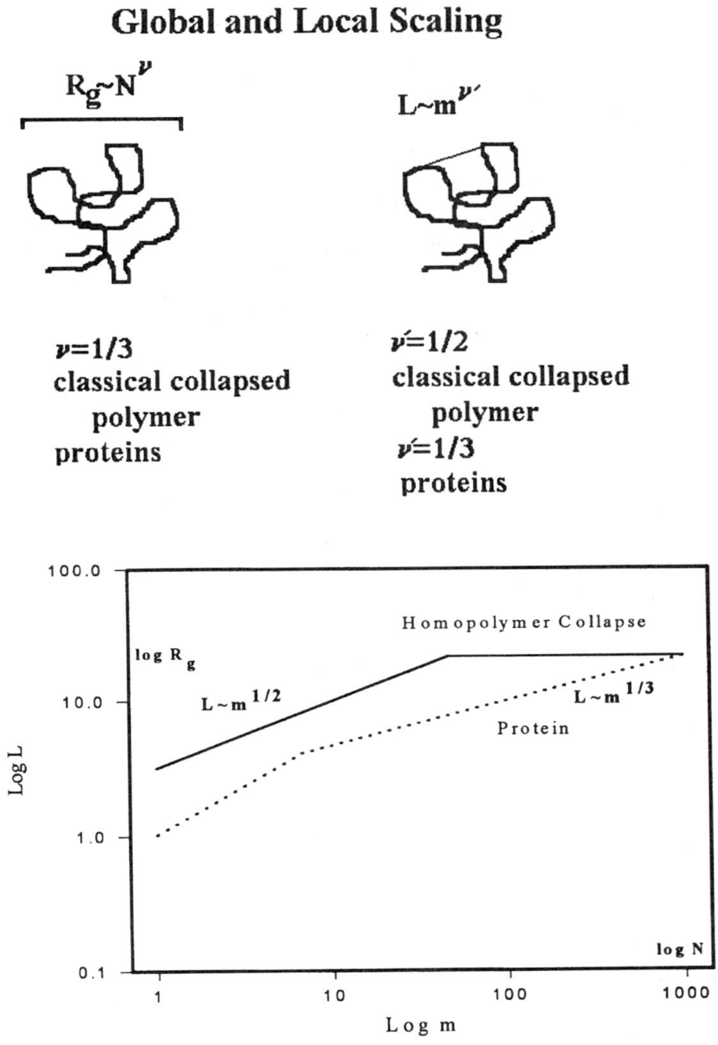

FIGURE 2.4 Representation of length scaling in polymers. (Top) Comparison of scaling of the radius of gyration and of internal lengths of the polymer. (Bottom) Plot of *log L* versus the logarithm of the number of units, *log m*. This plot shows how proteins differ from "classical" collapsed polymers. (Adapted from Dewey, 1995.)

crossover behavior between local and global scaling occurs at the point m^*, which is given by $m^* = N^{2/3}$ for a collapsed polymer. As can be seen from the experimental data in proteins (average behavior illustrated in Figure 2.4), a very different curve is obtained. As discussed above, the initial, short length, scaling regime ($m < 10$) may not be meaningful. However, the second regime ($m > 10$) behaves much differently than a conventional polymer. In the protein case, m scales identically with L as N scales with R_g. Thus, the end-to-end and the internal lengths both behave identically. Proteins are self-similar in this respect. These results show that

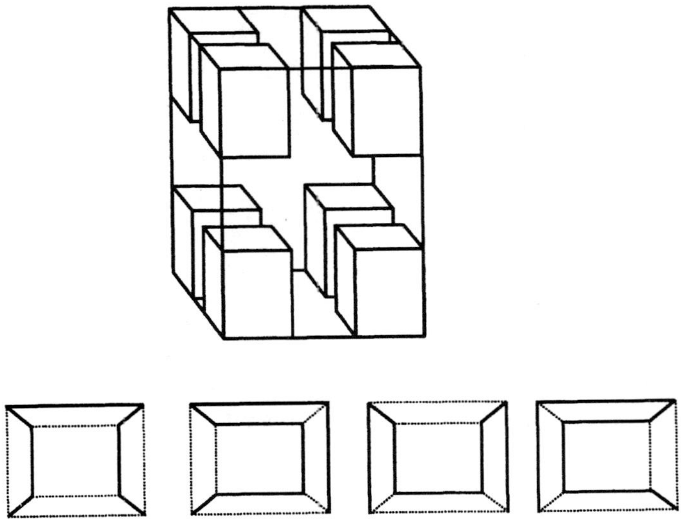

FIGURE 2.5 (Top) A second-generation cubic lattice created by combinations of walks shown in the bottom of the figure. Both the individual walks and the higher order constructs scale as a collapsed polymer. (Bottom): Example of Hamiltonian walks on a cubic lattice. The cube is shown in a two-dimensional projection; the bold line traces the path of the walk. (Adapted from Dewey, 1995.)

proteins behave very differently from classical collapsed polymers. This means that some theories of polymer physics cannot be appropriately adapted to describe protein structure.

Polymers that show compact local and global scaling have been referred to as "crumpled globules" (Grosberg and Khokhlov, 1989). A simple way to model a "crumpled globule" is through a mathematical structure known as a Hamiltonian walk. A Hamiltonian walk is a lattice walk in which every lattice point is visited once and only once, and no paths intersect. (Hamilton was originally interested in such walks on a dodecahedron.) Hamiltonian walks have received considerable attention from mathematicians, but have been used sparingly to describe collapsed polymers and glass transitions (Gordon et al., 1976; Malakis, 1976; Pechold and Grossmann, 1979). Figure 2.5 shows examples of Hamiltonian walks and how larger walks can be constructed from units of the smaller ones. For a cubic lattice, space-filling walks can be constructed in which $\nu = 1/3$ on all scales. A loose analogy can be made between the unique individual walks that make up the larger walk and secondary structural units of proteins. Because local scaling in proteins is identical to global scaling, the secondary structural units must also be units of collapsed polymer. Individual subsets of walks (or secondary structural units) can be pieced together to give a larger walk (or protein) with identical scaling properties. This analogy is not inappropriate considering the data in Table 2.1. As mentioned previously, the fractal dimension of the backbone is independent of the secondary structural content of the protein. This suggests that all classes of secondary structure formed in a protein are compatible with collapsed conditions, i.e., all secondary structural classes are Hamiltonian walks. A potential physiological advantage of having proteins form crumpled globules is that knot formation is prevented (Grosberg et al., 1993). Knots have never been seen in proteins, presumably as a result of the problems they create for protein synthesis and turnover.

Hamiltonian walk models of polymers (see Gordon et al., 1976; Malakis, 1976) originally arose as a limiting case of the Gibbs–DiMarzio theory of glass transitions (DiMarzio and Gribbs, 1958; Gibbs and DiMarzio, 1958). These models are simple limiting cases of lattice models that consider solvent and monomers to occupy lattice sites individually. In Flory–Huggins models of polymers, the Hamiltonian walk is achieved simply by letting the number of solvent molecules approach zero. However, the Flory–Huggins approximation is inaccurate because the number of configurations, Ω, is not properly computed. The starting point for statistical theories of polymers is the configurational free energy contribution given by $G = kT \ln \Omega$. If the number of Hamiltonian walks of n steps, H_n, can be enumerated and the following limit exists:

$$\lim_{n\to\infty} H_n^{1/n} = \omega \tag{2.18}$$

then the number of configurations is given by $\Omega = \omega^n$. For special lattice structures, H_n can be determined from recursion relationship generated by creating larger walks from units of small walks (see Figure 2.5).

Originally, Gibbs and DiMarzio (1958) explored models of chains that could assume *gauche* and *syn* configurations. Such models are readily adapted to consider heteropolymers that contain polar and nonpolar residues. Dill's lattice model (Dill, 1985) of protein stability employs a mean-field, hydrophobicity energy term that is analogous in form to the *gauche*/*syn* configurational energetics. With such energy terms, the Hamiltonian walk model is seen to show a coil–globule transition, similar in nature to the glass transition of Gibbs and DiMarzio. The coil–globule transition in proteins is marked by the favorable energetics of reducing the aqueous exposure of hydrophobic residues. This driving force is opposed by the great configurational restriction of collapsing the chain. To construct a statistical model of protein stability, Dill considers a hypothetical two-step process associated with protein folding. This process consists of random condensation followed by chain reconfiguration. Random condensation is essentially collapse of the polymer to a compact state. The chain reconfiguration step rearranges the hydrophobic residues along the sequence to minimize the configurational energy with respect to both polymer self-interactions and solvent interactions.

To calculate the random condensation contribution to the number of configurations, Dill employed a Flory–Huggins approximation to a Hamiltonian walk on a cubic lattice (Dill, 1985). This gives:

$$\Omega = \left(\frac{z}{a'}\right)^n \tag{2.19}$$

where z is the number of rotational isomeric states available to a peptide bond rotation. It was argued that z is, at most, 3.8. The value of a' is given to first approximation by e (= 2.7718). A slightly more sophisticated treatment gives (see Flory, 1982):

$$a' = \left(1 - \frac{2}{q}\right)^{-(q/2 - 1)} \tag{2.20}$$

where q is the coordination number of a lattice. Assuming a cubic lattice, $q = 6$ and a' is 2.25. Using these estimates and Eq. 2.19, the entropy of polymer collapse can be estimated. The number of configurations is further restricted by the chain reconfiguration steps. Including these terms, entropy values consistent with experimental values can be obtained. Because of the number of adjustable parameters, it is difficult to determine how well the Flory–Huggins approximation does in this case. Certainly, for a square lattice it does not do particularly well (see Gordon et al., 1976). It is also difficult to provide a quantitative test of the cubic lattice model of a protein. Nevertheless, the model provides a reasonable physical, albeit qualitative, picture of the folding transition in proteins.

2.3 Scaling Properties of Protein Surfaces

Since protein backbone conformations obey statistical laws, it is obvious to ask whether surface areas follow similar laws. Except for the photosynthetic and visual systems, the surface of a protein represents the first level of "communication" with its environment. The geometry of the protein surface will presumably have a significant effect on the nature of the solvation shell at the boundary. This in turn will influence recognition events with solutes – events that are key to a diversity of biochemical phenomena. The nature of the surface will also influence a range of physical properties, such as isotope exchange and fluorescence quenching by small molecules. The kinetics of isotope exchange in proteins is discussed in Chapter 5. The surface properties of proteins also play a crucial role in determining the thermodynamic stability of the final, folded state (see Chan and Dill, 1989), and to some extent dictate the structure.

There are a number of ways to determine the surface area of a protein and these various techniques all provide a similar statistical description of such surfaces. Typically, X-ray structures are examined and surfaces are generated from van der Waals radii (see Lewis and Rees, 1985; Fedorov et al., 1993) or isoenergetic contours (Pfeifer et al., 1985). Techniques for determining fractal surface dimensions fall into two main categories. In the first method, a "ball-rolling" algorithm is used, as illustrated in Figure 2.6. In this algorithm, a spherical probe of fixed radius, R, is rolled on the outside of the of the protein while maintaining contact with the van der Waals surface. The center of this sphere will sweep out a surface that is known as the "accessible surface." Because of the size of the sphere, contact will not be made with the entire van der Waals surface of the protein. The "contact surface" consists of the patches of the van der Waals surface that the probe actually rolled over. The "re-entrant surface" is also a patchwork that represents the inward surface of the probe when it is touching more than one atom. The contact and

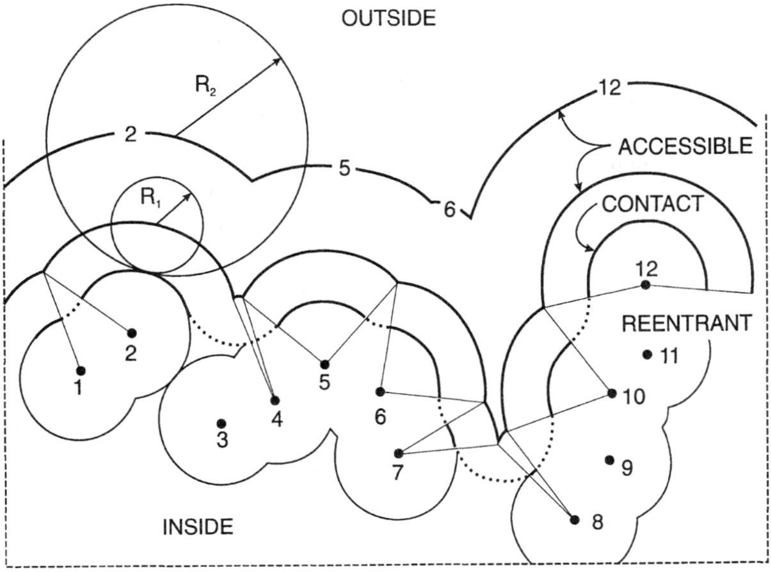

FIGURE 2.6 Representation of a protein surface and the definition of different surfaces from a ball-rolling algorithm. A cross-section of the van der Waals surface of a hypothetical protein is shown. Numbered dots show atomic centers. Surfaces created by rolling balls of two different radii are shown. (Figure reprinted, with permission, from Richards, 1977.)

re-entrant surface sum to make a continuous surface known as the "molecular surface." Despite these multiple definitions, in practice the accessible surface has been predominantly reported in the literature (see Richards, 1977).

Early on, it was recognized that the nature of the surface will depend on the size of the probe molecule. For convenience a probe approximately the size of a water molecule was often used, i.e., 1.4 or 1.5 Å. Before the advent of fractal geometry, the conundrum of surface areas dependent on the size of the measuring device had to be faced, as witnessed by Richards' statement that "The accurate calculation of even the static surface area is a complex geometrical problem" (Richards, 1977). The fractal resolution of this problem lies in the definition of the fractal surface dimension (see Eq. 1.5).

Using the ball-rolling algorithm, the fractal surface dimension, D_s, can be determined in two ways. In the first approach, the probe radius, R, is varied for a given protein, and the accessible surface area, A_S, is determined at each value of the radius. The surface area is then given by $A_s = N(R)\,R^2$, where $N(R)$ is the number of balls required to cover the surface. Recalling Eq. 1.5, one has $N(R) \sim R^{-D_s}$. The slope of a plot of log A_s versus log R is given by (Lewis and Rees, 1985):

$$\frac{d \ln A_s}{d \ln R} = 2 - D_s \tag{2.21}$$

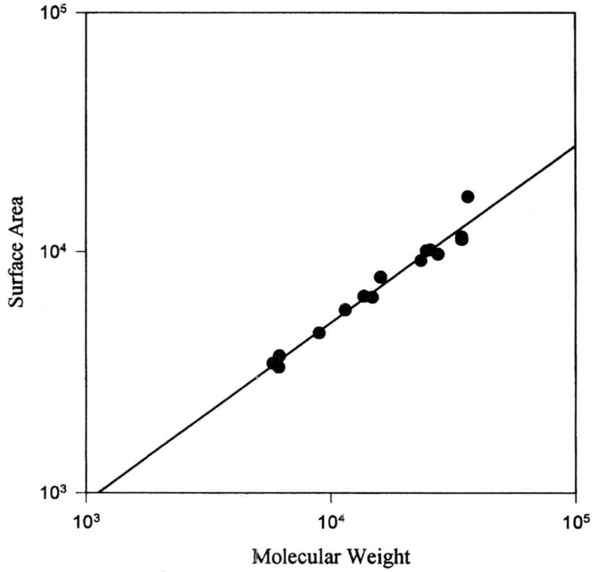

FIGURE 2.7 Log-log plot of the molecular weight versus surface area for 14 different proteins. As molecular weight is proportional to volume, this plot was used to determine the fractal surface dimension of 2.16. (Data from Chothia, 1975.)

In the data reported by Lewis and Rees (1985), these log–log plots did not show an extensive linear region. The value of D_s for the linear region was approximately 2.4.

In an alternate treatment, the protein surface was varied by considering different molecular weight proteins while the probe radius was held fixed (Dewey, 1993b, 1995). The volume of the protein is proportional to the molecular weight, M_W, and the following scaling relationship was used:

$$V \sim M_W \sim A_S^{D_S/3} \tag{2.22}$$

A log–log plot of molecular weight versus surface area is shown in Figure 2.7 (data for 14 different proteins, taken from Chothia (1975)). The slope of this plot gives $D_S = 2.16$.

In another method of determining the surface dimension, the coastline of the cross-sectional area of a protein is measured. This is achieved by cutting the protein with a series of parallel planes. A resulting cross-section of a protein is shown in Figure 2.8. The fractal dimension of the boundary of the protein appearing in a given cross-section is determined. This is usually done using a box covering algorithm (see Pfeifer et al., 1985) and the fractal dimension of the curve, D_c, is determine as in Eq. 1.3. The surface dimension is then given by:

$$D_s = D_c + 1 \tag{2.23}$$

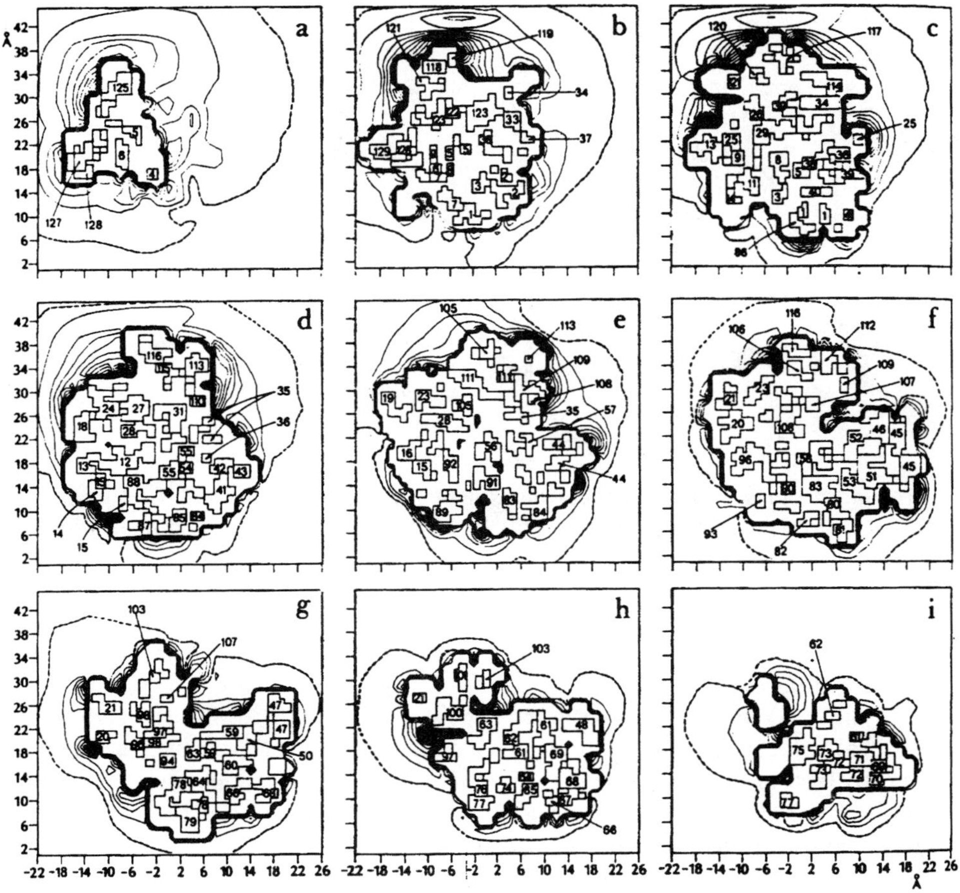

FIGURE 2.8 Consecutive cross-section of the protein, lysozyme, at a 4 Å spacing. Contours represent isoenegetic points as determined computationally. The innermost contour line marks the sharp transition from an attractive to a repulsive potential, and this line is considered the surface. (Figure adapted from Clements et al., 1977.)

Eq. 2.23 can be derived from the set theoretical results discussed in Chapter 1 (Eq. 1.15); its validity has been discussed previously (Falconer, 1990). However, see the discussion in Farin et al. (1985) for examples of failure. Typically, the values for all the cross-sections are averaged to give the final answer. In Table 2.2, the surface dimensions determined by these various methods are tabulated. Measurements have also been made of the fractal dimension of the silhouette of a protein (Farin et al., 1985). In these cases, relationships equivalent to Eq. 2.23 are much less obvious, and are not discussed further here.

Despite differences in computational techniques and proteins, the value of D_s typically falls in the range from 2.1–2.2. However, values as high as 2.4 have been observed. On the whole, these results indicate that proteins do not deviate greatly

TABLE 2.2 Fractal dimensions of protein surfaces*

Protein	Dimension	Reference
14 Proteins[1]	2.16	Janin (1976)
12 Proteins[1]	2.06	Teller (1976)
Lysozyme[2]	2.17	Pfeifer et al. (1985)
3 Proteins[3]	2.44	Lewis and Rees (1985)
11 Proteins[4]	2.11–2.13	Fedorov et al. (1993)

* Table adapted from Dewey (1993a).

[1] Data obtained from the slope of a log–log plot of surface area versus molecular weight.

[2] Calculations for an isoenergetic surface instead of a van der Waals surface.

[3] Data did not show an extensive linear region on a log–log plot of surface area versus probe radius.

[4] Data taken from the van der Waals surface of individual proteins.

from a smooth, spherical surface, which, of course, would have $D_s = 2$. This may not appear surprising, giving a picture of a protein that collapses into a globule with a smooth surface. This is also consistent with statistical mechanical models of polymer collapse (Dewey, 1993a). However, it is entirely possible to have a space-filling structure ($D = 3$) with a highly corrugated surface (D_s significantly greater than 2). It is conceivable that under some conditions a collapsed polymer could have a rough surface.

The experimental data consistently show a surface fractal dimension slightly greater than 2. Interestingly, the fractal dimension of a protein backbone is typically slightly less than 3. Is there any significance to these slight variations in both fractal dimensions? Recently, it was speculated that there is (Dewey, 1993b). The law of additivity of fractal codimensions (Falconer, 1990; Pietronero, 1988) can be used to relate the backbone dimension to the surface dimension. As discussed in Chapter 1 (Eq. 1.16), the intersection (space **C**) of two fractal sets (**A** and **B**) with fractal dimensions of $D_{\mathbf{A}}$ and $D_{\mathbf{B}}$, respectively, has a fractal dimension, $D_{\mathbf{C}}$, which is given by:

$$D_{\mathbf{C}} = D_{\mathbf{A}} + D_{\mathbf{B}} - E \tag{2.24}$$

where E is the Euclidean dimension of the pace in which the structures are embedded. The following question is now asked: "What must be the fractal dimension of a set (**B**) that intersects the polymer (set **A**) in such a manner as to produce a smooth surface, i.e., $D_{\mathbf{C}} = 2$?" Taking the protein fractal dimension to be $D_{\mathbf{A}} = 1/v = 1/0.35 = 2.86$, it is seen that $D_{\mathbf{B}} = 2.14$. The ball-rolling algorithm creates a smooth surface ($D = 2$) which can be associated with $D_{\mathbf{C}}$. Any surface that intersects the fractal polymer structure in such a manner as to create this smooth surface must by necessity have the dimension $D_{\mathbf{B}}$. This is the fractal dimension of the protein surface that underlies the smooth ball-rolling surface and explains why a fractal backbone dimension of slightly less than 3 gives a surface dimension slightly greater than 2.

In Chapter 5 the surface dimension will be used to determine scaling laws for the kinetics of hydrogen isotope exchange. In this model, isotope exchange occurs

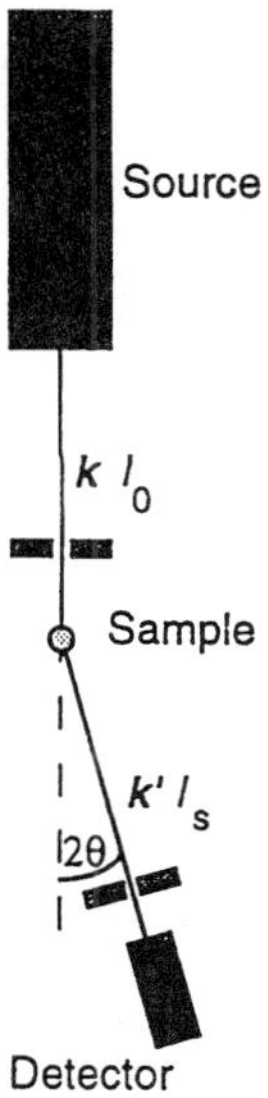

FIGURE 2.9 Schematic diagram showing the experimental set-up for measuring low-angle scattering. (From Schmidt, 1989. Reprinted by permission of John Wiley & Sons.)

as a result of concentration fluctuations at the protein surface. The surface dimension is used in a predictive fashion in a quantitative analysis of the kinetic data. This model allows X-ray crystal structures to be used to calculate kinetic scaling parameters.

2.4 Determining Fractal Dimensions by Scattering Techniques

2.4.1 Basic Concepts

Fractal dimensions of proteins have also been determined using low angle X-ray diffraction (Chen and Teixeira, 1986). Scattering techniques can be extremely useful in determining fractal dimensions of various kinds of aggregates (see Schmidt, 1989). A typical scattering arrangement is show in Figure 2.9. In such techniques the scattered intensity, $I(q)$, is measured as a function of the scattering vector, $\boldsymbol{q}$. This vector is varied by changing the collection angle, θ. The value of the scattering vector $\boldsymbol{q}$ is determined by:

$$\boldsymbol{q} = 4\pi\lambda^{-1} \sin\left(\frac{\theta}{2}\right) \tag{2.25}$$

where λ is the wavelength of the incident radiation. To determine a fractal dimension, one typically tries to make measurements that cover several orders of magnitude in $\boldsymbol{q}$. Power law behavior over a wide range of $\boldsymbol{q}$ provides confidence that scattering from a fractal is occurring. Often, to achieve this range of $\boldsymbol{q}$ values, X-ray and neutron scattering are supplemented with light scattering. In the Rayleigh–

Debye–Gans approximation, identical equations are obtained for all these scattering techniques. This regime occurs at angles such that $0.1 \leq \boldsymbol{q}l$, where l is a characteristic size of the scatterer. For X-ray scattering, this conditions is usually satisfied at very small scattering angles, $\theta < 7°$. Under these conditions sin $(\theta/2)$ in Eq. 2.25 is accurately approximated by $\theta/2$. For light scattering, much wider angles must be observed and this approximation does not hold. Typically, the largest values of $\boldsymbol{q}$ from light scattering will be 3–10 times smaller than the smallest values of $\boldsymbol{q}$ determined from X-ray or neutron scattering. The two techniques are complementary and are often both required to span a large scattering regime.

The measured intensity of the scattered radiation for a randomly oriented aggregate of N identical spherical scatterers is given by (see Schmidt, 1989; Vicsek, 1992):

$$I(\boldsymbol{q}) = NI_0(\boldsymbol{q}a)S(\boldsymbol{q}) \tag{2.26}$$

where a is the diameter of the spheres, $I_0(\boldsymbol{q}a)$ is the scattering intensity that reflects the interior structure of the sphere, and $S(\boldsymbol{q})$ reflects the structure of the aggregate. When considering scattering from a mass fractal, the variation of scattering with $\boldsymbol{q}$ is contained in $S(\boldsymbol{q})$ and $I_0(\boldsymbol{q}a)$ is essentially constant. However, the case of a surface fractal is just the opposite. A surface fractal such as a spherical object with a roughened surface will have a constant density in the interior and a relatively short fractal region on the surface. In this instance, $S(\boldsymbol{q})$ will be constant and I_0 $(\boldsymbol{q}a)$ will vary. In the present treatment, our focus will be on mass fractals.

The scattering intensity of an aggregated structure will be dominated by $S(\boldsymbol{q})$, and is given by:

$$S(q) = 1 + 4\pi \frac{N}{V} \int_a^l r^2\, g(r) \frac{\sin \boldsymbol{q}\mathbf{r}}{\boldsymbol{q}\mathbf{r}}\, dr \tag{2.27}$$

where l is the largest distance between a pair of scatterers within the aggregate and $g(r)$ is the pair correlation function. For a mass fractal the pair correlation function, $g(r)$, is given by a function of the form:

$$g(r) = r^{D-3} f\left(\frac{r}{l}\right) \tag{2.28}$$

where $f(r/l)$ is a cutoff function. It falls rapidly to zero when $r >> l$ and is constant for $r << l$. Inserting Eq. 2.28 into Eq. 2.27 and changing variables such that $z = qr$, one obtains:

$$S(\boldsymbol{q}) \sim \boldsymbol{q}^{-D} \int_0^\infty z^{D-2} f(z/\boldsymbol{q}\mathrm{l}) \sin z\, dz \tag{2.29}$$

The function $f(r/l)$ is such a slowly varying function that the resulting integral is virtually independent of $\boldsymbol{q}$ and, therefore, is merely a constant. From Eq. 2.29, it is seen that the scattering intensity will scale as:

$$I(\boldsymbol{q}) \sim \boldsymbol{q}^{-D} \tag{2.30}$$

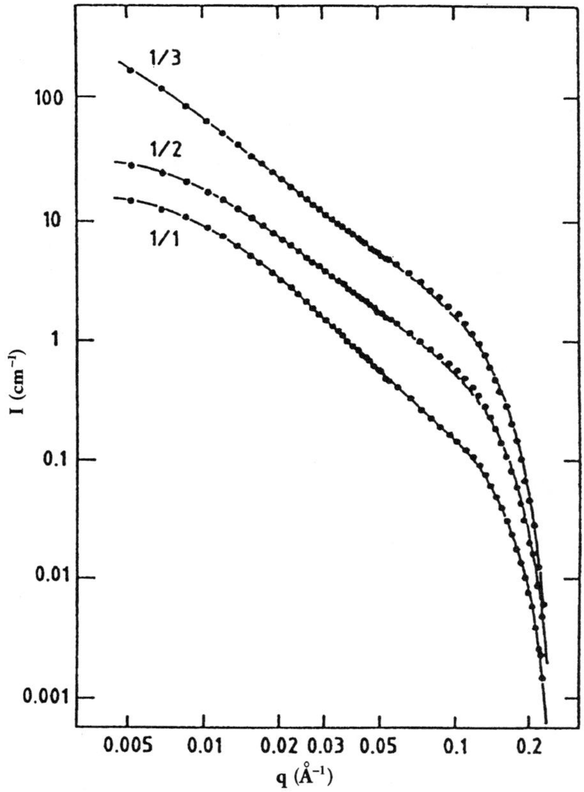

FIGURE 2.10 Scattering cross-section, $I(k)$, for bovine serum albumin at three different detergent concentrations versus q in a log–log plot. Solid lines are fits to a model that assumes a specific cutoff function. (From Chen and Teixeira, 1986, with permission.)

This scaling law has frequently been used to determine fractal dimensions from the log–log plot of I versus $\boldsymbol{q}$. It will be applicable in the regime $1/l \ll \boldsymbol{q} \ll 1/a$. For aggregates where the range from $1/l$ to $1/a$ does not cover several orders of magnitude, serious errors can occur in determining D in this manner. A more complete analysis that uses a specific function for $f(r/l)$ is useful in this situation. However, even in cases where a cutoff function is known, there can be statistical constraints on the accurately of D. Intrinsic limits on this accuracy can be computed. To determine the fractal dimension to within 5% error, the minimum number of data points, N_{min}, must satisfy the inequality $N_{min} > 42^M$, where M is the greatest integer less then D (Smith, 1988).

2.4.2 Low-Angle Scattering from Proteins

The low-angle scattering from native proteins in solution is of small amplitudes. However, when a detergent such as lithium dodecyl sulfate (LDS) is added, the amplitudes are greatly enhanced and the $\boldsymbol{q}$ dependence is easier to measure. Figure 2.10 shows a log–log plot of $I(\boldsymbol{q})$ versus $\boldsymbol{q}$ for the protein bovine serum albumin in the presence of three different concentrations of LDS (Chen and Teixera, 1986). Increased detergent concentration will progressively denature the protein and causes it to change from a compact structure to a random coil. The data in Figure 2.10 can be fitted accurately using a model in which detergent micelles cluster on the protein as a string of pearls. The analysis of this data is particularly interesting because it uses both the small and large $\boldsymbol{q}$ regimes and determines structural parameters from $S(\boldsymbol{q})$ and $I_0(\boldsymbol{q}a)$, respectively. The large $\boldsymbol{q}$ regime was treated as scattering from spherical particles. It provides evidence for the micellar structure in the string of pearls. From this regime a micellar radius of 18 Å was determined. This radius corresponds to the pearl size in our analogy.

The small $\boldsymbol{q}$ regime can be analyzed using a micelle–micelle pair correlation function, as in Eq. 2.28. The cutoff function, $f(r/l)$, was assumed to be an exponential. With this functional form, Eq. 2.29 can be integrated analytically and the resulting expression could be fitted accurately to the data (see solid lines in Figure 2.10). This fit provided a correlation length and the fractal dimension. It is important to note that, when analyzed in this manner, the values obtained for the fractal dimension were approximately 20% higher than those obtained using Eq. 2.30 and taking the slope of the linear region in Figure 2.10. This shows the system to be in a regime where the cutoff function is important.

As the detergent concentration is increased, the protein is progressively denatured. The observed fractal dimension decreases, presumably as a result of a globule–coil transition. At the lowest level of detergent used, partially denatured protein has a D value of 2.3. This puts a lower limit on the fractal dimension for the native protein. The nondenatured protein would presumably have a fractal dimension close to 3, consistent with the polymer collapse model. At the highest concentration of detergent used, the fractal dimension was determined to be 1.76. This is approaching the value of 5/3 expected for an excluded volume polymer and suggests that the protein has completely unraveled. While the fractal dimension of a protein in the presence of detergent is a rather specialized set of conditions, it is of some interest because of increased work on the nature of the denatured state. Also, gel electrophoresis of proteins is commonly performed in the presence of detergents. The structure of these protein–detergent aggregates is required for a deeper understanding of electrophoretic mobilities.

Scattering techniques can also be used to study protein–protein aggregation. Such aggregates occur as a result of heating, cooling, or freeze-drying. Often these phenomena have significant practical implications for the pharmaceutical industry. To date, there have been few structural studies on protein aggregation. Low-angle X-ray scattering has been used to investigate the aggregation of human transferrin upon freeze-drying (Congiu Castellan et al., 1993). This scattering showed only a limited scaling regime, and very low fractal dimensions were obtained from log–log plots. These dimensions ranged from 1.13 to 1.38. No interpretation was given for these results. In a separate kinetic scattering experiment, the heat-induced aggregation of the antibody immunoglobulin G (IgG) was monitored by changes in quasi-elastic light scattering with time (Feder et al., 1984; Jøssang et al., 1984;

Feder, 1988). From these data, the radius of gyration of the antibody aggregates was determined as a function of time. Using the Smoluchowski aggregation model, it was shown that this radius would increase as $R_g \sim (1 + \gamma t)^{1/D}$, where γ is a constant and D is the fractal dimension of the aggregate. These results gave the a fractal dimension of 2.56 for the protein aggregates. It is noteworthy that computer simulations of diffusion-limited aggregates in two-dimensions give a fractal dimension of 1.7, and in three-dimensions a fractal dimension of 2.5 (see Feder, 1988) is observed.

2.5 Scaling Laws for Membrane Proteins

The mechanisms by which membrane proteins transmit signals across the lipid bilayer remain a crucial problem in physical biochemistry. This phenomenon is key to a diversity of physiological functions. Examples of such "signaling" are the activation of hormone receptors, the coordination of vectorial transport of metabolites and the sensing of electrical potentials by ion channels. In each case, information is being transferred over the rather long distance of the membrane (~ 50 Å). For the above examples, this "information" is hormone concentration, chemiosmotic potential, and membrane potential, respectively. How a protein achieves this transmembrane communication is of vital importance. While each set of phenomena may have its specific mechanistic traits, it is probable that general features of transmembrane signaling will emerge as more detailed physical data become available. For example, one common structural motif is the utilization of seven membrane-spanning protein domains (see Dohlman et al., 1991). At this stage our knowledge of the microscopic details of membrane proteins remains so sketchy that most mechanistic models must be viewed as highly speculative (for a review of ion transport models, see Tanford (1983)).

In this section, a heuristic presentation is given of scaling laws associated with various physical parameters of membrane bound proteins. This treatment is similar in spirit to deGennes' analysis of adsorbed polymers (deGennes, 1979). One could readily adopt the more rigorous approach of Kosmas and Freed (1978), as the membrane problem has strong analogies to their treatment of confined polymers. Figure 2.11 shows a schematic representation of a membrane protein confined to a region of width δ and embedded in a membrane of thickness L. The two spatial coordinates in the plane of the membrane are unconfined.

First, consider the length of the protein parallel to the membrane, $R_{\parallel}$. This will be given by:

$$R_{\parallel} = R_f \, \Phi\left(\frac{R_f}{\Delta}\right) \tag{2.31}$$

where R_f is the root mean squared radius of the free or unconfined protein. For conditions of polymer collapse, $R_f \sim aN^{1/d}$, where d is the topological dimension. Following deGennes, Φ is a function of the dimensionless parameter R_f/Δ, and in the limit of strong confinement is assumed to scale as $\Phi(R_f/\Delta) \sim (R_f/\Delta)^m$, with m being

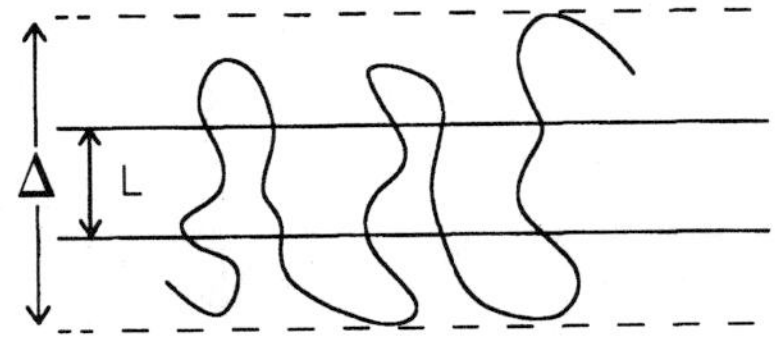

FIGURE 2.11 Schematic representation of a membrane-bound protein. The membrane thickness is L. As a result of the membrane association, the protein is confined to a slit of width Δ.

an exponent to be determined. As Δ becomes small, $R_{\|}$ will scale as a two-dimensional collapsed polymer, $R_{\|} \sim N^{1/2}$ and, therefore, using Eq. 2.31, $(m + 1)/3 = 1/2$. This gives the scaling law:

$$R_{\|} \sim a^{3/2} \left(\frac{N}{\Delta} \right)^{1/2} \tag{2.32}$$

Thus, the length along the membrane scales with N as a two-dimensional polymer.

$R_{\|}$ can be related to the membrane thickness, L, by a simple energy minimization procedure. The excess free energy of confinement for the polymer, ΔG, is given by the sum of two terms: the unfavorable entropy of confining the polymer in a slit of width Δ, and the favorable energy of interaction of the polymer with the membrane. The forms of these interaction terms are readily derived from scaling arguments. It is assumed that the unfavorable entropy of confinement will be a term linear in N. Again, consider the reduced variable defined as (R_f/Δ). In the limit of large Δ, there is no confinement and the entropy should decrease to zero. In the limit of small Δ, the confinement entropy will become infinite. Therefore, a scaling relationship for the entropy can be assumed, and it takes the form $\Delta S = -(R_f/\Delta)^P$. Because $\Delta S \sim N$, and for a collapsed polymer $R_f \sim N^{1/3}$, one finds that $p = 1/3$. The first contribution to the entropy will then be: $\Delta S = -(R_f/\Delta)^3 = N(a/\Delta)^3$.

The second term is the favorable energy of interaction between the polymeric units and the membrane. It is merely equal to the number of polymeric units in the membrane, N_M, times the average free energy of interaction per RT of a unit with the membrane, γ. The total excess free energy for a protein associating with a membrane will be given by:

$$\frac{\Delta G}{RT} = N \left(\frac{a}{\Delta} \right)^3 - N_M \gamma \tag{2.33}$$

where N_M is given by the fraction of units in the membrane times N. N_M is given simply by $N(L/\Delta)$. The distance Δ is unconstrained and is adjusted to minimize the free energy. Upon minimization of ΔG with respect to Δ, a relationship between Δ and L is obtained:

$$\Delta = a^{3/2} \left(\frac{3}{\gamma L} \right)^{1/2} \tag{2.34}$$

This allows the variable Δ to be eliminated from most expressions. This is important because Δ is not readily accessible experimentally, whereas L is known for a large number of systems. Because L is determined primarily by the lipid composition of the membrane, it is for the most part a constrained parameter. Using Eq. 2.34, one obtains:

$$\frac{\Delta G}{RT} \sim N\left(\frac{\gamma L}{3a}\right)^{3/2} \tag{2.35}$$

$$R_{\|} \sim N^{1/2}\left(\frac{\gamma L}{3}\right)^{1/4} a^{3/4} \tag{2.36}$$

The scaling behavior of the pair correlation function, $g(r)$, can also be demonstrated. This function scales as (deGennes, 1979):

$$g(r) \sim N\left(\frac{n}{r^3}\right) \tag{2.37}$$

where n is the number of residues within a radius r, with $r << R_f$. For an unconfined, collapsed polymer, $n \sim (r/a)^3$ and $g(r) \sim (N/a^3)$. Considering the membrane protein:

$$g(r) \sim N\left(\frac{n}{r_{\|}2r_{\perp}}\right) \tag{2.38}$$

where $r_{\perp}$ is the direction perpendicular to the membrane surface. Using Eq. 2.32, one obtains:

$$g(r) = \frac{N\Delta}{ar_{\perp}^3} = (N/a^2)\left(\frac{\Delta}{a}\right)r_{\perp}^{-1} \tag{2.39}$$

For a two-dimensional system $g_2 = N/a^2$, and in the limit of small Δ, $n \sim \Delta/a$. Therefore, for the membrane protein:

$$g(r) = g_2\,(r_{\|})\,g_1\,(r_{\perp}) \tag{2.40}$$

where g_1 represents a one-dimensional correlation function perpendicular to the membrane. In the plane of the membrane, the protein is seen to have a pair correlation function like a two-dimensional collapsed polymer.

The scaling results derived in this section provide a number of general relationships for membrane bound proteins. These results suggest, not too surprisingly, that in the plane of the membrane, the protein behaves as a two-dimensional collapsed polymer. Equation 2.34 is of some interest as it relates membrane properties, g and L, to the overall width of the molecule, δ. This simple connection may be of importance for membrane signaling. Experimentally, the membrane thickness is not easily varied. However, both lipid composition and protein exposure can change. This could result in changes in g, the free energy of interaction between a residue and the membrane. For instance, membrane proteins such as the adrenergic family of receptors, have hormone binding sites within the membranous region rather than in the aqueous region. Hormone–receptor binding energies can be quite large, easily of the order of 10 kcal/mol. Binding of hormone could produce conformational changes that expose different portions of the protein and, thereby, alter γ. This, in turn, will change the width Δ in accord with Eq. 2.34 to minimize the confinement energy. A change in Δ, especially an increase, could make the receptor more accessible for secondary recognition events. This interrelationship shows how properties in one region of the protein can influence those in others and, thereby, allow for long-range "communication." Both the width and interaction potential of the membrane limit the amount of protein exposed to the aqueous environment. Using Eq. 2.34, it is also seen that the membrane interaction parameter, γ, is likely to be small because $\gamma = 3a^3/\Delta^2 L$. The parameter a will be approximately a bond length and will be several angstroms. The values of Δ and L will be approximately the same as the thickness of membrane (~ 50–100 Å). This indicates that proteins can be confined to a membrane by fairly weak interactions. This is a consequence of the weaker confinement entropy for a collapsed polymer. For an excluded volume polymer the confinement entropy is proportional to $(a/\Delta)^{5/3}$, while for a collapsed polymer it is $(a/\Delta)^3$. As (a/Δ) is small, the difference in exponents has a dramatic effect. Consistent with this is the observation that the membrane has very little impact on the compactness of the protein (Eq. 2.40). These effects all support the view that membrane-bound proteins can exist without major structural alterations and will maintain similar secondary structural properties as soluble proteins.

2.6 Summary

In this chapter, scaling laws of polymer physics have been presented for three different polymeric states: excluded volume, ideal, and collapsed polymers. These laws have been related to the mass fractal dimension. Structural data for proteins have been examined, and it has been seen that proteins have fractal dimensions around 3 and are collapsed polymers. However, unlike classical collapsed polymers, proteins show identical local and global scaling. Implications for statistical mechanical models of proteins have been discussed. Protein surfaces also show similar fractal properties, and these have fractal dimensions of approximately 2.2. The use of scattering techniques to determine fractal dimensions has been discussed with particular reference to protein–detergent aggregates. Finally, scaling laws for membrane proteins have been derived by treating them as confined polymers.

References

Abad-Zapatero, C., and Lin, C. T. 1990. Statistical Descriptors for the Size and Shape of Globular Proteins. *Biopolymers* **29**, 1745–1754.

Chan, H. S., and Dill, K. A. 1989. Compact Polymers. *Macromolecules* **22**, 4559–4573.

Chan, H. S., and Dill, K. A. 1991. Polymer Principles in Protein Structure and Stability. *Ann. Rev. Biophys. Biophys. Chem.* **20**, 447–490.

Chan, H. S., and Dill, K. A. 1991. Polymer Principles in Protein Structure and Stability. *Ann. Rev. Biophys. Biophys. Chem.* **20**, 447–490.

Chen, S.-H., and Teixeira, J. 1986. Structure and Fractal Dimension of Protein–Detergent Complexes. *Phys. Rev. Lett.* **57**, 2583–2586.

Chothia, C. 1975. Structural Invariants in Protein Folding. *Nature* **254**, 304–308.

Chothia, C. 1984. Principles that Determine the Structure of Proteins. *Ann. Rev. Biochem.* **53**, 537–572.

Clementi, E., Ranghino, G., and Scardamaglia, R. 1977. Intermolecular Potentials: Interaction of Water with Lysozyme. *Chem. Phys. Lett.* **49**, 218–224.

Congiu Castellan, A., Barteri, M., Bianconi, A., Borghi, E., Cassiano, L., Castagnola, M., Della Longa, S., and LaMonaca, A. 1993. X-ray Small Angle Scattering of the Human Transferrin Protein Aggregates. *Biophys. J.* **64**, 520–524.

deGennes, P. G. 1975. Collapse of a Polymer Chain in Poor Solvents. *J. Phys. (Paris) Lett.* **36**, L55–L57.

deGennes, P. G. 1978. Collapse of a Flexible Polymer Chain II. *J. Phys. (Paris) Lett.* **39**, L-299–L-301.

deGennes, P. G. 1979. *Scaling Concepts in Polymer Physics*. Cornell University Press, Ithaca, NY.

Dewey, T. G. 1993a. Protein Structure and Polymer Collapse. *J. Chem. Phys.* **98**, 2250–2257.

Dewey, T. G. 1993b. Fractal Aspects of Protein Structure and Dynamics. *Fractals* **1**, 179–189.

Dewey, T. G. 1995. Fractal Dimensions of Proteins: What Are We Learning? *Hetergen. Chem. Rev.* **2**, 91–101.

Dill, K. A. 1985. Theory for the Folding and Stability of Globular Proteins. *Biochemistry* **24**, 1501–1509.

DiMarzio, E. A., and Gibbs, J. A. 1958. Chain Stiffness and the Lattice Theory of Polymer Phases. *J. Chem. Phys.* **28**, 807–818.

Dohlman, H. G., Thorner, J., Caron, M. G., and Lefkowitz, R. J. 1991. Model Systems for the Study of Seven-Transmembrane-Segment Receptors. *Ann. Rev. Biochem.* **60**, 653–688.

Falconer, K. 1990. *Fractal Geometry.* Wiley, Chichester, pp. 101–110.

Farin, D., Peleg, S., Yavin, D., and Avnir, D. 1985. Applications and Limitations of Boundary-Line Fractal Analysis of Irregular Surfaces: Proteins, Aggregates, and Porous Materials. *Langmuir* **1**, 399–407.

Feder, J. 1988. *Fractals.* Plenum Press, New York, pp. 36–37.

Feder, J., Jøssang, T., and Rosenqvist, E. 1984. Scaling Behavior and Cluster Fractal Dimension Determined by Light Scattering from Aggregating Proteins. *Phys. Rev. Lett.* **53**, 1403–1406.

Fedorov, B. A., Fedorov, B. B., and Schmidt, P. W. 1993. An Analysis of the Fractal Properties of the Surface of Globular Proteins. *J. Chem. Phys.* **99**, 4076–4083.

Flory, P. 1953. *Principles of Polymer Chemistry.* Cornell University Press, Ithaca, NY.

Flory, P. 1982. Treatment of Disordered and Ordered Systems of Polymer Chains by Lattice Methods. *Proc. Natl. Acad. Sci. USA* **79**, 4510–4514.

Freed, K. F. 1987. *Renormalization Group Theory of Macromolecules*. Wiley, New York.

Gibbs J. H., and DiMarzio, E. A. 1958. Nature of the Glass Transition and the Glassy State. *J. Chem. Phys.* **28**, 373–383.

Gordon, M., Kapadia, P., and Malakis, A. 1976. The Graph-Like State of Matter VII. The Glass Transition of Polymers and Hamiltonian Walks. *J. Phys. A* **9**, 751–769.

Grosberg, A. Y., and Khokhlov, A. R. 1989. *Statistical Physics of Macromolecules*. Nauka, Moscow.

Grosberg, A., Rabin, Y., Havlin, S., and Neer, A. 1993. Crumpled Globule Model of the Three-Dimensional Structure of DNA. *Europhys. Lett.* **23**, 373–378.

Isogai, Y., and Itoh, T. 1984. Fractal Analysis of Tertiary Structure of Protein Molecule. *J. Phys. Soc. Jpn.* **53**, 2162–2171.

Janin, J. 1976. Surface Area of Globular Proteins. *J. Mol. Biol.* **106**, 13–14.

Jøssang, T., Feder, J., and Rosenqvist, E. 1984. Heat Aggregation Kinetics of Human IgG. *J. Chem. Phys.* **120**, 1–30.

Kosmas, M. K., and Freed, K. F. 1978. On Scaling Theories of Polymer Solutions. *J. Chem. Phys.* **69**, 3647–3659.

Lewis, M., and Rees, D. C. 1985. Fractal Surfaces of Proteins. *Science* **230**, 1163–1165.

Malakis, A. 1976. Hamiltonian Walks and Polymer Configurations. *Physica* **84A**, 256–284.

Pechold, W. R., and Grossmann, H. P. 1979. Meander Model of Condensed Polymers. *Faraday Disc.* **68**, 58–77.

Pfeifer, P., Welz, U., and Wippermann, H. 1985. Fractal Surface Dimension of Proteins: Lysozyme. *Chem. Phys. Lett.* **113**, 535–540.

Pietronero, L. 1988. Fractals in Physics: Introductory Concepts. In *Order and Chaos in Nonlinear Physical Systems*, ed. S. Lundqvist, N. H. March, and M. P. Tosi. Plenum Press, New York, pp. 277–295.

Richards, F. M. 1977. Areas, Volumes, Packing, and Protein Structure. *Ann. Rev. Biophys. Bioeng.* **6**, 151–176.

Schmidt, P. W. 1989. Use of Scattering to Determine the Fractal Dimension. In *The Fractal Approach to the Chemistry of Disordered Systems*, ed. D. Avnir. Wiley, Chichester, pp. 67–79.

Smith, L. A. 1988. Intrinsic Limits on Dimension Calculations. *Phys. Lett. A* **133**, 283–288.

Tanford, C. 1983. Mechanism of Free Energy Coupling in Active Transport. *Ann. Rev. Biochem.* **52**, 379–410.

Teller, D. C. 1976. Accessible Area, Packing Volumes and Interaction Surfaces of Globular Proteins. *Nature* **260**, 729–731.

Vicsek, T. 1992. *Fractal Growth Phenomena*, 2nd edn. World Scientific, Singapore, pp. 74–81.

Wang, C. X., Shi, Y. Y., and Huang, F. H. 1990. Fractal Study of Tertiary Structure of Proteins. *Phys. Rev. A* **41**, 7043–7048.

3

Loops, Polymer Statistics, and Helix–Coil Transitions

As seen in the previous chapter, proteins follow scaling laws for collapsed polymers. These extremely compact structures have a mass fractal dimension of 3. The scaling laws of polymer physics were derived for extremely high molecular weight homopolymers that show no branching. Perhaps it is fortuitous that proteins have a scaling law similar to collapsed polymers? The diverse nature of amino acid side-chains allows for a number of different types of interaction between subunits. Consequently, most proteins show loop structures. Covalently attached loops can be formed as a result of disulfide bond formation between cysteine residues. Often, the protein will fold back on itself and form hydrogen bonds. The presence of such loop structures presents another question. Perhaps proteins are really extended polymers forming loops? Maybe the loop structures give an anomalously high fractal dimension? The answer to these questions can be found in cell renormalization studies of looped polymers (Family, 1982, 1984). This simple and elegant renormalization technique will be explored later in this chapter.

Initially, the scaling properties of loops are considered, along with their implications for biopolymer structure and function. We begin by introducing elementary concepts of polymer chain statistics with a comparison of linear and loop-forming chains. Because linear chains and ring polymers are in the same universality class, many of the scaling laws are identical. These approaches can also be used to investigate the "trail problem," the statistics of a linear polymer that doubles back on itself. Small cell renormalization group or real-space renormalization group (RSRG) theories can be used to explore this universality. Of all the renormalization techniques, RSRG is probably the easiest to visualize because the diagrams used have a direct physical correspondence. In these applications, one constructs all possible walks on a given lattice. The lattice is then decimated by removing lattice sites. Often, every other point is removed. The probability of finding a walk on the original lattice is then related to that of finding one on the decimated lattice. This gives rise to the renormalization equations and allows the determination of a fixed point. The decimation procedure allows one to determine how the system scales with size and yields the critical exponent. Surprisingly, small lattices can be used to obtain these exponents, and two specific examples will be treated in the present chapter.

After establishing the corresponding behavior of linear and loop-forming polymers, the question of the entropy of loop formation is treated. The probability of ring closure for a number of different topologies is presented. After discussing these theoretical results, experimental data on loop formation in proteins and nucleic acids are presented. Using these results, limitations on the size of enzymes and receptors are estimated. The binding sites of such proteins represent a trade-off between the number of favorable contacts and the unfavorable entropy of loop formation. In this section, we follow the arguments of deGennes (1990), suitably updated to account for the polymer collapse nature of proteins. Finally, the implication of loop thermodynamics on helix–coil transitions in nucleic acids is examined. After a short excursion into the statistical mechanics of helix–coil transitions, the crucial role of loop scaling laws in such order–disorder transitions is demonstrated.

3.1 Random Walks and Loops

Random walk problems have been studied extensively and have seen application to a wide range of phenomena. Polymers are often modeled as random walks, with the trajectory of the walk representing the polymer chain. In the present section, general features of random walks are discussed and special reference is made to the problem of loop formation. A central goal for most walk problems is to calculate the probability of a walk ending at a certain location after a given number of steps. For such problems, the probability distribution function is usually very difficult to obtain. For ideal and excluded volume polymers, the form of this probability function, $p_N(r)$, for a polymer of N units is given by (McKenzie, 1976):

$$p_N(r) = Ar^{\theta} \exp(-Br^{\delta}) \tag{3.1}$$

where $p_N(r)$ is the probability of finding the Nth unit at a location r from the first unit. A and B are constants, and the critical exponents, θ and δ, can be expressed in terms of the two universal critical exponents, ν and γ:

$$\vartheta = \frac{\gamma - 1}{\nu} \tag{3.2a}$$

$$\delta = (1 - \nu)^{-1} \tag{3.2b}$$

For an ideal polymer, $\gamma = 1$ and $\nu = 1/2$.

Interestingly, the form of the probability function in Eq. 3.1 is the same for ring-forming polymers when r is large ($r/N^{\nu} >> 1$) (Baumgartner, 1982; Prentis, 1982). The relationship in Eq. 3.2b also holds for rings. However, Eq. 3.2a is not universally valid, and values of θ will depend on specific loop topology (Redner, 1980). In the limit of short r ($r/N^{\nu} << 1$), the distribution function for loop-forming polymers has the form:

$$P_N(r) = AN^{-d\nu}\left(\frac{r}{N^{\nu}}\right)^{\theta} \sim N^{-\nu(d+\theta)} \tag{3.3}$$

where d is the Euclidean dimension in which the system is embedded. Usually, three different loops are considered: end-to-end closure (or a ring), closure between an end and an interior point (tadpole), and closure between two interior points.

In addition to Eq. 3.3, the scaling of the radius of gyration, R, with N is also useful in characterizing a walk. This scaling law takes the form:

$$R \sim N^{\nu} \tag{3.4}$$

where ν is a characteristic exponent. The scaling exponent ν is the same for linear and ring polymers. This is because these two systems are from the same universality class. As discussed in Chapter 1, for a mass fractal Eq. 3.4 can provide the fractal dimension, D, of the system, i.e., $\nu = 1/D$. The ratio of the squared radius of gyration for a self-avoiding ring and a linear polymer is a constant, given by (Prentis, 1982):

$$\frac{R^2_{\text{ring}}}{R^2_{\text{linear}}} = \frac{1}{2}\left[1 + \frac{13}{96}\,\varepsilon + O(\varepsilon^2)\right] \tag{3.5}$$

where $\varepsilon = 4 - d$. This ratio defines a universal constant for this system.

Another scaling law of particular importance relates the total number of possible walks (or configurations), Γ_N, to the size, N, of the system. This is given by:

$$\Gamma_N = \mu^N N^{\gamma - 1} \tag{3.6}$$

where μ is the average connectivity constant. For many problems, Γ_N is a rapidly increasing function of N. For percolation problems, it shows divergent behavior and allows for the study of phase transitions in surprisingly simple systems. Such percolation problems are discussed with reference to microdomain formation in biomembranes in Chapter 9.

In the loop problem, a law analogous to Eq. 3.6 also exists. Typically, a "reduction factor" or cyclization probability $R(N;i,j)$ is defined (see Chan and Dill, 1989, 1990). This is given by:

$$\mathrm{R(N;i,j)} = \frac{\Gamma_N\,(\mathrm{i,j})}{\Gamma_N} \sim |\mathrm{i}-\mathrm{j}|^{-\nu(\mathrm{d}+\theta)} \tag{3.7}$$

where i and j are the positions along the chain of the units that make contact. The number of units in the loop, n, is given by $n - 1 = i - j$. Equation 3.7 is useful in considering thermodynamic data on loop closure, since the logarithm of the reduction factor gives the entropy of loop formation as a function of the number of units in the loop. Thus, it is important to determine ν and θ for various loop topologies in order to understand the thermodynamics of loop formation in biopolymers.

3.2 Real Space Renormalization Groups

In many problems Eq. 3.1 is taken as an empirical law and the appropriate scaling behavior (Eqs 3.4 and 3.6) is determined by a variety of methods. These methods include computer simulations and renormalization group approaches (for some examples see Vicsek (1992). We now consider the determination of ν (and the fractal dimension) using a renormalization approach, which is known as the small cell RSRG (see Family, 1984; Creswick et al., 1992). This provides a simple and very visual application of renormalization concepts. These real-space renormalization techniques can also be applied to percolation problems, and are discussed again in Chapter 9.

A generating function, $G(K)$, is used to describe walks of all possible number of steps, N. This is given by:

$$G(K) = \sum_{N=1}^{\infty} \sum_{i} P_{N,i} K^N \tag{3.8}$$

where $P_{N,i}$ is the probability of creating the ith configuration for a walk of N steps, and K is the "fugacity" associated with a given step in the walk. The generating function is an analogue of the grand canonical partition function of statistical mechanics. It is mathematically similar to the "sequence generating function" that is used in our later discussion of the statistical mechanics of biopolymers. For the grand canonical partition function, K has a physical meaning related to the chemical potential of the system. In the present application, it is simpler to think of K as a mathematical device. It is an expansion variable the coefficients of which give the probability of an N-step walk. The probability of the ith configuration is given by the product of the probabilities of the individual, j steps, $P_{N,i}$ (J):

$$P_{N,i} = \prod_{j=1}^{N} P_{N,i}(J) \tag{3.9}$$

In the renormalization approach, the generating function for a decimated lattice is equated to that of the normal lattice. This is a result of the scale invariance of the problem. From this renormalization equation, the exponent ν is calculated. To make the connection with the critical exponent, the radius of gyration of the walk, R, is calculated using the generating function. This is given by:

$$R = \frac{\sum_N N^{\nu} K^N}{\sum_N K^N} \propto (K_C - K)^{-\nu} \tag{3.10}$$

where K_C is the radius of convergence of the sum. The proportionality on the right-hand side of Eq. 3.10 is made by using a common approximate relationship for gamma functions, $N^{\upsilon} = \lim_{N\to\infty} \Gamma(N+\nu)/\Gamma(N)$. This, in conjunction with an application of the binomial theorem, gives Eq. 3.10.

The renormalization group equations are derived by relating the fugacity of the decimated lattice, K', to that of the normal lattice. This is done in accordance with the weighting of individual steps in the walk, giving:

$$K' = \sum_{n} \sum_{i'} P_{n,i'} K^{n} \tag{3.11}$$

The double summation in Eq. 3.11 is best visualized by using the graphs for a specific problem. The double sum represents all possible walks that span the lattice. Note that there are walks with different numbers of steps. All of these are accounted for in the first summation. There will also be walks that have the same number of steps, but different probabilities associated with them. These are accounted for in the inner summation. The probabilities assigned to steps in the walk are dictated by the specific model of interest. For irreversible growth models, probabilities are assigned on the basis of equal likelihood of all sites. Thus, the number of neighbors at the growing end of the walk will determine the probability assigned to the next step. For problems that treat equilibrium properties, each step is assigned an equal probability, regardless of the number of available sites.

3.2.1 The Self-Avoiding Walk

The method is illustrated with a simple example of a self-avoiding walk (SAW), an example that serves as a good model for an excluded volume polymer. In this problem, we seek all walks that span a "lattice cell" in a self-avoiding fashion. In Figure 3.1 a 2 × 2 cell on a square lattice is considered. The walks that start in the lower-left corner and cross the cell vertically are depicted. In the decimated or renormalized 1 × 1 cell, all these walks will be equivalent to the single graph with one vertical step (see Figure 3.1). To model an excluded volume polymer, we deal with equilibrium configurations or walks, so each step in the walk is given a weight, K. The probability of a given graph is equal to K^n, where n is the number of steps in the graph. For the problem illustrated in Figure 3.1, Eq. 3.11 becomes:

$$K' = K^2 + 2K^3 + K^4 \tag{3.12}$$

The critical point, K_C, is determined by the condition that $K' = K$. This is determined numerically, and in this instance $K_C = 0.4656$.

To calculate the scaling exponent, the fugacity is expanded in a Taylor series about the critical point. This gives:

$$K' - K_c = \lambda\ (K - K_c) \tag{3.13}$$

where $\lambda = (\partial K'/\partial K)_{Kc}$. The scaling law (Eq. 3.10) for the decimated lattice now becomes:

$$b(K_c - K')^{-\nu} \approx (K_c - K)^{-\nu} \tag{3.14}$$

where b is the change in lattice spacing as a result of decimation. In this instance $b = 2$, since two spaces are converted into one. Equation 3.14 is a consequence of

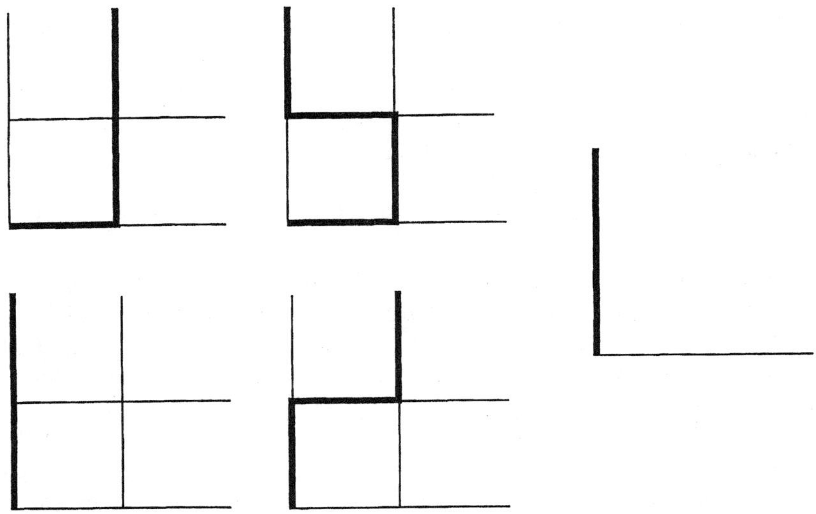

FIGURE 3.1 Graphs of all random walks that span a 2 × 2 lattice. Renormalization equates these graphs to the single graph (far right) that spans a 1 × 1 lattice.

the assumed scale invariance of the system. Using Eqs 3.13 and 3.14, the critical exponent is now calculated using:

$$\nu = \frac{1}{D} = \frac{\ln b}{\ln \lambda} \tag{3.15}$$

For the two-dimensional SAW problem, a value of ν of 0.7152 is obtained. The value predicted from Flory's theory is 3/4 (see Chapter 2). These renormalization results can be improved by considering larger cells. In such cases, configurations must be enumerated using a computer. Family (1984) treated the problem up to $b = 6$, and using an extrapolation procedure determined a ν of 0.746 ± 0.004. The approach is readily extended to three dimensions, but again a computer enumeration is required because of the dramatic increase in labor at the higher dimension. Extremely accurate results are also obtained for the SAW in three dimensions.

The small cell renormalization can provide a facile means of calculating the fractal dimension of complicated random walk problems. However, it is often difficult a priori to assess the level of accuracy of these results. Typically, the accuracy should improve as the cell size is increased. However, this improvement may not always be dramatic and can sometimes give the appearance of converging to a given value. Nevertheless, this method is computationally so simple that it is often useful in determining the qualitative behavior of the system. It has proven particularly powerful when investigating the dependence of a system on a particular parameter. Such problems occur in the investigation of sol–gel transitions (see Creswick et al., 1992) and in gel–fluid transitions in biomembranes (see Chapter 9). In the present chapter, it is seen to provide a convenient approach to problems regarding loop formation.

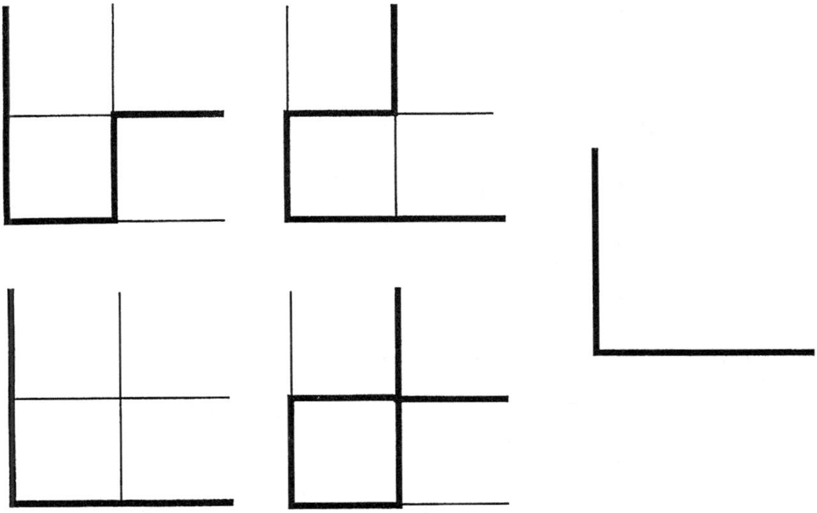

FIGURE 3.2 Graphs for ring-forming walks on a 2 × 2 lattice. Walks originate in the lower left corner and leave the cell at the top. The walk returns to the cell from the right-hand side and joints the starting point. The graph on far right is the renormalized, 1 × 1 lattice.

3.2.2 Ring Closure

The fractal dimension of a SAW that forms a ring is now considered. This problem is illustrated in Figure 3.2 where a 2 × 2 cell is shown. The walk originates in the lower left corner and leaves the cell at the top. It returns to the cell from the right and joins its starting point. There are four such ring-closure walks for a 2 × 2 cell. Note that the cell with fugacity, K^6, has two point contacts. This walk renormalizes to a single contact in the 1 × 1 cell and is, therefore, allowed. In this specific model, a ring is formed when two points are common to the walk. If two bonds (the line between points) are taken as common, very different results are obtained. The renormalization equation associated with Figure 3.2 is (Family, 1982):

$$K'^2 = K^4 + 2K^5 + K^6 \tag{3.16}$$

The fixed point is $K_C = 0.62$ and $\nu = 0.79$ ($D = 1.27$). When a 3 × 3 cell is used a slightly lower ν is obtained. However, renormalizing a 3 × 3 cell to a 2 × 2 cell gives $\nu = 0.75$, in agreement with Flory's result for a linear polymer. Family verified that a ring and a linear polymer are of the same universality class by considering a two-parameter renormalization where different fugacities were associated with spanning and returning configurations. The flow diagram associated with the recursion relations shows that rings and linear SAWs reached the same critical point, indicating that they are in the same universality class. Thus, the ring polymer will have the same fractal dimension ($1/\nu$) as a linear polymer. These results parallel the very different approach of Prentis (1982) and Baumgartner (1982), who essentially established the same behavior using Eq. 3.5.

These results on simple rings are readily extended to more complicated loop topologies. The problem of a SAW that has a single contact in its interior can also

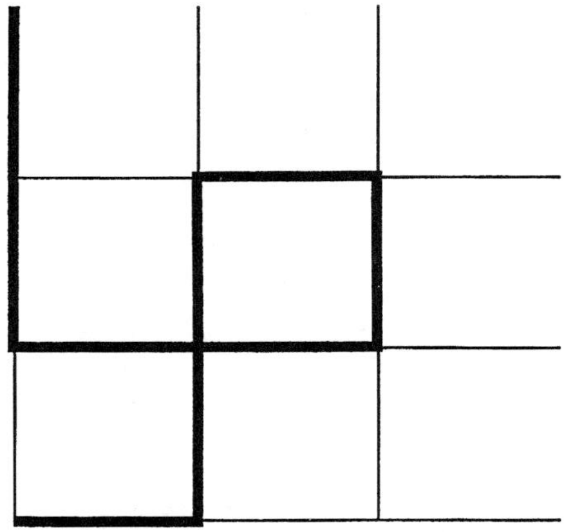

FIGURE 3.3 An example of a loop-forming walk that spans the 3 × 3 lattice. Determining the statistics of this type of walk is sometimes referred to as the "trail" problem.

be treated with real-space renormalization techniques. This is sometimes referred to as the "trail" problem. Figure 3.3 shows an example of a loop-forming SAW that spans a 3 × 3 cell. (There are no loop-forming configurations that span a 2 × 2 cell). The fractal dimension of the SAW when interior loop-forming configurations are allowed can be determined. When all spanning configurations are considered, the renormalization equation for a 3 × 3 cell going to a 1 × 1 cell is (Family, 1982):

$$K' = K^3 + 3K^4 + 9K^5 + 5K^6 + 9K^7 + 4K^8 + 8K^9 + 2K^{10} \tag{3.17}$$

This gives a value of ν of 0.72, which is very close to the Flory result. A two-parameter renormalization can again be used to generate a flow diagram. This shows that polymers with and without loops converge to the same fixed point. Once more, this demonstrates that they are in the same universality class. Interestingly, if the polymer is allowed to fold back on itself so that two bonds overlap, a very different result is obtained. In this instance, it scales as an ideal, or theta, polymer.

These results show that loop formation will not affect the fractal dimension of an excluded volume polymer. Thus, it is extremely unlikely that the high fractal dimension of 2.8 obtained for a large data set of proteins (see Chapter 2) (Dewey, 1993) is a result of loops in such polymers. Independent of this argument, one might not anticipate loop effects on the fractal dimension (or v) of collapsed polymers. In such cases, interactions between units are so favorable that the polymer becomes close packed. Since multiple contacts are numerous within the polymer, the concept of a loop loses some of its meaning in this context.

As a word of caution, in the biochemical literature a loop region in a protein is defined as a disordered region that occurs between regions of well-defined secondary structures. In a recent, interesting study (Thomas, 1990), the end-to-end lengths of such regions were determined from a large set of X-ray data. Although there is significant scatter in the data, it is seen that this subset of protein structures scaled as a theta, or ideal, polymer. Thus, these disordered regions or "loops" give a local structure consistent with the Gaussian statistics of ideal polymers. Nevertheless, when proteins are viewed as global structures and units are not taken in isolation, the structures appear as collapsed polymer.

3.3 The Entropy of Loop Formation

We now turn our attention to the thermodynamics of loop formation in biopolymers. These entropic effects are important for the conformational stability of both proteins and nucleic acids. As mentioned earlier, the entropy of loop formation is obtained from the reduction factor (Eq. 3.7), the ratio of the number of loop configurations to linear configurations. The entropy, S, of a loop of n units has the general form:

$$S(n) = -c \ln n \tag{3.18}$$

where c is a constant. From Eq. 3.7, it is seen that $c = \nu\,(d + \theta)$. Originally, simple theoretical arguments by Flory (Poland and Scheraga, 1966) and by Jacobsen and Stockmayer (see Chan and Dill, 1989) predicted a value for c of $d/2$. This result appears ingrained in the biophysical literature. However, more recent theoretical results show that the value of c is much larger and will change with loop topology. This is a result of changes in θ. Before considering these theoretical results, we examine experimental data on the thermodynamic effects of loop formation in proteins and nucleic acids.

3.3.1 Loop Formation in Proteins

It is not uncommon for proteins to form covalent loops as a result of disulfide bridges between the side-chains of cysteine residues. This has the curious effect of stabilizing the protein structure as a result of destabilizing the denatured state (Johnson et al., 1978; Lin et al., 1984; Ueda et al., 1985; Pace et al., 1988). Protein stability is determined by the difference in free energy between the native, conformational state and the denatured state. The native state is usually a highly specific, folded conformation. This restriction in number of configurations results in a large unfavorable entropic contribution in this state. Often, such native states are confined to a limited number of configurations, even when the disulfide bridge forming a loop is broken. The denatured state, on the other hand, has considerable configurational freedom. The imposition of a loop structure on this random state now has a large unfavorable entropic impact. Thus loop structures destabilize the denatured state much more than the native state. This results in increased protein stability. This effect has been explored experimentally for a number of systems.

Stability studies have been performed on a variety of proteins with disulfide linkages by studying thermal and urea denaturation. Perhaps the most elegant of these investigations was the site-directed mutagenesis studies on ribonuclease T_1 (Pace et al., 1988). In this work, the ribonuclease was genetically engineered so that

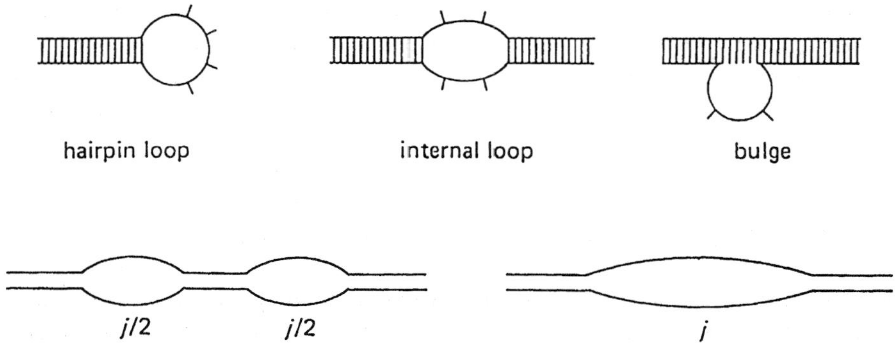

FIGURE 3.4 Examples of different loop topographies possible in nucleic acids. (Top left) Hairpin loop formed by self-complementary regions of single strands. (Top middle) Interior loops formed by noncomplementary regions or "melted" regions of double strands. (Top right) Bulge loops formed by noncomplementary regions in one strand of a complementary double strand. (Bottom) Comparison of a double strand with two loops of $j/2$ units to a double strand with a single loop of j units. The single-loop structure is favored entropically.

it contained zero, one, and two loops formed by disulfide bridges. Care was taken in placing these mutations so that the native state was not disrupted by the replacement of natural amino acids with loop-forming cysteines. In addition to investigating these mutant proteins, Pace et al. compiled thermodynamic data from other systems. The focus of the interpretation in this work was not on the value of the exponent, c, but rather on the pre-exponential term associated with Eq. 3.18. These results were analyzed assuming a value of c of 3/2 (the Flory result). If these mutation results are analyzed with c as an adjustable parameter, a much higher value ($c = 2.41$) gives the best fit to the experimental data (Chan and Dill, 1989). Thus, the protein data are not in agreement with the early, simple theoretical results. Nevertheless, they do emphasize the importance of loop effects in protein stability.

3.3.2 Loop Formation in Nucleic Acids

Nucleic acids also form structures in which the entropy of loop formation influences conformational stability. Because of the double stranded nature of many nucleic acids, a wide range of loop topologies is possible. These are illustrated in Figure 3.4, where structures known as hairpins, interior loops, and bulge loops are shown. Hairpins result from self-complementarity in single-stranded structures. Nucleic acids are composed of purines and pyrimidines that "complement" each other by forming hydrogen bonded complexes called Watson–Crick base pairs. Thus a specific sequence must find its complement to form a helix. When the nucleotide bases are complementary and hydrogen bond with each other, the chain can fold back on itself and form a double helical region. The bases in between the complementary regions are forced into a loop structure. Bulge loops are formed in double-stranded structures when a noncomplementary region occurs within one of the two complementary strands. The interior loop can occur as a result of noncomplementary regions in both halves of an otherwise complementary double strand. The lengths of the loop region for each strand need not be identical. The interior loop is important in considering the "melting" or strand separation behavior of a double strand at high temperature. Short double-stranded polynucleotides will, at high temperature, fray from the ends and separate. For longer double strands, the melting is not from the ends, but rather from the growth of interior loops.

To recognize the importance of loop entropy in dictating how double helical DNA melts, consider the situation illustrated in Figure 3.4. The figure shows a double helix with one interior loop of size j. Compare this with the other double helix with two interior loops of half the size, $j/2$. The single-loop structure has a more favorable entropy than two loops with comparable number of loop bases. This entropy difference between the single loop and the double loop is given by:

$$\Delta S = 2c \ln (j/2) - c \ln j = -c \ln 2 + c \ln (j/2) \tag{3.19}$$

As can be seen, this difference is always positive for $j > 4$. This positive entropy difference contributes a negative term to the free energy. This indicates that a single large loop is thermodyamically more stable than two smaller loops. Since loops can be very large in DNA, easily 100 to 1000 bases, a situation is created where a single large loop will be dominant equilibrium structure. Although this effect is countered by other energy contributions, it contributes very strongly to the cooperativity of helix–coil transitions in nucleic acids. A detailed statistical mechanical treatment of such effects is discussed in the next section.

The energetics of loop formation in nucleic acids has been investigated using synthetic oligonucleotides of specific base sequences. The thermodynamics of double helix formation can be determined for a given complementary sequence by thermal denaturation. Single strands that contain the same complementary sequence can then be constructed with an intervening noncomplementary sequence. By varying the length of this intervening sequence, the loop size is varied. Usually it is assumed that the energetics of the double helix is the same regardless of the loop structure in the rest of the molecule. Thus the "excess" thermodynamic parameters in the loop-forming sequence are due to the energetics of the loop. For short oligonucleotides this interpretation provides empirical laws for the free energy of loop formation that take the form (Bloomfield et al., 1974):

$$\Delta G_{\text{loop}}\,(75^\circ\text{C}) = A_0 + 0.9 \ln \left(\frac{j}{A_1} \right) \tag{3.20}$$

where the units are kilocalories per mole of loop formed. Surprisingly, these empirical laws do not change with the different loop topologies shown in Figure 3.4. The prelogarithmic factor, after correcting for proportionality constants, corresponds to a c value of 2.59, slightly higher than the value of 2.41 determined from protein data. Although it is unlikely that the three loop structures in Figure 3.4 will all have the same c value, the quality of the data and the assumptions of the thermodynamic model preclude a more accurate determination of this parameter.

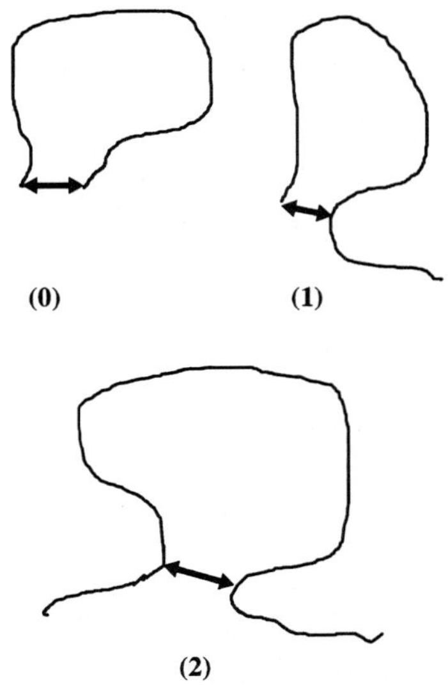

FIGURE 3.5 Different loop topographies used in theoretical calculations: (0) ring-forming polymer; (1) closure from one end of a loop (tadpole); (2) an internal loop with two ends.

3.3.3 Comparison with Theory

It is interesting to compare the experimental values of c with those obtained from theoretical models. As mentioned earlier, initial theoretical work suggested that $c = d/2$ for ring closure. A more sophisticated theoretical analysis by Fisher (1966) gave c values of 1.46 and 1.75 in two and three dimensions, respectively. At the time of Fisher's work it was recognized that excluded volume effects between the loop and the "tails" of the polymer might have a significant effect. Later work confirmed this. Three different exponents have been calculated. They are: loop formation by the entire polymer (c_0); a loop at the end of a long polymer chain (c_1); and a loop in the middle of a long chain (c_2). These three cases are illustrated in Figure 3.5. Unfortunately, these cases do not correspond with those in Figure 3.4, which are relevant to nucleic acids. Nevertheless, they will be of importance in considering protein structure. Numerical enumerations have provided values for each of these constants. The value for end-to-end closure, c_0, is now generally accepted to be 11/6 for SAWs in two dimensions and 23/12 for three dimensions. Redner (1980) has tabulated values of c_1 from a variety of approaches. There is some variation in these values, but typically they are 2.13 for two dimensions and 2.12 for three dimensions. An exact enumeration of lattice walks was used by Redner (1980) to determine c_2, and values of 2.95 and 2.16 are obtained in two and three dimensions, respectively. These computational results are essentially in agreement with those obtained from a renormalization group approach known as an ε expansion (des Cloizeaux, 1980). Recall that $\varepsilon = 4 - d$ and that $c = \nu(d + \theta)$. The ε expansion gives $\theta_0 = \varepsilon/4 + 9\varepsilon^2/128$, $\theta_1 = \varepsilon/2 - 3\,\varepsilon^2/8$, and $\theta_2 = \varepsilon - 15\,\varepsilon^2/32$. Although this expansion agrees favorably with most computational results, higher order terms in the two-dimensional ε expansion give results that differ significantly from the computational result of 2.95. In a smaller scale enumeration, Chan and Dill (1990) obtained values of 3.17 and 2.42 for c_2 in two and three dimensions, respectively. Thus, excellent (and probably fortuitous) agreement is obtained with the experimental results for proteins ($c_2 = 2.41$).

Interestingly, end-to-end closure is more probable in two than in three dimensions, while loop formation in the middle of a polymer is more probable in three dimensions. In the first case, the polymer finds it more difficult to return to the origin when the dimension is increased. (A four-dimensional walk has virtually no probability of returning to the origin.) For loop formation in the middle of the chain, there is the added problem of the "tails" getting in the way and blocking loop closure. This problem becomes more severe in compact spaces, making loop formation less probable at lower dimensions. As mentioned above, the nucleic acid results for a variety of loop topologies give a c value of 2.59. One should bear in mind that this number was determined using a specific thermodynamics model of nucleic acid interactions and from a limited data set. Consequently, the experimental value may not be highly accurate. Nevertheless, it is a very high value that has profound implications for the nature of the helix–coil transition. For $c > 2$, theoretical arguments show that the system can support a first-order phase transition. This has not been observed experimentally.

3.4 Constraints on the Size of Enzymes and Receptors

A typical enzyme or receptor will bind small molecules that are substrates or ligands in a reversible fashion. For enzyme–substrate binding this equilibrium can be represented as:

$$\mathrm{ES} \leftrightarrow \mathrm{E} + \mathrm{S}$$

where E is enzyme, S is the substrate, and ES is the enzyme–substrate complex. Biochemists typically discuss the strength of the complex using the dissociation constant defined as $K_d = [\mathrm{E}][\mathrm{S}]/[\mathrm{ES}]$, where the brackets represent concentrations. This dissociation constant is usually in units of molarity. Although there is considerable variation in this parameter, a typical dissociation constant might be 10^{-6}M for an enzyme–substrate complex. Hormone receptors generally have a greater affinity for their ligands, and a typical binding constant would be roughly 10^{-9}M. The binding energy for these two examples are 8.2 and 12.4 kcal/mol at 300 K, respectively. The binding site for an enzyme or receptor can consist of a range of amino acid side-groups, each capable of interacting with the ligand in a specific fashion. While there is extreme variability in the nature and extent of these ligand–protein interactions, a number of crude rules of thumb apply. Hydrogen bonding, hydrophobic interactions, and salt bridges (electrostatic) are types of ligand contacts that can contribute 2–10 kcal/mol per interaction (Fersht, 1985). Mutagenesis and binding studies suggest that most interactions will fall in the lower end of this range. Dissecting the binding of a ligand into specific pairwise interactions is fraught with difficulties. Often the strength of a given interaction is compromised to enhance other aspects of the binding. Bearing in mind these difficulties, one can estimate the number of "contacts" in a typical enzyme or

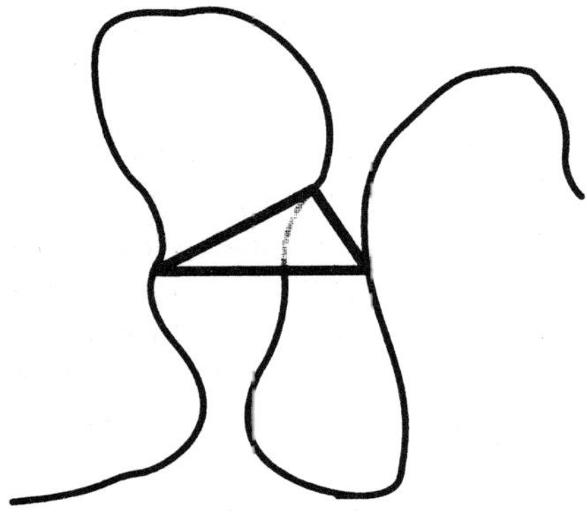

FIGURE 3.6 Schematic representation of the loop structures required to bring specific amino acids together to form an active site of an enzyme.

receptor. It is not unrealistic to have three or four specific interactions that must occur at the binding site to create a functional protein. Typically these interacting sites are on distant residues, so that two or three loops must be created to position these sites in a confined region. The confinement at the binding site and the loop structure is shown in Figure 3.6.

Given this constraint, deGennes (1990) presented an insightful analysis on the minimum size requirement for an enzyme or receptor. Unfortunately, this analysis was based on the loops being Gaussian (or ideal) coils, an assumption inconsistent with the scaling laws discussed in this and the previous chapter. In this section, deGennes' arguments are recreated with the appropriate Hamiltonian walk behavior of the protein inserted into the model. This analysis is based on the probability of loop formation. First, we consider the distribution function, $\rho(x,y,z)$, for finding a polymer configuration of N units starting at the origin and ending at the point (x,y,z). In the previous treatment (deGennes, 1990), this was given by:

$$\rho(x,y,z) = \Omega(2\pi)^{-3}\, P(x,y,z) \tag{3.21}$$

where Ω is the number of possible configurations and $P(x,y,z)$ is taken as the Gaussian probability function of Eq. 2.5.

Following a Flory–Huggins treatment (Flory, 1982) (see also Chapter 2), the number of configurations was taken as $\Omega = (20/e)^N$, because there are 20 different amino acids, each one with a different configuration. However, this is clearly an overestimate as a given position on the chain will often be insensitive to substitution by another amino acid. As was demonstrated in the previous chapter, globular proteins with a variety of secondary structures are consistent with Hamiltonian walk conditions. Therefore, the number of configurations in Eq. 3.21 should really be the number of possible Hamiltonian walks. This is given by:

$$\Omega = n_H^N \tag{3.22}$$

where the constant, n_H, is determined within the Flory–Huggins approximation as:

$$n_H = \left(\frac{z-1}{e}\right) \tag{3.23}$$

where, again, z is the connectivity of the lattice. For a cubic lattice, $z = 6$, it was shown that the Flory–Huggins approximation is surprisingly good (Pande et al., 1994). Therefore, a physically more realistic approach gives the much smaller value of $\Omega = (5/e)^N$.

In addition to correcting Ω for polymer collapse, the function $P\,(x,y,z)$ must also be suitably changed. A simple modification of Eq. 2.5 would be to alter the expression for R_0 (Eq. 2.6). It should now be $R_0 = N^\nu\ a$, where $\nu = 1/D = 1/3$. However, this correction alone is not sufficient to give proper scaling behavior for a collapsed polymer. As discussed earlier in this chapter (Eq. 3.1), a more general form of the probability function is:

$$P(r) \sim \left(\frac{r}{R_0}\right)^{\theta} \exp\left\{-B\left(\frac{r}{R_0}\right)^{\delta}\right\} \tag{3.24}$$

where θ and δ are critical exponents and B is a constant. As it turns out, the form of the exponential is unimportant, because for loop formation $r << R_0$, and because the exponent δ is approximately 1. For our present purpose, we need only to know θ, and we use values determined from the numerical simulations discussed in the previous section.

With the above considerations and explicitly introducing the proportionality constants, the distribution function for loop formation, $\rho_{\text{loop}}(r)$, is now given by:

$$\rho_{\text{loop}}\ (r) = (2\pi)^{-3}\left(\frac{5}{e}\right)^N\left(\frac{3}{2\pi a^2 N^{2\nu}}\right)^{d/2}\left(\frac{r}{aN^{\vartheta}}\right)^{\theta} \tag{3.25a}$$

$$= (2\pi)^{-3}\left(\frac{5}{e}\right)^N\left(\frac{3}{2\pi a^2}\right)^{3/2} N^{-\nu(d+\theta)}\left(\frac{r}{a}\right)^{\theta} \tag{3.25b}$$

The exponent, $\nu(d + \theta)$, is equal to c, and we will assume an interior loop and use the c_2 value of 2.16 determined numerically by Redner (1980). From Redner's work, θ_2 is found to be 0.67, and the r/a term will not strongly influence the final result. Following deGennes, a value of 4 Å is used for the length, a, of an amino acid in a protein. The distance r will represent the size of binding pocket. For simplicity, we assume a value of 4 Å for this as well, and the (r/a) term is eliminated.

Equation 3.25 represents a probability density. To calculate the probability, the volume available to the protein in configurational space, w, must be estimated. This volume is the variation in an active site locus that would still result in a functional enzyme. Previously, this was taken as a the volume of thermal fluctuations in the end-to-end distances in the loop (deGennes, 1990). The spatial variation was assumed to be 0.2 Å in each of three directions and the angular variation was taken

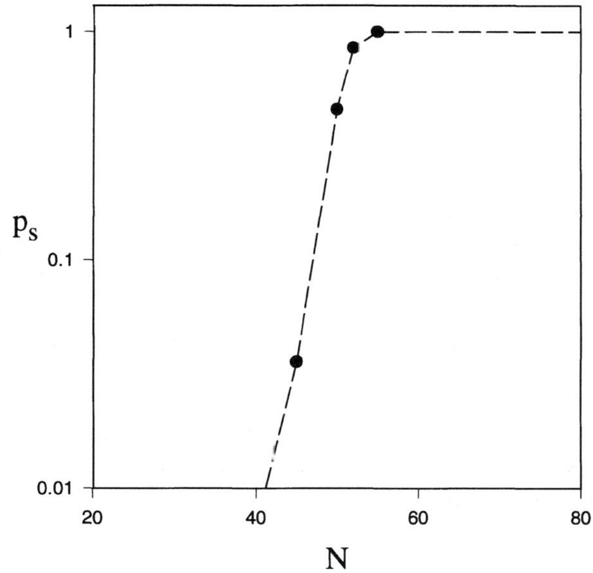

FIGURE 3.7 Plot of the probability of a successful loop orientation, p_s, versus the number of amino acids in the loop, N. The minimum size for a successful loop is approximately 50 units.

as 1/10 rad or 6°. Consequently, w was estimated as 8×10^{-6} Å^3 rad^3. It is assumed that configurations are evenly distributed and are at low density. Consequently, a Poisson distribution is used. The probability of no points in configuration space falling into the volume w is now given by:

$$p_f = \exp\{-\rho w\} = \exp\left\{-\frac{1.663 \times 10^{-10}}{N^{2.16}}\left(\frac{5}{e}\right)^N\right\} \tag{3.26}$$

where p_f is the probability of failure of a proper configuration being found. The probability of success, p_s, of forming a properly folded loop is then $p_s = 1 - p_f$, and these values are plotted for different values of N in Figure 3.7. As can be seen, the smallest number of amino acids that could form a loop with a precise end-to-end contact is approximately 50. This number is to be compared with deGennes' number of 13 obtained for a Gaussian coil. It has been argued that the smallest loop size will be independent of the number of loops in the structure. Consequently, a lower limit on the size of an enzyme can be calculated.

To create an enzyme with three contacts at the active site, two loops are required. Using the above arguments, the smallest possible enzyme should be approximately 100 amino acids. There are few enzymes that have less than 100 amino acids. Ribonuclease and lysozyme are examples of small enzymes, being composed of 124 and 129 amino acids, respectively. Examination of the active sites of a number of enzymes shows that amino acids participating in catalysis are frequently well separated along the sequence. A separation of approximately 50 residues is not uncommon. However, considerably smaller and larger separations are observed, and there is wide variability in the "loop structures" that form an active site. No attempts have been made to characterize the distribution of such structures. However, from a survey of "textbook enzymes" the estimate of 50 residues between contacts is not unreasonable.

3.5 The Statistical Mechanics of Biopolymers

During the late 1950s and 1960s, considerable effort went into developing theoretical models of the helix–coil transition in biopolymers (see Poland and Scheraga, 1970). Collectively this work represents one of the more successful applications of statistical mechanics in molecular biophysics. Currently there is renewed interest in this problem (Qian and Schellman, 1992). This is largely due to advances in synthetic techniques that can provide specific sequences for both polynucleotides and polyamino acids. With these sequences, it is now possible to test a number of postulates underlying the theoretical models. In addition to these fundamental considerations, there has been considerable interest in RNA structure and stability (Turner et al., 1988). Even relatively short RNA sequences can show a rich diversity of looped structures. With renewed attention to such problems, it is important to re-examine the previous statistical mechanical models. It is also important to reconsider these models in the light of the computational work on loop entropies discussed in the previous section. The result of this work bears on the issue of first-order phase transitions in biopolymers.

While a wide variety of biopolymer models have been explored, most efforts have focused on two specific structures: the alpha helix used to describe polyamino acids, and the double helix used to describe polynucleotides. Helix–coil transitions in these two models are illustrated in Figure 3.8. In both cases, the units of the polymer exist in an extremely compact, helical conformation. The alpha helix is characterized by hydrogen bonding between the peptide linkage of every fourth residue. The hydrogen of the amide group bonds to the carbonyl oxygen. These hydrogen bonds are close to being parallel to the helix axis. For the double helix model, helical segments hydrogen bonds between Watson–Crick base pairs on opposing strands. This bonding is perpendicular to the helix axis. In both cases, the helical regions create extremely compact structures, perhaps the most compact structure possible. However, because of their high orientational order, at intermediate length these structures are rigid rods. The energetics governing such structures are a complex interplay of a variety of forces. A common "textbook" mistake is to assume that the hydrogen bond energy holds them together. While hydrogen bonding is important, no single contribution can account for the overall stability of alpha or double helices. Because of the "zipper" nature of both structures, they undergo thermal transitions that are highly cooperative. Typically, a biopolymer can "melt" from a predominantly helical structure to a predominantly coiled structure over the range of a few degrees. While the coil state is often assumed in theoretical models to be that of an excluded volume polymer, there is

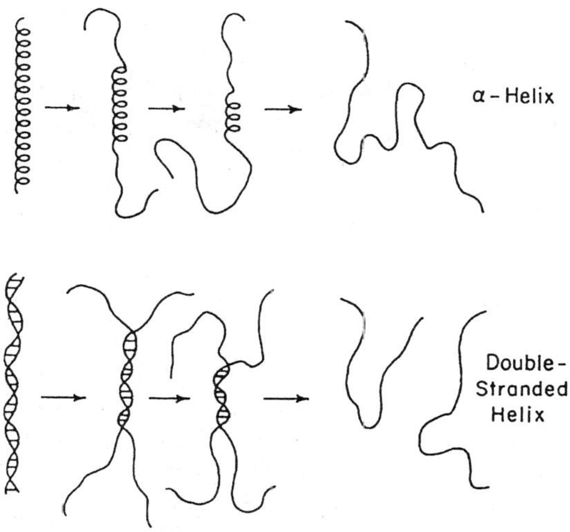

FIGURE 3.8 Helix–coil transitions in an alpha helix and a double stranded nucleic acid The alpha helix has hydrogen bonds between every fourth amino acid residue. These bonds are approximately parallel to the helix axis. The double strand has complementary nucleic acid bases that hydrogen bonds between strands. These hydrogen bonds are perpendicular to the helix axis. In both cases helix formation is favored by nearest-neighbor interactions, making helix–coil transitions highly cooperative.

little experimental information to support this contention. The steepness of this order–disorder transition has intrigued experimentalists and theoreticians alike, and has resulted in extensive work on this subject over the past 30 years.

There are a variety of statistical mechanical techniques used to investigate order–disorder transitions in biopolymers. They fall into three main categories: combinatoric, transfer matrix, and sequence generating functions. Each mathematical approach has its appeal and utility. In the current development, sequence generating functions will be used. This approach provides a facile connection with the multifractal formalism to be discussed in the next chapter. Also, recent advances have given this method a wider range of applicability (Dewey, 1994). In the majority of models, a unit of the biopolymer can exist in one of two states: helix or coil. Statistical weights are then assigned to these states depending on their occurrence in the sequence. Lifson developed the sequence generating function technique to calculate the partition function for a number of statistical problems involving biopolymers (Lifson, 1964; Litan and Lifson, 1965). This method is an extension of Guggenheim's generalized divergent partition function (Guggenheim, 1939), which is discussed in a rigorous fashion by Hill (1956). This approach has appeared in a variety of applications, including one-dimensional gases (Runnels, 1965) and other combinatorial problems (Temperley, 1956).

3.5.1 Sequence-Generating Functions

Helix–coil (or order–disorder) transitions in linear chain biopolymers can be described by the symbolic sequence representing the state of each unit in the polymer. The helix–coil sequence of a biopolymer is determined by a state (h or c) associated with each unit of the polymer. For a given polymer, the state can be specified by giving the length of alternating sequences of type h and c, in the order of their occurrence along the chain. Thus, a given microscopic state is determined by:

$$i_0, j_1, i_1, j_2, i_2, \ldots, j_s, i_s, j_0 \tag{3.27}$$

where the i and j terms are the lengths of sequences containing c and h units, respectively. If the polymer begins with an h unit then $i_0 = 0$, and likewise if it ends in a c then $j_0 = 0$. All sequences, excluding the ends, contain at least one unit. This means that the total length of sequences, $2s$, cannot exceed the length in the polymer, n. Also, the sum of the lengths of the sequences must be n:

$$\sum_{\sigma=0}^{s} (i_\sigma + j_\sigma) = n \qquad 0 \le s \le (n/2) \tag{3.28}$$

To determine the partition function, a statistical weight, $w(n;i_0, \ldots, j_0)$, is associated with each microscopic state and a sum is made over all microstates. A widely used model is to give each sequence, i_σ and j_σ, within the microstate a weight, u_{i_σ} for c sequences and v_{j_σ} for h sequences. Such models assume that there are no long-range interactions between sequences. The weight for a given microstate is now given by:

$$w(n;\, i_0, \ldots, J_0) = \prod_{\sigma=0}^{s} u_{i_\sigma} v_{i_\sigma} \tag{3.29}$$

The specific values of u_{i_σ} and v_{j_σ} will depend on the model used and the nature of the interactions. The restriction, $u_0 = v_0 = 1$, must hold for end sequences of zero length, i.e., $i_0 \neq 0$ and $j_0 \neq 0$. The partition function, $Z(n)$, for the polymer is the sum over all microscopic states:

$$Z(n) = \sum_k w(n;\, i_0, \ldots, j_0) = \sum_{s=0}^{n/2} \sum_{\{i_\sigma, j_\sigma\}} \prod_{\sigma=0}^{s} u_{i_\sigma} v_{i_\sigma} \tag{3.30}$$

where k represents a microscopic state of the system and the first sum runs over all possible states. In the second sum, $\{i_S, j_S\}$ indicates that the summation is over all possible lengths of sequences subject to the restriction in Eq. 3.28 with n and s being held fixed. The difficulties imposed by this restriction are circumvented by inverting the summations.

In Lifson's original work, the partition function was evaluated using a divergent moment generating function, Ξ, which is a function of an undetermined variable, x:

$$\Xi(x) = \sum_{n=0}^{\infty} Z(n)\, x^{-n} \tag{3.31}$$

It is noted that the free energy, $-kT \ln Z(n)$, is proportional to the length, n, of the polymer. Thus, a parameter x_1 can be defined such that:

$$Z(n) = x_1^n \tag{3.32}$$

The crux of the method is that $\Xi(x)$ can be calculated directly, and its divergent properties are now used to determine x_1 and the partition function. For $x > x_1$ the series converges, but in the limit of $x \to x_1$ it diverges. Equation 3.31 can be simply represented as follows, using sequence generating functions:

$$\Xi(x) = \frac{U_0 V_0}{1 - UV} \tag{3.33}$$

where U and V are the sequence generating functions for coils and helices, respectively. They are defined as:

$$U(x) = \sum_{i=1}^{\infty} u_i x^{-i} \tag{3.34}$$

$$V(x) = \sum_{i=1}^{\infty} v_i x^{-i} \tag{3.35}$$

Similar sums over i_0 and j_0 define $U_0(x)$ and $V_0(x)$, except that these summations run from $i = 0$ to ∞. The terms U and V can also differ from U_0 and V_0 when specific end effects are considered. The specific forms associated with u_i and v_i are determined from the model of interest and usually result in simple analytic forms for $U(x)$ and $V(x)$. For Eq. 3.33 to diverge:

$$U(x)\, V(x) - 1 = 0 \tag{3.36}$$

The largest root of this equation gives x_1 and, therefore, provides a direct determination of the partition function.

In a later version of this technique, Lifson changed the definition of the generating function, so that it was given by an ascending rather than inverse power series in x (Litan and Lifson, 1965). Part of the motivation for the new definition was that the generating function now becomes analogous to the grand canonical partition function. Indeed, it is sometimes referred to as the "quasi-grand canonical partition function." This definition lends itself to a closer association with standard statistical mechanical relationships. The main drawback with the method is that the method of "steepest descent" (Fowler, 1929; Schroedinger, 1989) is used to evaluate the resulting contour integral. This requires that the number of units, n, is large. Thus, the method cannot handle finite length polymers. However, this problem is readily remedied, by returning to the original definition (Eq. 3.25) of the generating function (Dewey, 1994). In the present application, the generating function is nothing more than a mathematical device for determining the partition function. There is no physical reason for making the association with the grand canonical partition function. Bearing this in mind, Cauchy's integral equation is applied to Eq. 3.31. This gives:

$$Z(n) = (2\pi i)^{-1} \oint_C \Xi(x)x^{n-1}\, dx \tag{3.37}$$

where the contour C encompasses the singularities of the integrand. The evaluation of the partition function using Eq. 3.37 is a powerful technique and has a comparable utility to the popular "matrix method" (see Poland and Scherega, 1970). In the case of self-complementary polymers that can form loops, the sequence generating function is actually preferred to the matrix method (Gô, 1967; Wiegel, 1983).

The validity (and utility) of Eq. 3.37 is readily demonstrated using a simple calculation. The matrix method provides partition functions of the general form (see Applequist, 1963):

$$Z(n) = \sum_{r=1}^{\rho} C_r \lambda_r^n \tag{3.38}$$

where C_r is a constant and λ_r is a root of a determinant of rank ρ. Taking this as a general representation of a partition function, the sequence generating function can be "back-calculated" as follows:

$$\Xi(x) = \sum_{n=0}^{\infty} Z(n)x^{-n} = \sum_{n=0}^{\infty} \left(\sum_{r=1}^{\rho} C_r\lambda_r^n \right) x^{-n} \tag{3.39a}$$

$$= \sum_{r=1}^{\rho} C_r \left[\sum_{n=0}^{\infty} \left(\frac{\lambda_r}{x} \right)^n \right] = \sum_{r=1}^{\rho} \frac{C_r x}{x - \lambda_r} \tag{3.39b}$$

Using Eq. 3.37, it is seen that:

$$Z(n) = (2i\pi)^{-1} \oint_C \sum_{r=1}^{\rho} \frac{C_r x}{x - \lambda_r} x^{n-1}\, dx = \sum_{r=1}^{\rho} C_r \lim_{x \to \lambda_r} \frac{(x - \lambda_r)x^n}{x - \lambda_r} \tag{3.40a}$$

$$= \sum_{r=1}^{\rho} C_r \lambda_r^n \tag{3.40b}$$

Thus, application of Eq. 3.37 recovers the original partition function and confirms the validity of the approach. In the next section, two simple problems are considered and compared with results obtained by combinatoric and matrix approaches.

This demonstrates that the above approach is valid for all values of n, i.e., finite polymers. This recent extension of the sequence generating function technique (Dewey, 1994) greatly enhances its utility. For some problems, such as loops within loops (see Gô, 1967), it is clearly the method of choice.

3.5.2 Calculations on the Alpha Helix

In this section the Bragg–Zimm model of the alpha helix is treated with the sequence generating function approach. After solving for the partition function, a variety of parameters of experimental relevance are calculated. In the Bragg–Zimm model (Zimm and Bragg, 1959), the coils are given statistical weights of $u_i = 1$, and the helical units have weights such that $v_1 = \sigma^{1/2}$ and $v_i = \sigma s^{i-2}$ for $i \geq 2$, where σ is the weight associated with the "boundaries" of the helical region and s is the weight associated with each helical unit. The term s is a Boltzmann factor that contains the energetics of the favorable neighbor interactions within the helix. The boundary effect is an unfavorable term because it is the start of a helical segment and the first unit does not benefit from a favorable association with the preceding unit in the coil. Nevertheless, this unit must adopt the restricted configuration of a helix (as opposed to a random coil). Consequently, σ is a Boltzmann factor dominated by this unfavorable entropy of configuration. Experimental values for σ are typically in the range 10^{-2} to 10^{-4}. With these assignments the sequence generating functions are:

$$U = \frac{1}{x-1} \tag{3.41a}$$

$$V = \frac{s\sigma}{x-s} \tag{3.41b}$$

In the Bragg–Zimm model the end sequences have slightly different generating functions. As one does not usually start with a helical region on the end, $V_0 = 1$ and the coil generating function is given by:

$$U_C = \frac{x}{x-1} \tag{3.42}$$

Combining these results gives:

$$\Xi(x) = \frac{x(x-s)}{(x-1)\ (x-s) - s\sigma} = \frac{x(x-s)}{(x-\lambda_0)\ (x-\lambda_1)} \tag{3.43}$$

where λ_0 and λ_1 are the roots of the quadratic in the denominator of the middle expression in Eq. 3.37. By expressing Eq. 3.43 in terms of partial fractions, as in Eq. 3.39b, the partition function is determined by Eq. 3.40. Thus:

$$\Xi(x) = \frac{C_0 x}{x-\lambda_0} + \frac{C_1 x}{x-\lambda_1} \tag{3.44a}$$

where

$$C_0 = \frac{\lambda_0 - s}{\lambda_0 - \lambda_1} \tag{3.44b}$$

$$C_1 = \frac{s - \lambda_1}{\lambda_0 - \lambda_1} \tag{3.44c}$$

The partition function is identical to the Bragg–Zimm model and is valid for finite n. It is given by:

$$Z(n) = \frac{\lambda_0^n(\lambda_0 - s) - \lambda_1^n(\lambda_1 - s)}{\lambda_0 - \lambda_1} \tag{3.45}$$

This result is identical to the one obtained with the transfer matrix technique (Applequist, 1963).

Once the partition function has been determined, it can be used to calculate a number of different quantities (see Poland and Scherega, 1970). Experimentally, the most accessible quantity is the fraction of helical units, Θ_h, per polymer. This is often measured by the extent of change of a spectroscopic parameter. This parameter is given by:

$$\Theta_h = \frac{1}{N}\frac{\partial \ln Z_N}{\partial \ln s} \tag{3.46}$$

This parameter gives the entire "melting" curve for the helix–coil transition. The transition temperature is defined as the point where $\Theta_h = 1/2$. The scaling behavior about this transition determines the order of the transition. A first-order phase transition will have a discontinuity in Θ_h at the transition temperature. The sharpness of the thermal transition is an important parameter and is readily determined by taking the derivative of Eq. 3.40 with respect to temperature. Commonly measured experimental parameters are the transition temperature and the slope of Θ_h versus T at the transition point.

Another parameter of interest is the number average length of a helical segment, L_h, and this is given by:

$$L_{\rm h} = \frac{\Theta_{\rm h}}{\Theta_{\rm hc}} \tag{3.47a}$$

where

$$\Theta_{\rm h} = \frac{1}{N}\frac{\partial \ln Z_N}{\partial \ln \sigma} \tag{3.47b}$$

For most models Eqs 3.46 and 3.47 provide tractable expression for analyzing the melting behavior of a biopolymer and are central to the analysis of experimental data. These expressions will also dictate phase behavior and scaling exponents.

3.5.3 The Finite Perfect Matching Double Strand

Turning now to the nucleic acid double helix, the perfect matching double strand is treated. The development by Litan and Lifson (1965) is followed throughout. This model consists of helical regions with sequence generating function $V(x)$ and weights $v_i = s^i$, and loop regions with sequence generating function $U(x)$ and weights $u_i = i^{-C}$, where c is the exponent associated with the loop entropy. As in the alpha helix model, the statistical weight for a coil is taken to be 1. The entropy of the resulting internal loop is accounted for by the factor i^{-c} in accordance with Eq. 3.18. A cooperativity parameter, σ_0, is associated with the boundary between the loop and helical regions. Including end effects, the generating functions are then given by:

$$V = V_0 = \frac{s}{x - s} \tag{3.48a}$$

$$U(x) = \sigma_0 \sum_{i=r}^{\infty} x^{-i} i^{-c} \tag{3.48b}$$

$$U_0 = \frac{x}{x - 1} \tag{3.48c}$$

The quasi-grand partition function is now represented as a partial fraction, giving:

$$\Xi(x) = \frac{sx^2}{(x-1)^2 \left(x - s - s\sigma_0 \sum_{i=1}^{\infty} x^{-i} i^{-c}\right)} = \sum_r \frac{C_r x}{x - \lambda_r} \tag{3.49}$$

where

$$C_r = \frac{\lambda_r s}{(\lambda - 1)^2 \left(1 + s\lambda_r^{-1}\sigma_0 \sum_{i=1}^{\infty} \lambda_r^{-i} i^{1-c}\right)} \tag{3.50}$$

and the values of λ_r are determined from the roots of Eq. 3.51.

$$x - s - s\sigma_0 \sum_{i=1}^{\infty} x^{-i} i^{-c} = 0 \tag{3.51}$$

The summation over r in Eq. 3.49 runs over all the roots of Eq. 3.51. Once the roots have been determined, Eq. 3.50 can be used to calculate the partition function as given in Eq. 3.38.

To solve Eq. 3.51, the transformations $x = 1 + \varepsilon$ and $s = 1 + \delta$ are made and the region where ε and δ are small is considered. The summation in Eq. 3.51 is approximated by the form derived by Fisher (1966), (see also Poland and Scheraga, 1970) to give the following algebraic equation:

$$\varepsilon + \sigma' \varepsilon^{c-1} - (\delta + \sigma') = 0 \tag{3.52}$$

where $\sigma' = \sigma_0 \sum_{i=1}^{\infty} i^{c-1}$. Using Eq. 3.46, the melting behavior of the double strand can be explored. It is seen that:

$$\Theta_h = \frac{1}{1 + \frac{\sigma'}{(2-c)} (\varepsilon^{c-2} - 1)} \tag{3.53}$$

From scaling arguments, two different types of behavior are seen. From previous arguments (Fisher, 1966); Poland and Scheraga, 1970), the parameter $(\delta + \sigma') \propto (T - T_M)$ for T below the transition temperature, T_M. For $c < 2$:

$$\Theta_h \propto (T - T_M)^{\frac{2-c}{c-1}} \tag{3.54}$$

The fraction of the helix varies continuously for $c < 2$, and no first-order phase transition exists. However, for $c > 3/2$, the slope of the melting curve will be infinite and higher order transitions exist. For the case of $c > 2$, the following scaling is seen:

$$\Theta_h \propto (T - T_M)^{2-c} \tag{3.55}$$

Thus, in this case, the fraction of helix diverges as it approaches the transition temperature, and a true, first-order phase transition is observed. Similar results were obtained by Dewey (1994) within the context of the Yang–Lee theory of phase transitions.

The initial observations of the "melting" of DNA over a very small temperature range generated considerable excitement. At the time, it was thought that a true phase transition could be occurring in a one-dimensional system. This spurred theoretical efforts on phase transitions in one-dimensional systems, and it was soon established that two conditions had to be met for a first-order phase transition to occur (Poland and Scheraga, 1966a,b). Either the system had to be infinite or the interactions between units must be infinite in range. As discussed in the previous section, loop formation can support phase transitions in linear polymers. This observation is not in conflict with the early theoretical work because, strictly speaking, loop formation models are not one dimensional. When incorporating loop statistics (via the exponent c) into a model, the three-dimensional structure of the loop enters into the argument.

The formation of a loop in double strand models gives the appearance of long-range interaction between helical regions. Two opposing entropic forces are involved here. The favorable combinatorial entropy of alternating helical and coil regions is counterbalanced by the unfavorable loop entropy of the coil regions. This results in the helical regions not behaving independently, and gives the appearance of long-range interactions. Interestingly, as the double strand approaches its "melting" transition, the number of helical regions does not change sharply, but rather the length of these regions changes. The determination of the parameter, c, is crucial to assessing the possibility of phase behavior in this system. The topological considerations implicit in a choice of c values build features of higher dimensional models into this problem. As was seen in the previous section, for loop formation in the middle of a polymer, c_2 has experimental and theoretical values greater than 2. Thus, first-order phase transitions would be expected for infinitely long polymers. Such an abrupt transition has, of course, not been observed. This presumably is the result of the finite length of the polymers observed in experimental settings (see Dewey, 1994).

3.6 Summary

In this chapter, the self-similarity of polymer loops has been explored using applications of real-space renormalization group theory. These applications employ a decimation procedure for renormalizing walks on a lattice. These procedures allow critical exponents to be calculated. Similar approaches will appear in Chapters 7 and 9. Using renormalization arguments, it has been seen that loops are in the same universality classes as extended polymers. The scaling exponents associated with the entropy of loop formation have been discussed with respect to loop topology. Results from computer simulations and renormalization theory have been compared with experimental results for proteins and nucleic acids. The basic models governing the statistical mechanics of helix–coil transitions in biopolymers have been introduced. These statistical methods and models will be employed in the following chapter on multifractals. It has been seen how the entropy of loop formation can play an important role in determining the sharpness of order–disorder transitions in biopolymers.

References

Applequist, J. 1963. On the Helix–Coil Equilibrium in Polypeptides. *J. Chem. Phys.* **38**, 934–941.

Baumgartner, A. 1982. Statistics of Self-Avoiding Polymers. *J. Chem. Phys.* **76**, 4275–4280.

Bloomfield, V. A., Crothers, D. M., and Tinoco, Jr., I. 1974. *The Physical Chemistry of Nucleic Acids*. Harper & Row, New York, pp. 343–353.

Chan, H. S., and Dill, K. A. 1989. Intrachain Loops in Polymers: Effects of Excluded Volume. *J. Chem. Phys.* **90**, 492–509.

Chan, H. S., and Dill, K. A. 1990. The Effects of Internal Constraints on the Configuration of Chain Molecules. *J. Chem. Phys.* **92**, 3118–3135.

Creswick, R. J., Farach, H. A., and Poole, Jr., C. P. 1992. *Introduction to Renormalization Group Methods in Physics*. Wiley, New York, pp. 69–107.

deGennes, P. G. 1990. *Introduction to Polymer Dynamics*. Cambridge University Press, Cambridge, pp. 16–26.

des Cloizeaux, J. 1980. Short Range Correlation between Elements of a Long Polymer in a Good Solvent. *J. Phys.* **41**, 223–238.

Dewey, T. G. 1993. Protein Structure and Polymer Collapse. *J. Chem. Phys.* **98**, 2250–2257.

Dewey, T. G. 1994. Order–Disorder Transitions in Finite Biopolymers: A Sequence Generating Function Approach. *J. Chem. Phys.* **100**, 7648–7654.

Family, F. 1982. A Direct Renormalization Group Study of Loops in Polymers. *Phys. Lett.* **92A**, 341–344.

Family, F. 1984. Polymer Statistics and Universality: Principles and Applications of Cluster Renormalization. *AIP Proc.* **109**, 33–72.

Fersht, A. 1985. *Enzyme Structure and Mechanism*. W. H. Freeman, New York, pp. 293–308.

Fisher, M. E. 1966. Effect of Excluded Volume on Phase Transitions in Biopolymers. *J. Chem. Phys.* **45**, 1469–1473.

Flory, P. J. 1982. Treatment of Disordered and Ordered Systems of Polymer Chains by Lattice Methods. *Proc. Natl. Acad. Sci. USA* **79**, 4510–4514.

Fowler, R. H. 1929. *Statistical Mechanics*. Cambridge University Press, Cambridge, pp. 14–31.

Gô, M. 1967. Statistical Mechanics of Biopolymers and Its Application to the Melting Transition of Polynucleotides. *J. Phys. Soc. Jpn* **23**, 597–608.

Guggenheim, E. A. 1939. Grand Partition Functions and So-Called "Thermodynamic Probability". *J. Chem. Phys.* **7**, 103–107.

Hill, T. L. 1956. *Statistical Mechanics*. McGraw-Hill, New York, p. 59.

Johnson, R. E., Adams, P., and Rupley, J. A. 1978. Thermodynamics of Protein Cross-Links. *Biochemistry* **17**, 1479–1484.

Lifson, S. 1964. Partition Functions of Linear-Chain Molecules. *J. Chem. Phys.* **40**, 3705–3710.

Lin, S. H., Konishi, Y., Denton, M. E., and Scheraga. H. A. 1984. Influence of an Extrinsic Cross-Link on the Folding Pathway of Ribonuclease A. Conformational and Thermodynamic Analysis of Cross-Linked ($Lysine^7$-$Lysine^{41}$)-Ribonuclease A. *Biochemistry* **23**, 5504–5512.

Litan, A., and Lifson, S. 1965. Statistical-Mechanical Treatment of Multistranded Polynucleotide Molecules. *J. Chem. Phys.* **42**, 2528–2532.

McKenzie, D. S. 1976. Polymers and Scaling. *Phys. Rep.* **27**, 1–88.

Pace, C. N., Grimsley, G. R., Thomson, J. A., and Barnett, B. J. 1988. Conformational Stability and Activity of Ribonuclease T_1 with Zero, One, and Two Intact Disulfide Bonds. *J. Biol. Chem.* **263**, 11 820–11 825.

Pande, V. S., Joerg, C., Grosberg, A. Y., and Tanaka, T. 1994. Enumerations of the Hamiltonian walks on a cubic sublattice. *J. Phys. A: Math. Gen.* **27**, 6231–6236.

Poland, D., and Scheraga, H. A. 1966a. Phase Transitions in One Dimension and the Helix-Coil Transition in Polyamino Acids. *J. Chem. Phys.* **45**, 1456–1463.

Poland, D., and Scheraga, H. A. 1966b. Occurrence of a Phase Transition in Nucleic Acid Models. *J. Chem. Phys.* **45**, 1464–1469.

Poland, D., and Scheraga, H. A. 1970. *Theory of Helix–Coil Transitions in Biopolymers*. Academic Press, New York.

Prentis, J. J. 1982. Spatial Correlations in a Self-Repelling Ring Polymer. *J. Chem. Phys.* **76**, 1574–1583.

Qian, H., and Schellman, J. A. 1992. Helix–Coil Theories: A Comparative Study for Finite Length Polypeptides. *J. Phys. Chem.* **96**, 3987–3994.

Redner, S. 1980. Distribution Functions in the Interior of Polymer Chains. *J. Phys. A: Math. Gen.* **13**, 3525–3541.

Runnels, L. K. 1965. One-Dimensional Lattice Gas. *J. Chem. Phys.* **43**, 212–216.

Schroedinger, E. 1989. *Statistical Thermodynamics*. Dover, New York, pp. 5–21.

Temperley, H. N. V. 1956. Combinatorial Problems Suggested by the Statistical Mechanics of Domains and Rubber-Like Molecules. *Phys. Rev.* **103**, 1–16.

Thomas, D. J. 1990. The Entropic Tension of Protein Loops. *J. Mol. Biol.* **216**, 459–465.

Turner, D. H., Sugimoto, N., and Freier, S. M. 1988. RNA Structure Prediction. *Ann. Rev. Biophys. Biophys. Chem.* **17**, 167–192.

Ueda, T., Yamada, H., Hirata, M., and Imoto, T. 1985. An Intramolecular Cross-Linkage of Lysozyme. Formation of Cross-Links between Lysine-1 and Histadine-15 with bis(Bromoacetamide) Derivatives by a Two-Stage Reaction Procedure and Properties of the Resulting Derivatives. *Biochemistry* **24**, 6316–6322.

Vicsek, T. 1992. *Fractal Growth Phenomena.* World Scientific, Singapore, pp. 48–97.

Wiegel, F. W. 1983. Conformational Phase Transitions in a Macromolecule: Exactly Solvable Models. In *Phase Transitions and Critical Phenomena*, **Vol. 7**, ed. C. Domb and J. L. Liebowitz. Academic Press, London, pp. 105–151.

Zimm, B. H., and Bragg, J. K. 1959. Theory of the Phase Transition between Helix and Random Coil in Polypeptide Chains. *J. Chem. Phys.* **31**, 526–535.

4

The Multifractality of Biomacromolecules

The multifractal formalism provides a method of characterizing complex structures and dynamics (for reviews see Pieteronero et al., 1988; Stanley and Meakin, 1988; Tél, 1988). It has been used to investigate a diversity of phenomena such as diffusion-limited aggregation, dielectric breakdown, and turbulence. This formalism allows the association of fractal dimensions with specific, configurational subsets of a given statistical problem. There are two general types of problem that have been treated. In the first case, a single, complicated physical structure is considered and fractal dimensions of subsets of this structure are determined using box-counting algorithms (see Meisel et al., 1992). This approach has been successfully used to characterize growth probabilities in aggregation models (see Vicsek, 1992). The second case investigates the multifractal nature of an ensemble of structures or trajectories (Evertsz and Lyklema, 1987). For random walk problems, subsets of trajectories can be partitioned into equivalence classes with identical fractal dimensionalities. Evertsz and Lyklema developed this approach to define a generalized mean square displacement exponent for a specific linear growth model. This application is closely related to the multifractal description of tree structures used to characterize generalized Cantor sets and dynamical processes (Halsey et al., 1986; Sarkar, 1987; Feigenbaum et al., 1986, 1989).

The present chapter explores multifractal descriptions of biomacromolecules. We begin with a discussion of general features of multifractals and how they originate in physical systems. Two situations that give rise to multifractals are discussed. These are random multiplicative processes and random walks on random lattices. The latter situation results in a convolution of two narrow probability functions, which results in a broad distribution. As an example of a probability distribution on a disordered network, we consider simple binomial distributions on Cantor sets. We then move on to a discussion of order–disorder configurations of synthetic biopolymers. Models for both the alpha helix and the double helix can be represented as "trajectories" on a binary tree and, therefore, can be represented as random multiplicative processes. From this representation, the intrinsic multifractal character of helix–coil transitions can be seen. The sequence generating function method discussed in the Chapter 3 will be used to develop the multifractal formalism for these models. The double helix model is especially interesting as it shows a phase transition in the multifractal spectrum. After discussing simple biopolymers, the more complicated issue of multifractality in proteins is considered. Solvent accessibilities of protein side-chains, derived from X-ray structures, can be analyzed by a "box-counting" algorithm and show multifractal behavior. These multifractals provide a diagnostic test for proper protein folding, an important consideration when considering protein structure prediction algorithms. Although proteins are very different from model biopolymers such as the alpha helix, it is anticipated that similar processes will contribute to their multifractal nature. It will be argued that proteins, also, have an intrinsic multifractal nature due to the convolution of probability functions. These arguments go to the heart of the origin of multifractals. The Appendix to this chapter gives mathematical details showing how to relate multifractal spectra to parameters for simple nearest neighbor models. Also, the connection between sequence generating functions and the more popular transfer matrix methods is made in that section.

4.1 Origins of Multifractality

Later in this chapter, we show that a box-counting analysis of protein solvent accessibilities leads to a multifractal spectrum for each protein. To understand the implications of such observed multifractality, one must address the question of the very nature of multifractals. Indeed, this has been the topic of recent research (see Stanley, 1991), and there is, at this time, no single answer to this question. However, two phenomena are known to give rise to multifractal behavior. They are: random multiplicative processes and processes that involve the convolution of two probability functions. As will be seen, order–disorder configurations in biopolymers can be represented by binary trees. These binary trees can be generated by random multiplicative processes, and have interesting multifractal properties. Using a spin glass analogy, it is seen that similar multifractal behavior can occur in protein structures. Proteins are even more complicated because a probability convolution appears in addition to the random multiplicative nature of the configuration.

4.1.1 Random Multiplicative Processes

To approach the question of the origin of multifractals in random multiplicative processes, it is helpful to consider a biophysical example. Often, two-state models can be used to characterize protein conformational transitions. Two-state models abound in biochemistry. Such models may be used to describe transitions from the native to the denatured state of a protein or nucleic acid. They also appear in enzyme kinetics where "inactive" and "active" forms of a native enzyme structure are invoked. These two-state models are often successful in dealing with the macroscopic kinetic and thermodynamic behavior of the system. However, they belie the simplicity of the system. The work of Frauenfelder et al. (1991) suggests that a single conformational state does not exist, but rather a conformation will have many substates (see Chapter 6). There may be a hierarchy of states within states. To proceed from one conformation to another requires a movement across an extremely complicated potential energy surface. Let us, hypothetically, consider how such a transition might occur. The probability of a conformational change happening during a given time interval may be associated with a range of events. This overall transition probability requires the determination of the probabilities of each of these events. First, the protein must be in a given conformational substate (with probability, P_1). Additionally a specific side-chain, say a tyrosine, must be in a specific rotameric configuration (with probability, P_2) during the time interval. There might be a requirement for a peptide bond to be in a specific vibrational state. This will have probability P_3. A multitude of independent events must coincide to "kick" the protein from one conformation to another. The probability of the transition, $\langle P \rangle$, is then given by:

$$\langle P \rangle = \prod_{i=1}^{k} P_i \tag{4.1}$$

Within this hypothetical model, it is seen that protein conformational changes are random multiplicative processes, i.e., that they are represented by a product of independent probabilities. In Chapter 6, we look at the nonexponential kinetic behavior of such a multiplicative model.

Random multiplicative processes appear in a number of interesting settings. Such diverse and colorful examples as the size of crushed rocks, scientific productivity, and the distribution of wealth have all been implicated as random multiplicative processes (see West and Shlesinger, 1989). Despite the ubiquity and importance of multiplicative processes, the mathematics of such behavior is not well understood. This is in contrast to the situation of random additive processes. Sums of random variables are as equally pervasive in natural phenomena. However, such processes are more readily handled using the central limit theorem. In the limit of large numbers of random variables, sums of these variables can be represented by the familiar Gaussian probability distribution function. Unfortunately, there is no equivalent of the central limit theorem for random multiplicative processes, and this is why multifractal descriptions are required.

One obvious trick is to take the logarithm of Eq. 4.1, so as to convert the product into a sum:

$$\ln \langle P \rangle = \sum_{i=1}^{k} \ln P_i \tag{4.2}$$

Now, if P_i can be treated as a random variable, why not treat *ln* P_i as a random variable as well? Application of the central limit theorem to Eq. 4.2 leads to a distribution known as a log–normal distribution, i.e., a Gaussian in which the random variable is *ln P*. While this trick works for some cases, the argument is not always valid (for a clear discussion of the failure of the log–normal distribution see Redner, 1990). For the log–normal distribution, the probability function, $\langle P \rangle$, has a narrow distribution the moments of which obey $\langle P^q \rangle = a_q \langle P \rangle^q$, where a_q is some nonsingular function in q. However, other pathological cases exist and the multifractal formalism can be used to characterize such cases. Multifractal distributions do not show simple moment relationships. This signifies a loss of characteristic scales for the system. This results in an extremely broad distribution. For multifractals the moments are such that:

$$\langle P^q \rangle \sim \langle P \rangle^{\tau(q)} \tag{4.3}$$

where $\tau(q)$ is the mass exponent (see Chapter 1). The multifractal spectrum is obtained from Legendre transformations involving $\tau(q)$.

Thus, multifractals can result from random multiplicative processes, and these processes are characterized by distributions given in Eq. 4.3. There are two major differences between the behavior of multiplicative and additive random processes. For multiplicative processes, a rare event (small P_i) can dominate the distribution, while for additive processes, rare events have very little impact. The result is that multiplicative processes have distributions with "long tails." Also, in multiplicative processes short-range correlations can have a strong impact on the product in Eq. 4.1. Additive processes are insensitive to short-range correlations because correlated pairs are often distributed as if they were a single random variable.

4.1.2 Random Walks on Random Media

Multifractal behavior can also arise in cases where a distribution results from a convolution of two other distributions. This situation occurs in problems involving random walks on random structures (Havlin and Bunde, 1989; Bunde et al., 1990). Two probability functions enter into such problems: one to characterize the structural disorder of the random substrate and one to treat the random nature of the walk on such a substrate. To calculate the behavior of the entire system, an average must be performed over all random substrate structures. The resulting probability of the walk is a convolution of the two probability functions associated with the problem. Even for two relatively narrow distributions, the convolution can result in a broad logarithmic distribution. In such cases, the convolution exhibits multifractal behavior.

In the random walk problem, the two distributions are $P(l,t)$ and $\Phi(l,r)$, where t is time, r is a Euclidean distance, and l is a "chemical" or polymer contour distance, i.e., distance within the substrate. The dynamics of the problem is associated with the probability of moving the walker along the random structure, $P(l,t)$. For a normal diffusion problem, the dynamics of the walker are represented by the Gaussian distribution:

$$P(l,t) = \mathrm{P}(0,t)\exp\left(\frac{-l^2}{4Dt}\right) \tag{4.4}$$

where D is the diffusion constant. The probability density, $\Phi(l,r)$, represents the distribution of the substrate or lattice in Euclidean space. This is the probability of finding two sites separated by a chemical distance l while being separated by a Euclidean distance, r. For a substrate formed by a random walk $\Phi(l,r)$ is also Gaussian and is given by:

$$\Phi(r,l) = \Phi(0,l) \exp\left(\frac{-r^2}{4\Delta l}\right) \tag{4.5}$$

where Δ is a constant.

The multifractal spectrum for a random walk on a random substrate is generated by the convolution integral:

$$\langle P^q \rangle = \int_0^\infty \Phi(l,r) P^q(r,t)\, dr = \langle P \rangle^{\tau(q)} \tag{4.6}$$

It can be shown (Stanley, 1991) that $\tau(q) \sim q^{1/3}$ for $q > 0$. This gives the nonlinear dependence that makes the probability distribution multifractal. While each probability function in the convolution may have a narrow distribution, the resulting convolution gives a logaritmically broad distribution. These convolution problems present a direct analytic theory of multifractal behavior. Later in this chapter, spin glass models of protein structure are discussed. These show similar probability convolutions and can produce multifractals. As a simple example of how the underlying substrate can impart multifractal behavior to a random distribution, the case of a binomial distribution on a fractal lattice, is considered in the next section.

4.2 Multifractal Behavior of the Generalized Cantor Set

We now consider the case of a simple binomial distribution that is spatially associated with a Cantor set. In a sense, this is a discrete version of the continuos distribution problem of Eq. 4.6. This problem mimics growth problems, such as diffusion-limited aggregation, where growth probabilities are distributed unevenly over a disordered structure. This scenario also has relevance for nonlinear dynamic systems. In chaotic systems, trajectories are confined to a fractal structure known as a "chaotic attractor." Points in phase space will fall on this attractor. If one asks how frequently a given region of the attractor is visited, it is found that there is an uneven probability distribution. This situation creates a multifractal as a result of a probability distribution on a fractal.

Presently, we consider the simple example of a binomial distribution superimposed on a Cantor set and calculate the multifractal parameters. We consider the Cantor set in Figure 1.3. Initially the line segment is partitioned into two segments. For a binomial distribution, we assign a weight of p to the first segment and a weight of $(1 - p)$ to the second one. In the second generation, each of these segments is broken into two and again assigned weight of p and $(1-p)$. The second generation will yield four segments with weights of: p^2, $p(1-p)$, $(1-p)p$, and $(1-p)^2$. The process is repeated and ultimately the binomial distribution is spread out over the Cantor dust.

The multifractal parameters for this distribution can be calculated by returning to Eq. 1.19:

$$Z(q) = \sum_i \mu_i^q = \delta^{-\tau} \tag{4.7}$$

where μ_i is the probability density and δ is the box size. For a single iteration, one has three intervals of which the leftmost has probability p, the middle one has probability of zero, and the rightmost has probability $(1-p)$. Normalizing the interval size to unity, Eq. 4.7 becomes:

$$Z(q) = p^q + (1-p)^q = \left(\frac{1}{3}\right)^{\tau} \tag{4.8}$$

giving:

$$\tau = \frac{\ln\,[p^q + (1-p)^q]}{\ln 3} \tag{4.9}$$

Recalling Eqs 1.24 and 1.25, the f versus a spectrum can be obtained.

For the case of a uniform distribution on the Cantor set, $p = 1/2$ and:

$$\tau = \frac{1}{\ln 3}\left\{\ln\left[2\cdot\left(\frac{1}{2}\right)^q\right]\right\} = (1-q)\,\frac{\ln 2}{\ln 3} = (1-q)D \tag{4.10}$$

where D is the fractal dimension of the Cantor set. The other parameters are given by:

$$\alpha = -\frac{d\tau}{dq} = D \tag{4.11a}$$

$$f = \tau + q\alpha = D \tag{4.11b}$$

From Eq. 4.11, the multifractal spectrum is a delta function at $\alpha = D$ with height $f = D$. This delta function occurs for uniform distributions regardless of the substrate. A uniform distribution on a regular deterministic fractal reflects only the dimension of the fractal and has no multifractal character. In the above example, if $p \neq 0.5$ then Eqs 4.10 and 4.11 are not so simple. In these instances, a broad multifractal distribution exists rather than a single dimension. For simple models the breadth of the multifractal spectrum is related to the uneven nature of the distribution. The magnitude of the spectrum will reflect the underlying fractality of the support. These consideration will be important when discussing multifractal spectra derived from protein structural parameters.

4.3 Multifractality of Order–Disorder Transitions in Biopolymers

As discussed in Chapter 1, the multifractal approach is closely related to the partition function of statistical mechanics and provides a method for defining "generalized" thermodynamic quantities. Indeed, many of the techniques of statistical mechanics are readily adapted to the calculation of multifractal spectra. The sequence generating function technique described in Chapter 3 for the statistical mechanics of biopolymers is now adapted to the multifractal formalism. Sequence-generating functions and their associated partition functions (Dewey, 1994a,b) can be used to determine the multifractal spectrum for a variety of one-dimensional systems. Our present emphasis is on order–disorder transitions in biopolymers of the alpha helix and the perfectly matched double stranded helix. When discussing order–disorder transitions, it is important to distinguish the behavior of synthetic polypeptides from proteins. Homopolymers have been investigated that can exist in a pure alpha helical state. Such polymers will give the highly cooperative transitions that are the subject of this section.

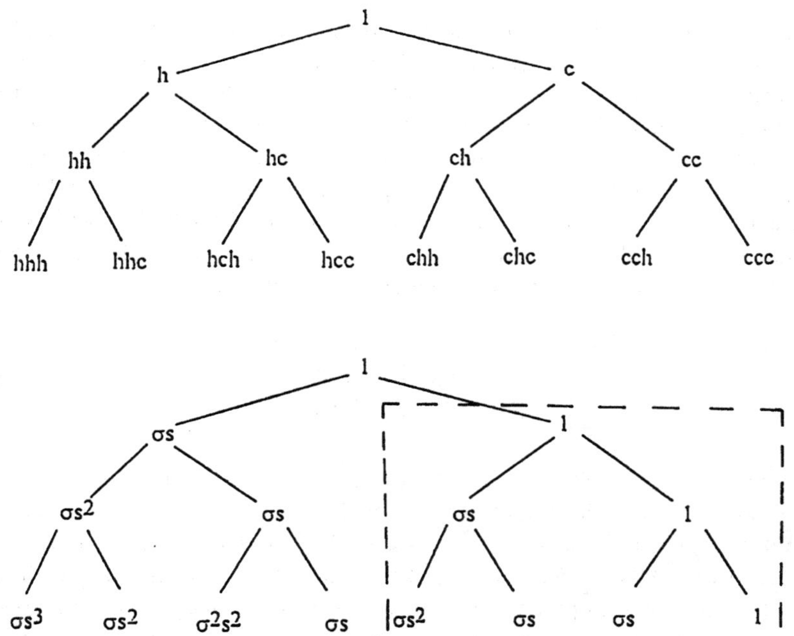

FIGURE 4.1 Binary tree structure for the Bragg–Zimm model of an alpha helix. Units in the linear polymer can exist in one of two states: h (helix) or c (coil). The top tree shows the binary generation of all possible configurations. The bottom tree shows the statistical weights associated with each sequence. Each coil unit gets a weight of 1, each helix a weight of s and each coil-helix boundary provides a weight of σ. Self-similar regions of the weighted tree originate from branches that have a weight of 1. The box encloses a region that is self-similar with the entire tree.

Proteins are usually a complicated mixture of secondary structures, and cannot be described by such simple statistical models. The multifractal behavior of proteins is treated in the subsequent section.

As seen in Chapter 3, most theories consider the units of a linear polymer as being in one of two states, helix (ordered) or coiled (disordered). The possible states of a biopolymer can then be represented as a binary tree structure (Figure 4.1, top) where each generation represents an additional unit added to the polymer. Statistical mechanical models associate a Boltzmann weight to each unit of the sequence. In Figure 4.1, the weights used in the Bragg–Zimm model of the alpha helix are superimposed on the "sequence tree." This model was discussed in Chapter 3; it is a binary model with one-step memory. The reason that the memory is one-step is that the probability of the new unit depends only on the nature of the preceding unit. Double helix models that form loops do not have this property and have long-term memory. Even for the simple Bragg–Zimm weighting scheme, self-similar branching of the tree occurs. The region enclosed in the box in Figure 4.1 is self-similar, demonstrating a fractal statistical weighting of sequences.

4.3.1 Sequence-Generating Functions and Multifractals

In this section, the connection between binary sequences and the multifractal formalism of Evertsz and Lyklema (1987) is made. Essentially, the multifractal formalism is a way of classifying the branches of the tree according to the probability of their occurrence. Similarly weighted branches will belong to the same "equivalence class." Each equivalence class has its own "singularity strength" and "generalized dimension" associated with it. A probability p_k is assigned to the kth microstate (or branch of a tree). The kth state will be some specific sequence, $\{i_0, \ldots, j_0\}$, of helix and coil states (see Chapter 3 for the notation used in this section). A singularity strength, α_k, is defined as:

$$\alpha_k = \lim_{n \to \infty} \left[\frac{\ln p_k}{\ln (1/\Omega)^n} \right] \tag{4.12}$$

where Ω is the number of possible states that a unit can assume. For the helix–coil problem, $\Omega = 2$ (h and c). The formalism may be extended to the more general case where Ω is unrestricted by methods described previously (Lifson, 1964; Dewey, 1994c). The factor $(1/\Omega)^n$ enters Eq. 4.12 as the weight of a microstate in which no preference is given for h or c states. This corresponds to the weight of a symmetric random walk in the "trajectory" formalism and is the counterpart of the box size in the box-counting formalism of multifractals. This term is essentially a normalization factor for embedding a given microstate in the set of all microstates. In constructing the multifractal spectrum, the probabilities of a given sequence are used rather then the statistical weights, w, which are Boltzmann factors.

The set of all microstates can be subdivided into equivalence classes with the same value of α. A different fractal dimension, $f(\alpha)$, can be found for each of these subsets. This is done by first defining a generalized dimension, d_q:

$$d_q = \lim_{n \to \infty} \left[\frac{-\ln \Gamma_n(q)}{(q-1)\ln \Omega^n} \right] \tag{4.13}$$

where

$$\Gamma_n(q) = \sum_k p(n;i_0, \ldots, j_0)^q = \sum_k \left\{ \frac{w(n;i_0, \ldots, j_0)^q}{Z(n,1)^q} \right\} \tag{4.14a}$$

$$= \sum_{s=0}^{n/2} \sum_{(i_\sigma, j_\sigma)} \prod_{\sigma=0}^{s} \left(\frac{u_{i_\sigma}}{x^{*i_\sigma}} \right)^q \left(\frac{v_{j_\sigma}}{x^{*j_\sigma}} \right)^q \tag{4.14b}$$

where now the summation of k runs over all microstates in the equivalence class. The conversion from probabilities to Boltzmann factors is made in Eq. 4.14a by dividing the weights by the normal partition function, now designated as $Z(n, 1)$. In Eq. 4.14b the partition function is represented as x^{*n}, and u_{i_σ} and v_{j_σ} are the Boltzmann weights for coil sequences of length i_σ and for helix sequences of length j_σ, respectively. The dimension, d_q, is the analog of the generalized dimension, D_q, found in the box-counting algorithm.

To obtain the multifractal spectrum, the fractal dimension, $f(\alpha(q))$, for each singularity, $\alpha(q)$, is calculated. This is done using (Evertsz and Lyklema, 1987):

$$z(\alpha(q)) = \Omega^{q\alpha(q) - (q-1)d_q} \tag{4.15}$$

$$\alpha(q) = \frac{d}{dq}\{(q-1)d_q\} \tag{4.16}$$

In the Evertsz and Lyklema notation, $z(\alpha)$, is the counterpart of $f(\alpha)$. The main computational task is centered on calculating, $\Gamma_n(q)$. Once this has been done, the straightforward manipulations of Eqs 4.13, 4.15, and 4.16 provide the multifractal spectrum.

The sum in Eq. 4.14b can be handled as the sum in the partition function of Chapter 3 (see Dewey, 1994c). Thus, a generalized treatment of sequence generating functions gives:

$$Z(n,q) = (2i\pi)^{-1} \oint_c \Xi(q,z) z^{n-1}\, dz \tag{4.17}$$

where the generalized moment generating function is

$$\Xi(q,x) = \sum_{n=0}^{\infty} Z(n,q)x^{-n} \tag{4.18a}$$

$$= U_0(q,x)V_0(q,x) \sum_{s=1}^{\infty} [U(q,x)V(q,x)]^s$$

$$= U_0(q,x)V_0(q,x)\,[1 - U(q,x)V(q,x)]^{-1} \tag{4.18b}$$

with the generalized sequence generating functions as

$$U(q,x) = \sum_{i=1}^{\infty} u_i^q x^{-i} \tag{4.19}$$

$$V(q,x) = \sum_{i=1}^{\infty} v_i^q x^{-i} \tag{4.20}$$

The moment generating function, $\Xi(q,x)$, and the generalized "partition" function, $Z(q,n)$, are extensions of the normal definitions given in Chapter 3. The generalized partition function will be of the form, $Z(q,n) = x^*(q)^n$ and the probability weighted function is given by $\Gamma_n(q) = [x^*(q)/x^*(1)^q]^n$. Now the generalized dimension, d_q, is:

$$d_q = \frac{-\ln x^*(q)}{(q-1)\ln 2} + \frac{q \ln x^*(1)}{(q-1)\ln 2} \tag{4.21}$$

The second term on the right-hand side of Eq. 4.21 results from the introduction of $Z(n,1)$ in Eq. 4.14a.

As a simple example, consider again the two state model with no nearest neighbor interactions (see Chapter 3). The sequence generating functions are:

$$U(q,x) = \frac{u^q}{(x - u^q)} \tag{4.22a}$$

$$V(q,x) = \frac{v^q}{(x - v^q)} \tag{4.22b}$$

Equation 4.17 is solved by obtaining the roots, $x^*(q)$, of the following equation:

$$U(q,x)V(q,x) - 1 = 0 \tag{4.23}$$

For this example, $Z(q,n) = x^{*n}(q) = (u^q + v^q)^n$. The generalized dimension now becomes:

$$d_q = \frac{-\ln (u^q + v^q)}{(q-1)\ln 2} + \frac{q \ln (u+v)}{(q-1)\ln 2} \tag{4.24}$$

Noting that the probability of a u unit occurring in a sequence can be defined as $p = u/(u+v)$, Eq. 4.24 gives the expected result for a binomial multiplicative process (Feder, 1988):

$$d_q = \frac{-\ln [p^q + (1-p)^q]}{(q-1)\ln 2} \tag{4.25}$$

Thus, the sequence generating function gives results equivalent to those obtained with other very different approaches.

4.3.2 Alpha Helical Model

The multifractal nature of the distribution of configurations in an alpha helix is treated within the framework of the Bragg–Zimm model (Zimm and Bragg, 1959). Essentially, we are examining the mathematics behind the self-similar branching of the binary tree in Figure 4.1. To do this, the sequence generating functions, Eqs 4.19 and 4.20, are given the specific form, $u_i^q = u^{iq}$ and $v_i^q = \sigma^q s^{iq}$. The parameters u and v represent the weight of a coil and helix unit, respectively, and boundary interaction results in the weight, σ. The generating functions are:

$$U(q,x) = \frac{u^q}{(x - u^q)} = (x-1)^{-1} \tag{4.26a}$$

$$V(q,x) = \frac{\sigma^q s^q}{(x - s^q)} \tag{4.26b}$$

For this situation, the integral in Eq. 4.17 is evaluated using Eq. 4.23 and gives:

$$x^*(q) = \frac{(1+s^q)}{2} + \frac{\{(1-s^q)^2 + 4\sigma^q s^q\}^{1/2}}{2} \tag{4.27}$$

From Eqs 4.13, 4.15, and 4.16, expressions for the various multifractal parameters can be derived, and give:

$$\alpha(q) = -\left(\frac{\ln \sigma}{2 \ln 2}\right)\left(\frac{\sigma^{q/2}}{1+\sigma^{q/2}}\right) + \frac{\ln(1+\sigma^{1/2})}{\ln 2} \tag{4.28}$$

$$f(q) = z(q) = (1+\sigma^{q/2}) \exp\left[-\frac{q \ln \sigma}{2}\left(\frac{\sigma^{q/2}}{1+\sigma^{q/2}}\right)\right] \tag{4.29}$$

Most studies of order–disorder transitions in biopolymers focus on the transition region. For the alpha helix model, the midpoint of the transition is defined as the point where the number of helical units equals the number of coil units. This occurs in the Bragg–Zimm model when $s = 1$. The cooperativity parameter, σ, determines the steepness of the transition as the temperature is changed. This parameter is temperature independent and typically has values of 10^{-2}–10^{-4} for polypeptides. Figure 4.2 shows the multifractal spectrum for the alpha helix at the transition temperature ($s = 1$) for different values of σ. As can be seen from the figure, the curves become broader as σ decreases, i.e., as the cooperativity increases. The breadth of these curves reflects the wider range of sequences that contribute to the distribution. A simple manifestation of this effect is also seen in standard statistical mechanical results. The average length of a helical segment, L_h, and the mean squared length, L_h^2 is calculated from the

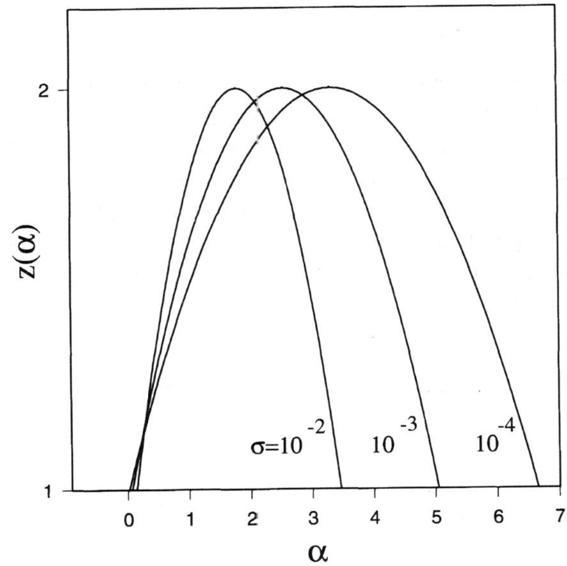

FIGURE 4.2 The multifractal spectra, $z(\alpha)$ versus α, for the alpha helix at the transition temperature ($s = 1$). The parameter $z(\alpha)$ is equivalent to $f(\alpha)$ in some notations. Three different values of the cooperativity parameter, σ, are shown. As σ decreases the distribution becomes broader.

partition function. The variance in length at the transition point, $s = 1$, is given by (Poland and Scheraga, 1970):

$$\Delta^2 = \overline{L_h^2} - \overline{L_h}^2 = \left(\frac{x^* s}{(x^* - s)^2} \right)_{s=1} = \sigma^{-1} + \sigma^{-1/2} \tag{4.30}$$

Thus, as σ decreases the distribution of helical lengths gets larger, indicating a broader range of sequences is being populated. This results in broader multifractal spectra.

The scaling indices α can be understood by reference to Figure 4.1. The multifractal spectrum is an ordering of the sequences in the binary tree according to their singularity strengths. For instance, at $s = 1$, the sequences, hhh . . . hhh and ccc . . . ccc have a probability associated with them of p_1^n, which is given by $1/x^{*n}$. Observing that $x^* = 1 + \sigma^{1/2}$, these sequences are seen to have a singularity strength at the minimum α ($f = 0$). This is given by: $\alpha_{min} = d_\infty = \ln(1 + \sigma^{1/2})/\ln 2$. The sequences, hchc . . . hchc and chch . . . chch, on the other hand, have a probability of $1^{n/2}\sigma^{n/2}/x^{*n}$ giving the maximum α with $\alpha_{max} = d_{-\infty} = \ln(1 + \sigma^{1/2})/\ln 2 - \ln \sigma/2 \ln 2$. These latter sequences are the most improbable because the unfavorable hc boundary interaction has been maximized. The multifractal spectrum orders the sequence tree via a folding or kneading process. In this application, a particularly simple folding occurs: the tree is folded in half, over top of itself. The sequences such as hhh . . . hhh and ccc . . . ccc are next to each other on the extreme left of the spectrum. Sequences such as hchchc . . . hchchc are now at the extreme right of the spectrum.

Changes in the multifractal properties of the system can be followed as it moves through the transition point. This is done by considering the behavior for three cases: $s < 1$, $s = 1$, and $s > 1$. It proves more convenient to monitor the generalized dimensions for these cases. Thus, we consider plots of d_q versus q for these conditions. From Eq. 4.21, one obtains the following results:

$$d_\infty = \frac{\ln\,[1 + \sigma s/(1 - s)]}{\ln 2} \qquad s < 1,\ s >> \sigma \qquad (4.31a)$$

$$= \frac{\sigma^{1/2}}{\ln 2} \qquad s = 1 \qquad (4.31b)$$

$$= \frac{\ln\,[1 - \sigma/(1 - s)]}{\ln 2} \qquad s > 1,\ s << 1/\sigma \qquad (4.31c)$$

and

$$d_{-\infty} = \frac{-\ln s\sigma}{2 \ln 2} + \frac{\ln\,[1 + \sigma s/(1 - s)]}{\ln 2} \qquad s < 1,\ s >> \sigma \qquad (4.32a)$$

$$= \frac{\sigma^{1/2}}{\ln 2} - \frac{\ln \sigma}{2 \ln 2} \qquad s = 1 \qquad (4.32b)$$

$$= \frac{\ln s}{2 \ln 2} - \frac{\ln \sigma}{2 \ln 2} + \frac{\ln\,[1 - \sigma/(1 - s)]}{\ln 2} \qquad s > 1,\ s << 1/\sigma \qquad (4.32c)$$

Using these expressions, the d_q versus q curves in Figure 4.3 were generated. As can be seen, the shapes of these curves are not dramatically affected by changes in the value of s. The average length of a helical stretch, L_h, and the variance, Δ^2, changes significantly through the transition point. However, the relative variance, Δ^2/L_h^2, does not change as dramatically. For s values of 0.9, 1.0, and 1.1, the relative variance changes as 0.89, 0.97, and 0.99, respectively. This suggests that there is a degree of symmetry between h and c states in the problem. For $s < 1$ the helix is the favored state, while for $s > 1$ the coil becomes the favored state. The same sequences exist in both states, so now an hhh . . . sequence at $s < 1$ will have a comparable weighting to a ccc . . . sequence at $s > 1$. This results in the similarity between the curves in Figure 4.3 for $s = 0.9$ and $s = 1.1$.

4.3.3 Perfect Matched Double Helix

In this section, the problem of a perfectly matched double stranded polymer is considered (see Chapter 3 for a more complete description of the model). This

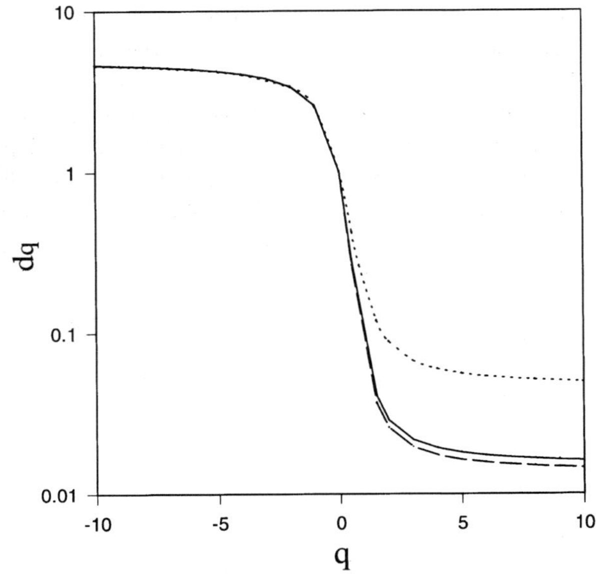

FIGURE 4.3 Plot of the generalized dimension, d_q, versus q for the alpha helix as the system proceeds through the transition temperature. Three different values of s are considered: s = 0.9 (–), 1.0 (. . .), and 1.1 (–––). The plot is semilogarithmic, to enhance the difference between the curves.

problem is somewhat more interesting then the alpha helix case and illustrates the strength of the sequence generating function approach over more traditional transfer matrix methods. In this case, the polymer no longer has one-step memory as in the previous example of the alpha helix. The long-range effects of the loop entropy create a situation where the sequence has long-term memory. From Chapter 3, recall that the entropy of a loop is given by:

$$S(N) = - c \ln N \tag{4.33}$$

where c is a constant and N is the number of members in the loop. The value of c depends on loop topology and the dimension of the embedding space (see Chapter 3). This entropy term effectively provides a long-range interaction, as all N members of the loop contribute to it. Because the perfectly matched double strand cannot simply be described by nearest neighbor effects, matrix techniques prove cumbersome.

To explore the multifractal behavior of this model, the treatment of Litan and Lifson (1965) is followed. The helical units are given Boltzmann weighting factors of s, so that $v_i = s^i$. The boundary factor, σ, is associated with the coil regions, and the weight of the loop is given, in accordance with Eq. 4.33, by $u_i = \sigma i^{-c}$. Thus, the sequence generating functions are now:

$$U\ (q,x) = \sigma \sum_{i=1}^{\infty} x^{-i}\, i^{-qc} \tag{4.34a}$$

$$V(q,x) = \sum_{i=1}^{\infty} x^{-i} s^{qi} = \frac{s^q}{x - s^q} \tag{4.34b}$$

The multifractal behavior is calculated from $x^*(q)$, which is obtained from the roots of Eq. 4.23. One must obtain the solution to:

$$x - s^q - s^q \sigma^q \sum_{i=1}^{\infty} x^{-i} i^{-qc} = 0 \tag{4.35}$$

With this solution, and using Eq. 4.21, the multifractal properties for the double strand are calculated.

Only the behavior of Eq. 4.35 in the interesting regime near the transition point, $s = 1$, is considered here. Following Fisher (1966), we make the transformations, $x = 1 + \varepsilon$ and $s = 1 + \delta$. With these transformations and using Fisher's asymptotic approximation for the summation, Eq. 4.35 can be greatly simplified. For the case of $q > 0$, it becomes:

$$\varepsilon + \sigma^q \varepsilon^{qc-1} = q\delta + \sigma^q \tag{4.36}$$

The generalized dimension is now given for small ε by:

$$d_q = \frac{-\varepsilon(q)}{(q-1)\ln 2} + \frac{q\varepsilon(1)}{(q-1)\ln 2} \tag{4.37}$$

where $\varepsilon(q)$ is the solution to Eq. 4.36 for a given value of q.

It is instructive to first consider the solution to Eq. 4.58 for the case $q = 1$. For small ε, there will be two different, approximate solutions depending upon the c value. These are given by:

$$\varepsilon(1) = \left[\frac{(\delta + \sigma)}{\sigma}\right]^{1/(c-1)} \qquad c < 2 \tag{4.38a}$$

$$\varepsilon(1) = \delta + \sigma \qquad c > 2 \tag{4.38b}$$

The implications of these solutions for the physics of the system have been explored extensively (Fisher, 1966; Poland and Scheraga, 1966, 1970). When $c > 2$, the system can exhibit a true first-order phase transition. In the context of the double strand model, this means that the fraction of the polymer in the helical state can abruptly drop to zero at a given temperature. Thus a pure coil phase will exist at high temperatures. Equation 4.38 represents behavior of the partition function at temperatures slightly below the transition temperature. From Eq. 4.38, it is seen that the generalized dimension will be different for $c < 2$ compared with $c > 2$ as a result of the $\varepsilon(1)$ term in Eq. 4.37. It is interesting to compare the d_q versus q curves for these two cases. When $c > 2$, the phase transition manifests itself as a discontinuity in d_q. This is because Eq. 4.36 has two different solutions depending on the value of qc. In a manner similar to Eq. 4.38, there is one solution for $qc < 2$ and a different

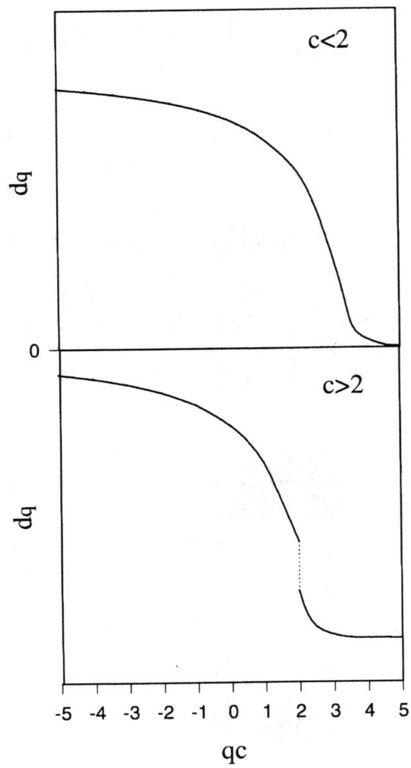

FIGURE 4.4 Plots of the generalized dimension, d_q, versus qc, for the perfect matched double stranded helix. The value of q is the moment of the multifractal expansion, and c is the exponent characterizing the loop entropy. Plots are general representations of the behavior of the system. The bottom curve shows the phase transition.

one for $qc > 2$. The discontinuity in d_q versus q occurs at $qc = 2$ and is illustrated in Figure 4.4. Values of the generalized dimension at this discontinuity are given by:

$$d_{2/c} \approx \frac{\sigma^{2/c}}{(1 - 2/c)\ \ln 2} \qquad 2/c \leftarrow q \tag{4.39a}$$

$$d_{2/c} \approx \frac{1}{(1 - 2/c)\ \ln 2} \qquad q \rightarrow 2/c \tag{4.39b}$$

where in Eq. 4.39a q approaches $2/c$ from the right, and in Eq. 4.39b q approaches from the left. One also can obtain the limit $d_\infty = \sigma/\ln 2$. For $c < 2$, one still has two different solutions: one at $qc > 2$ and one at $qc < 2$. However, in this case no discontinuity occurs, a consequence of there being no phase transitions for $c < 2$. The curves for $q < 0$ can also be obtained. In this case, the summation in Eq. 4.35 is more easily handled. This results in a single simple solution that does not have any discontinuities. The details of this calculation are omitted for brevity. The resulting generalized dimensions will have curves that appear as in Figure 4.4 for the two conditions $c < 2$ and $c > 2$.

The phase behavior exhibited by this system is similar to that seen in chaotic systems that exhibit intermittency (Csordás and Szépfalusy, 1989; Beck and Schlögl, 1993). In these cases, dynamical phase transitions occur as a result of anomalous scaling of sequence probabilities. For intermittent systems, the orbit remains in a subregion of phase space for an extended period of time. It can become almost periodic for this time and this results in long-time correlations in the symbolic sequences. One finds a specific sequence for which the probability decreases slower than exponentially with increasing length, N. Typically, one has $P(\text{sequence}) \sim N^{-\alpha}$, where α is a positive constant. This behavior is analogous to the probability weighting for a sequence of units within a loop. This analogy suggests that sequence generating function methods may find utility in the analysis of intermittent systems.

4.4 Protein Structure and Multifractals

We now turn to the more difficult problem of multifractals in proteins. This section is considerably more empirical than the previous one. An analysis is presented of a protein structural parameter, the solvent accessibilities of the amino acid side-chains, that reveals the multifractal nature of proteins. The solvent accessibility of an amino acid side-chain can be measured using the ball-rolling techniques described in Chapter 2. These accessibilities are determined from X-ray crystallographic data and usually a single ball radius is used. The most convenient representation is the fractional solvent accessibility. This is defined as the surface area of the side-chain in the protein divided by the surface area of a fully exposed amino acid side-chain in a tripeptide. This profile provides a sequential representation of a parameter that reflects the packing of the side-chains in the three-dimensional structure of the protein. In Figure 4.5, the solvent accessibilities of the amino acid side-chains are shown as a function of position along the protein backbone for the protein myoglobin. In Chapter 8, the hydrophobicity profile of the protein is introduced. This sequential array reflects the chemical composition of the protein.

Two questions are asked regarding such profiles. First, are there any correlations in this seemingly random pattern? Secondly, if there are correlations, what are the implications for the structure of the protein? There are a wide variety of methods for analyzing correlations in "noisy data." Previously, a Hurst analysis was used to demonstrate long-range correlations among Debye–Waller factors at different positions along a protein chain (Dewey, 1993). The present focus is to develop the multifractal formalism for analyzing sequential data (Balafas and Dewey, 1995; Strait and Dewey, 1995). One reason for choosing this particular approach is that the multifractal spectrum is a reflection of a hierarchical structure (as in a binary tree). Ultimately, one would hope to relate the multifractal properties of proteins to other hierarchical models such as the conformational substate model proposed by

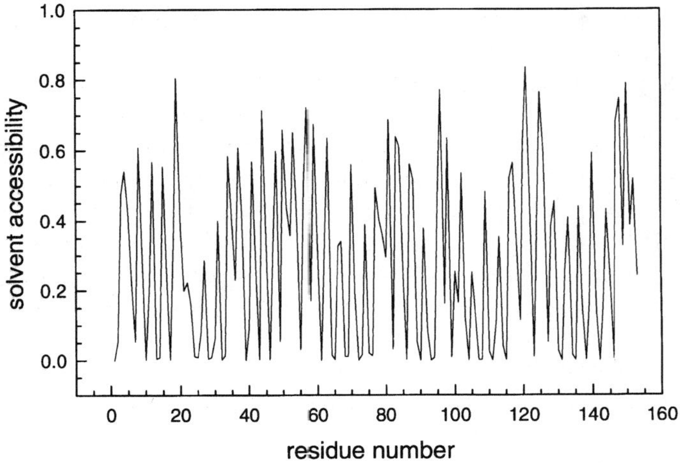

FIGURE 4.5 Solvent accessibility as a function of residue number for the protein myoglobin (153 residues). A value of 1 represents full exposure of the amino acid side-chain to the solvent; a value of 0 corresponds to the amino acid being completely buried in the protein. (Data from Balafas and Dewey, 1995.)

Frauenfelder et al. (1991). As will be seen, the multifractal spectra of the solvent accessibilities can also be used as a diagnostic test for proper protein folding. This is potentially of great utility in testing structure prediction algorithms.

4.4.1 Algorithm for Calculating Multifractal Spectra

A generalized box-counting method similar to that described in Chapter 1 is used to analyze sequential data (Figure 4.5) such as the solvent accessibilities. This method is used to derive the "generalized" dimensions associated with the shape of the curve. These generalized dimensions provide information on the hierarchical nature of the "noise" in Figure 4.5. To proceed with the details of the method, consider the solvent accessibilities as a sequence of numbers of length n:

$$\{x_i: i = 1, \ldots, n\} \tag{4.40}$$

where n is the number of amino acids in the protein. The solvent accessibilities are ordered along a linear array according to their respective positions in the amino acid sequence. This sequence is "covered" with boxes of a defined length and the value(s) of the accessibilities within a given box is assigned to the box. Initially, one starts with boxes of size 1 and covers a single amino acid residue. The value of the accessibility of that residue is assigned to its respective box. In the second iteration, the box size is doubled, giving two residues per box. The sum of the two accessibilities covered is now assigned to the respective box. The procedure is repeated with increasing box sizes. The following illustration shows how the box-counting algorithm is implemented for a linear data array. Boxes of size $\delta = 1, 2$, and k are illustrated below:

x_1	x_2	x_3	...	x_n
μ_1	μ_2	μ_3	...	μ_n

$x_1 + x_2$	$x_3 + x_4$	$x_5 + x_6$	...	$x_{n-1} + x_n$
μ_1	μ_2	μ_3	...	$\mu_{n/2}$

...

$\sum_{1}^{k} x_i$	$\sum_{k+1}^{2k} x_i$	$\sum_{2k+1}^{3k} x_i$	...	$\sum_{n-k}^{n} x_i$
μ_1	μ_2	μ_3	...	$\mu_{n/k}$

where x_i is the value of the solvent accessibility for the ith residue along the chain, and μ_j is the sum in the jth box:

$$\mu_j (\delta) = \sum_{i=1}^{\delta} x_{i+\delta j} \tag{4.41}$$

A function, $Z_q(\delta)$, is now defined that provides the qth moment of the measure:

$$Z_q (\delta) = \sum_j \mu_j^q (\delta) \tag{4.42}$$

The scaling Ansatz is made such that (Feder, 1988):

$$Z_q(\delta) = \lim_{\delta \to 0} \delta^{-\tau(q)} \tag{4.43}$$

where $\tau(q)$ is identical to the exponent described in Eq. 4.7. This exponent is obtained from the initial or limiting slope of a log $(Z_q(\delta))$ versus log (δ) plot. This procedure is represented as:

$$\tau(q) = - \lim_{\delta \to 0} \frac{\ln (Z_q(\delta))}{\ln (\delta)} \tag{4.44}$$

The singularities of the measure are again characterized by the exponent, α. This parameter is related to τ according to the relation:

$$\alpha(q) = - \frac{d}{dq} \tau(q) \tag{4.45}$$

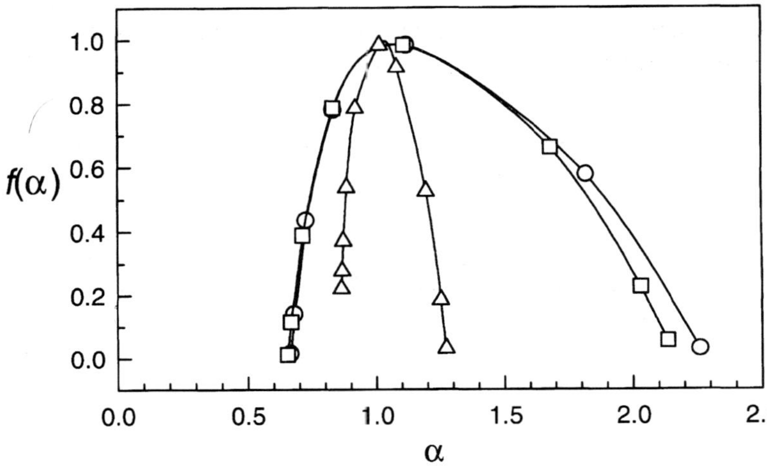

FIGURE 4.6 Multifractal spectra of hexokinase using the wrap-around algorithm: forward sequence (○) and reversed sequence (□). The multifractal spectrum for data derived from a random number sequence having a length identical to hexokinase (374 residues) is also shown (△). Note the narrowness of the random number spectrum compared to that of the protein. (Data from Balafas and Dewey, 1995.)

Substitution of Eq. 4.44 into Eq. 4.45 yields

$$\alpha(q) = -\lim_{\delta \to 0} \frac{\sum_j \mu_j^q \ln(\mu_j)}{Z_q(\delta) \ln(\delta)} \tag{4.46}$$

Similarly, $\alpha(q)$ is obtained from the limiting slope of a plot of $\Sigma_j \mu_j^q \ln(\mu_j)/Z_q(\delta)$ versus $\ln(\delta)$. Now, the multifractal spectrum, $f(\alpha)$ versus α, can be calculated according to $f(\alpha) = q\alpha(q) + \tau(q)$.

Thus, a multifractal spectrum is generated for the profile by calculating $Z_q(\delta)$ at a fixed q value and varying δ. From the two linear regressions, $f(\alpha)$ and α are determined. The entire spectrum is then generated by varying q. Both positive and negative integer values of q are used. For a given box size, δ, there may not be sufficient residues in the last box to complete the sequence. To circumvent this problem, an initial segment of the sequence was appended to the end of the sequence to fill the last box (Balafas and Dewey, 1995). Thus, the algorithm uses a wrap-around method to create periodic boundary conditions. As a check on the error introduced by this procedure, periodic boundary conditions were applied for sequences taken in both the forward (amino → carboxy) and reverse (carboxy → amino) directions. This changes the content of the appended sequence. The wrap-around algorithm produced essentially the same spectra, regardless of the direction of the sequence. A typical comparison of forward and reverse sequence spectra is shown in Figure 4.6.

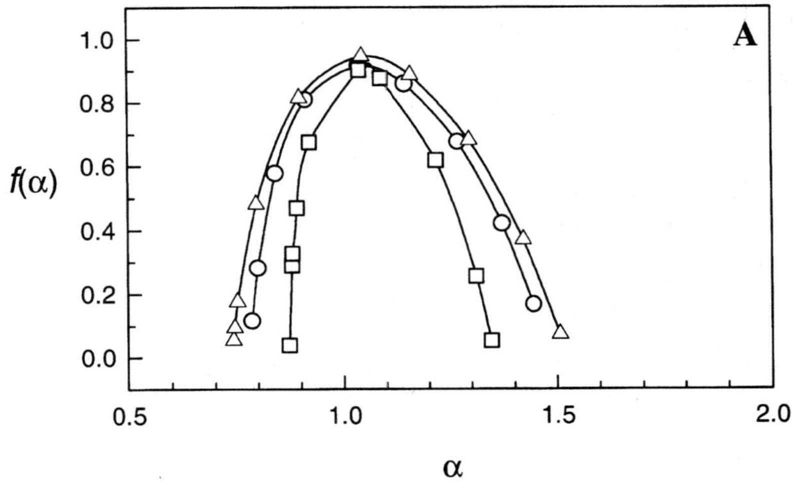

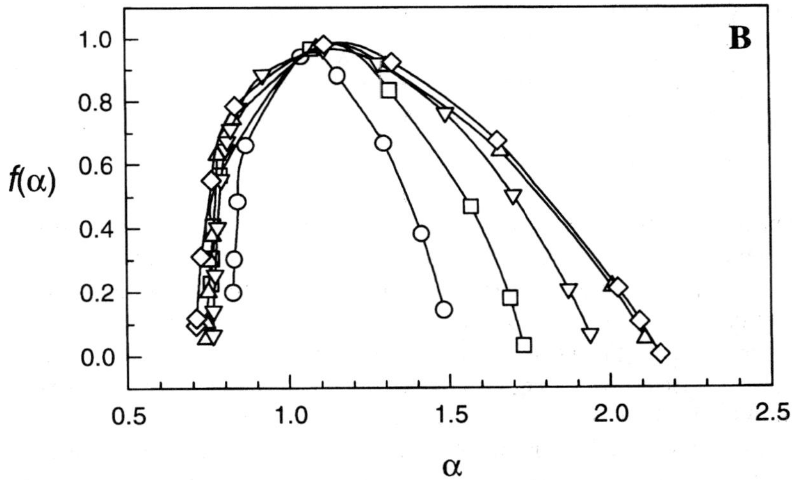

FIGURE 4.7 Multifractal spectra of solvent accessibilities. (A) Alpha class: (○) cytochrome c; (□) calcium binding parvalbumin; (△) myoglobin. (B) Beta class: (○) plastocyanin; (□) α-lytic protease; (△) concanavalin A; (▽) elastase; (◇) acid proteinase. (Data from Balafas and Dewey, 1995.)

4.4.2 Multifractal Spectra of Solvent Accessibilities from Proteins

Using the algorithm given in the previous section, the multifractal spectra of the solvent accessibilities of a number of proteins were examined (Balafas and Dewey, 1995). Figure 4.6 shows the spectrum for the protein hexokinase. As can be seen, a simple convex spectrum is observed that can be characterized by its intercepts, α_{min} and α_{max}. Note that all spectra have a maximum at $f = 1$, reflecting the dimensionality of the support, i.e., the linear array. As a comparison, the spectrum for a random sequence was generated, and this is also shown in the figure. The random number sequences have narrower spectra than the protein accessibility

TABLE 4.1 Summary of the results of the multifractal analysis of proteins*

Protein class	Protein	Length	α_{min}	α_{max}	$\sigma_p(00)$	$\sigma_p(11)$
FeS	Ferredoxin	54	0.88	1.12	0.54	0.46
	Ferredoxin	98	0.76	1.53	0.59	0.35
Alpha	Cytochrome c	103	0.79	1.47	0.58	0.36
	Calcium binding parvalbumin	107	0.87	1.36	0.55	0.39
	Myoglobin	153	0.73	1.52	0.60	0.35
Beta	Plastocyanin	99	0.82	1.51	0.57	0.35
	α-Lytic protease	198	0.75	1.73	0.60	0.30
	Concanavalin A	237	0.73	2.12	0.60	0.23
	Acid proteinase	330	0.71	2.13	0.61	0.23
Alpha/beta alternate	Flavodoxin	138	0.75	2.49	0.59	0.18
	Adenylate kinase	194	0.71	2.04	0.61	0.24
	Carboxypeptidase A	307	0.7	2.88	0.62	0.14
Alpha/beta segregate	Papain D	212	0.7	2.46	0.62	0.18
	Actinidin	218	0.7	2.4	0.62	0.19
	Carbonic anhydrase B	261	0.75	2.53	0.60	0.17
	Thermolysin	316	0.7	2.93	0.62	0.13

* $\sigma_p(01)$ and $\sigma_p(10)$ can be obtained from Eqs A4.9 and A4.13.

sequences of corresponding length. Thus, the breadth of the multifractal spectrum of the protein can be attributed to nonrandom effects within the data sequence.

In Figure 4.7, similar spectra for a number of different proteins are shown. These particular spectra are for alpha-helix-rich proteins (Figure 4.7A) and beta-sheet-rich proteins (Figure 4.7B). The spectra are generally asymmetric about the maxima, with the region of large α values (right-hand side) being much broader. The region to the left-hand side of the maximum is determined by the positive moments of the distribution and is dominated by high probability sequences of accessibility values. The region to the right of the maximum is due to the negative moments and, consequently, is dominated by low probability sequences. Much of the broadness of the entire spectra is due to these low probability sequences. The values of α_{min} and α_{max} for a number of different proteins are given in Table 4.1.

Ideally, one can use α_{min} and α_{max} to extract information regarding the underlying physics of this multifractal behavior. Such approaches are described in the Appendix to this chapter and in Chapter 7. It is difficult to make this connection because relatively simple multiplicative models can fit the spectra. Models are not easily discriminated by their ability to reproduce the spectra. However, these results do show that the underlying phenomenon does not require a complicated model to describe its multifractal behavior. To test the validity of the multiplicative model, one must provide an interpretation of the scaling parameters from a physical model. For instance, one could relate the scaling factors to the Bragg–Zimm model of an alpha helix in as described in the Appendix. Using these results, one obtains physically unrealistic parameters for the cooperativity parameter. Not too surprisingly, the helix–coil model of an alpha helix is seen not to be applicable to proteins.

In Chapter 7, the multifractal analysis of hydrophilicity profiles is discussed. As will be seen, these profiles follow an even simpler model than the solvent accessibility profiles. Also, in Chapter 7, the connection is made between the multifractal approach and encoded walks. Encoded walks have been used to examine sequence correlations in both nucleic acids and proteins.

An interesting and very useful comparison can be made between the multifractal behavior of sequences generated from native proteins and improperly folded proteins. Computer simulations can be used to generate improperly folded proteins (Novotny et al., 1984). This is done by selecting two proteins with a commensurate number of amino acid residues but dissimilar structures. The amino acid side-chains of one protein are then placed in sequence on the backbone of the other, and vice versa. The incorrectly folded structures were then energy minimized using a geometry optimization algorithm. This minimization procedure optimizes the local structure of the side-chains, while not affecting the tertiary structure of the protein. The final "product" is a misfolded protein that is not grossly incorrect.

A number of different protein pairs were used to generate misfolded protein structures. After computing the solvent accessibility profiles for both the native and misfolded pairs, the multifractal spectra were calculated. An example of such a pair of spectra is shown in Figure 4.8 for the elastase–concanavalin A swap-pair. The notable feature of the multifractal spectra of the misfolded structures is that they are much narrower than the properly folded proteins. The reduced width of the spectrum ($\Delta\alpha$) for a number of misfolded proteins is shown in Table 4.2.

TABLE 4.2 Comparison of multifractal parameters for native and misfolded proteins

Protein	α_{min}	α_{max}	$\Delta\alpha$[1]
Elastase	0.76	1.96	1.20
Misfold (concanavalin A frame)	0.78	1.71	0.93
Concanavalin A	0.74	2.14	1.40
Misfold (elastase frame)	0.76	1.72	0.96
Carboxypeptidase A	0.7	2.88	2.2
Misfold (thermolysin frame)	0.7	2.06	1.4
Adenylate kinase	0.71	2.04	1.33
Misfold (α-lytic protease frame)	0.71	1.76	1.05
Flavodoxin	0.75	2.49	1.74
Misfold (myoglobin frame)	0.75	1.70	0.95
Carboxypeptidase A	0.7	2.88	2.2
Misfold (Arabinose-binding protein frame)	0.74	1.96	1.22
Arabinose-binding Protein	0.77	2.21	1.44
Misfold (carboxypeptidase A frame)	0.70	2.26	1.56
Ferredoxin	0.76	1.53	0.77
Misfold (Plastocyanin frame)	0.83	1.40	0.57
Plastocyanin	0.82	1.51	0.69
Misfold (ferredoxin frame)	0.81	1.42	0.61

[1] $\Delta\alpha = \alpha_{max} - \alpha_{min}$.

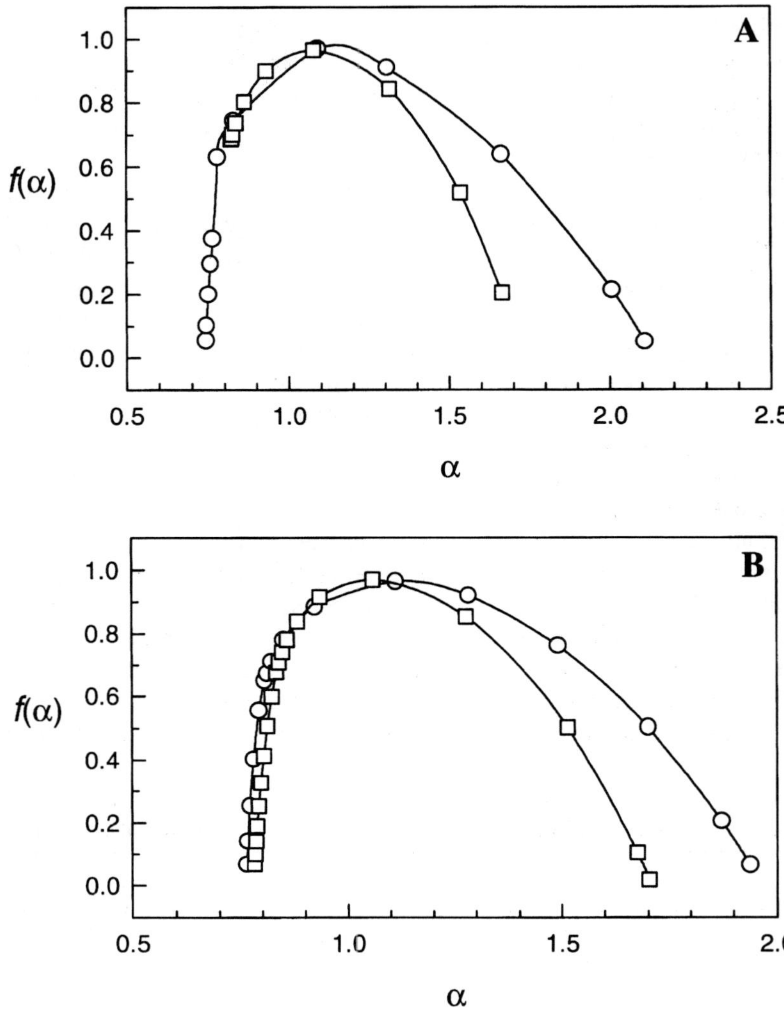

FIGURE 4.8 Multifractal spectra of solvent accessibilities of improperly folded proteins and the corresponding properly folded structure. (A): (○) concanavalin A; (□) misfolded concanavalin A on an elastase frame. (B): (○) elastase; (□) misfolded structure elastase on a concanavalin A frame. Note that misfolded proteins give narrower spectra than properly folded proteins. (Data from Balafas and Dewey, 1995.)

Visualization of the improperly folded proteins via computer graphics reveals that the structures are not closely packed, containing many open regions. These open spaces destroy the alternating regions of high and low solvent accessibilities of the sequence found in properly folded proteins. They introduce more randomness and less correlation into the sequence data. The narrowing of the spectrum is an observation that exists for all the misfolded proteins except one, arabinose-binding protein. In this case, the protein has a large binding cleft (Richardson, 1981), and thus a rather open structure. Improperly folding the arabinose-binding protein sequence on the carboxypeptidase A frame produced a well-packed structure. Therefore, the multifractal spectrum of arabinose-binding protein is narrower than that of the misfolded structure.

An important problem in biochemistry is the development of algorithms to predict protein structure from sequence information (Fetrow and Bryant, 1993; Blundell et al., 1987). As competing algorithms are developed, one would like to assess their relative performance on protein sequences the structures of which are not known. The multifractal approach potentially provides a diagnostic tool for evaluating the performance of such algorithms. It is anticipated that the best algorithm will generate the widest multifractal spectrum. Additionally, it is important to compare structures determined from nuclear magnetic resonance (NMR) and X-ray data. Often these structures are not the same, and it is not clear whether the difference is due to environmental effects or to failings of the technique. This algorithm provides a tool for a quantitative comparison. Since a correct structure is expected to have a broader spectrum than an incorrect one, the multifractal spectra may help sort out the causes of different X-ray and NMR structures.

4.4.3 Origins of Multifractality in Proteins

The previous section showed that the solvent accessibility of amino acid side-chains is distributed in a multifractal fashion along the length of the polymer. This distribution is consistent with that generated by a random multiplicative process with one-step memory. However, the physical implications of this are not so obvious. Returning to our original arguments (Section 4.1), there are at least two potential sources for the observed multifractality: a random multiplicative process and a convolution of probability functions. Both these factors may contribute to the multifractality of proteins.

As discussed in Section 4.3.2, the Bragg–Zimm model of the alpha helix exhibits multifractal behavior (Dewey, 1994a,b). This model can be mapped into the 2×2 P model used in the Appendix to analyze the experimental multifractal spectrum. Such simple helix–coil models are inadequate for describing protein structure because they give unrealistic physical parameters and thermodynamics. Nevertheless, they can serve as a starting point for other statistical approaches. Recent work has made the analogy between proteins and spin glasses (Bryngelson and Wolynes, 1987, 1989; Wolynes, 1992). Helix–coil models are a form of the Ising spin model. Similarly, spin glass models can be adapted to capture the long-range order and intrinsic disorder in proteins. This is done by using a spin Hamiltonian with random spin–spin coupling factors. Wolynes and coworkers have exploited these analogies to model protein structure and folding. To determine the energy of the system, averages over these random variables must be performed in a specific way. One method of doing this is the replica approach developed by Edwards (see Chan and Dill, 1991). This method results in a convolution of a probability function with the partition function. This convolution will have features of Eq. 4.6. With the spin glass analogy, multifractality in proteins can be attributed to two sources. First, the Ising problem, even in one dimension, has an intrinsic fractal nature (Bak and Bruinsma, 1982; Bruinsma and Bak, 1983). This can be put in the context of a random multiplicative process as done in Section 4.3. In addition to this effect, the averaging of coupling constants introduces a convolution of probability functions, a second potential source of multifractal behavior. More work is required to put these ideas on a quantitative footing.

Finally, we turn to the intriguing question of why misfolded structures show narrower multifractal spectrum than do properly folded structures. In plots of $f(\alpha)$ versus α, the region to the left of the maximum corresponds to positive moments ($q > 0$). This region is dominated by common events or sequences. In this region there are not strong differences between the native and misfolded proteins. The region to the right of the maximum is influenced by negative moments ($q < 0$). It is dominated by rare events or sequences. The misfolded spectrum shows large changes in the right-hand region. These changes make the misfolded protein appear closer to a random sequence. A properly folded protein will have rare (or nonrandom) combinations of exposed residues. The multifractal spectrum shows great sensitivity to correlated packing of side-chains as reflected in this right-hand region.

4.5 Summary

The origins of multifractality have been explored in this chapter with specific reference to biopolymers. Multifractals are probability distributions the moments of which are characterized by an infinite number of fractal dimensions. Two common sources of multifractal behavior are random multiplicative processes and convolutions of distributions. As a simple example, we have considered a binomial distribution on a Cantor set and shown how multifractal properties arise from such a random distribution on a disordered substrate. We then turned to a discussion of synthetic and natural biopolymers that show well-defined order–disorder transitions. Typically, these are helix–coil transitions and can occur in nucleic acids as well as polypeptides. Using the statistical mechanics of helix–coil models, the population distribution could be represented as a random multiplicative process that has multifractal behavior. Because of the diversity of secondary structural units and the intrinsic disorder, proteins offer a more complicated milieu for multifractals. Analysis of a structural parameter called the solvent accessibility show that the packing of protein side-chains is distributed in a multifractal fashion. The origin of this behavior is attributed to both random multiplicative and convolution effects, as demonstrated in spin glass models of proteins.

Appendix: Extraction Procedures and Comparison with Transfer Matrix Methods

Transfer matrix techniques have figured prominently in the statistical thermodynamics formulation of dynamic processes. These methods provide a facile means of extracting information on a dynamical process via the multifractal formalism. In this section we discuss the connection between transfer matrix techniques and sequence generating function. Because of the heavy use of transfer matrix methods, it is important to establish the correspondence between them and sequence generating functions. The motivation for using sequence generating functions over the transfer matrix is that they are more amenable for handling sequences with memory. We also demonstrate how extraction techniques can be used to relate the features of the multifractal spectrum to model parameters. Such extraction procedures are crucial in proceeding from experimental data to theoretical models. While this section is more mathematical in nature, it does provide useful results for analyzing multifractal spectra.

Recently, a general description of three types of multifractal model was presented using transfer matrix methods (Chhabra et al., 1989). Each of these three approaches can be used to associate a multiplicative process with a singular measure. These three approaches are known as the P, L, and LP models. The P model is perhaps the most common and, for our present purpose, the most natural. In this approach, the measure is covered with boxes of equal length and the probability in each box is allowed to vary. The variation of these probabilities with box size then reveals the complexity of the measure. In the L model, one requires that each box encompass a region of equal probability. Thus, the box sizes must vary, while the probabilities remain fixed. The probability in each box is then reduced in a fixed fashion and the resulting adjustment of the length of the boxes again reflects the complexity of the measure. The LP model is a hybrid model in which both the box length and probability are adjusted in a prescribed manner. In principle, each approach will provide identical information on the measure and the method of choice is dictated by the computational ease for a specific problem. For each of these transfer matrix formulations, a corresponding sequence generating function approach can be developed. For sequences in biopolymers, the P model is the most natural.

In the following development, the P model as described by Chhabra et al. (1989) is used. A partition function is defined as:

$$\Gamma_n(q) = \sum_{i=1}^{N} P_i^{(n)} = N^{-\tau} \tag{A4.1}$$

where N is the number of boxes in the box-counting algorithm and $P_i^{(n)}$ is the probability in the ith box in the nth generation. Assuming the boxes are created from a tree structure of n generations, then $N = \Omega^n$ where Ω was defined previously as the number of possible states in a generation. The exponent τ is related to the generalized dimension, d_q, by:

$$\tau = (q-1)d_q \tag{A4.2}$$

To develop the transfer matrix formalism it proves convenient to take the ratio of the partition function at two successive generations (n and $n+1$). This gives:

$$\frac{\Gamma_{n+1}(q)}{\Gamma_n(q)} = \frac{\sum_{i=1}^{\Omega^{n+1}} (P_i^{(n+1)})^q}{\sum_{i=1}^{\Omega^n} (P_i^{(n)})^q} = \Omega^{-\tau} \tag{A4.3}$$

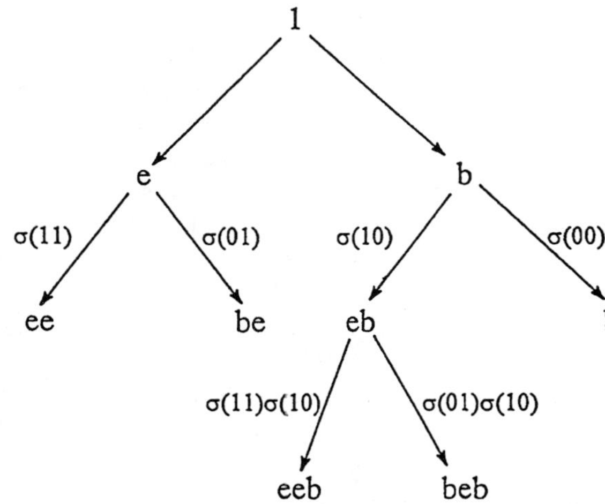

FIGURE A4.1 A schematic diagram showing a portion of a binary tree that corresponds to a random multiplicative process. The multifractal spectrum of the protein solvent accessibility data cold be analyzed with such a process. Sequences of exposed (e) or buried (b) units in the protein are generated by a random process. The probability of a given step is determined by the preceding unit. The Feigenbaum scaling functions, $\sigma_p(ij)$, are associated with the steps indicated in the tree.

Scaling information can be readily extracted using a scaling function, σ_p, defined as:

$$\sigma_p\,(\varepsilon_{n+1}, \ldots, \varepsilon_0) = \frac{P(\varepsilon_{n+1}, \ldots, \varepsilon_0)}{P(\varepsilon'_n, \ldots, \varepsilon'_0)}\, \delta_{\varepsilon'_n \varepsilon_n}, \ldots, \delta_{\varepsilon'_0 \varepsilon_0} \tag{A4.4}$$

where $P_i^{(n)}$ is now explicitly represented as $P(\varepsilon_n, \ldots, \varepsilon_0)$, where $\varepsilon_n, \ldots, \varepsilon_0$ is the string of states ε that make up the ith box in the nth generation of the tree. The delta functions in Eq. A4.4 insure that the probability ratio contains the same initial sequence. This measure is an analog of the Feigenbaum scaling function. Substituting Eq. A4.4 into A4.3 gives:

$$\sum_{(\varepsilon_{n+1}, \ldots, \varepsilon_0),\,(\varepsilon'_n, \ldots\, \varepsilon'_0)} \delta_{\varepsilon_n, \varepsilon'_n} \cdots \delta_{\varepsilon_0, \varepsilon'_0} [\sigma_p(\varepsilon_{n+1}, \ldots, \varepsilon_0)]^q\,]P(\varepsilon'_n, \ldots, \varepsilon'_0)]^q$$

$$= \lambda(-\tau) \sum_{(\varepsilon_n, \ldots, \varepsilon_0)} [P(\varepsilon_n, \ldots, \varepsilon_0)]^q \tag{A4.5}$$

where again the delta functions assure the similarity of the first n elements of the two sequences and $\lambda\,(-\tau)$ is the leading eigenvalue of the transfer matrix, **T**:

$$\mathbf{T}_{\varepsilon_{n+1}, \ldots, \varepsilon_0,\,(\varepsilon'_n, \ldots, \varepsilon'_0)} = \sigma_p^q\,(\varepsilon_{n+1}, \ldots, \varepsilon_0)\, \delta_{\varepsilon_n, \varepsilon'_n}, \ldots, \delta_{\varepsilon_0,\, \varepsilon'_0} \tag{A4.6}$$

The transfer matrix is related to the exponent τ through the largest eigenvalue, $\lambda(-\tau) = \Omega^{-\tau}$. This provides the association between the probabilities of a given model and the multifractal spectrum.

It is helpful to consider an example of this formalism and the 2×2 P model will be treated. Despite the simplicity of this model, it can be used in to generate the multifractal spectrum for a variety of mappings. This model can treat the two-scale Cantor set and tree structures with one time-step memory. Figures A4.1 and 4.1 illustrate a binary random multiplicative process. In Figure A4.1 the scaling factors, σ_{ij}, are given explicitly. In the 2×2 model, each unit in the sequence can exist in one of two states. For heuristic purposes, the context of a protein with exposed or buried units is considered. In Figure A4.1, the probability of an e or b is determined by the preceding unit, so the model has one-step memory. These binary trees are hierarchical structures, as the early branches dictate the region in which the sequence finally appears.

The multifractality of such binary trees has been investigated extensively. They give multifractal spectrum (f or z versus α) that can be characterized by three independent parameters, α_{min}, α_{max}, and f_{max}. An alternative (and equivalent) representation is to plot d_q versus q. For these plots, three points again dominate the curve. They are: d_0, d_∞, and $d_{-\infty}$. These parameters can be derived from the probabilities associated with each branch of the tree (see Chhabra et al., 1989). The connection between these probabilities and the multifractal spectrum is outlined here. For the model shown in Figure A4.1, there are four such scaling factors: $\sigma_p(00)$, $\sigma_p(10)$, $\sigma_p(11)$, and $\sigma_p(01)$. These scaling factors are defined as:

$$\sigma_p(ij) = \frac{P(ij)}{P(j)} \tag{A4.7}$$

where $P(ij)$ is the probability of an i unit following a j unit, and $P(j)$ is the probability of a j unit existing. The product of the scaling factors associated with each branch gives the probability for a specific configuration or sequence.

The transfer matrix for the 2×2 binary model is now given by

$$\mathbf{T} = \begin{bmatrix} \sigma_p^q(00) & \sigma_p^q(01) \\ \sigma_p^q(10) & \sigma_p^q(11) \end{bmatrix} \tag{A4.8}$$

where two conservation equations hold:

$$\sigma_p(00) + \sigma_p(10) = 1 \tag{A4.9a}$$

$$\sigma_p(01) + \sigma_p(11) = 1 \tag{A4.9b}$$

Equations A4.9a and A4.96 are a result of the conservation of probability in each binary splitting of the tree. Thus, for the 2×2 model there are three independent parameters, two from the transfer matrix and the parameter, Ω, determining the splitting ratio. These three values ultimately determine the shape of the $f(\alpha)$ versus α and of the d_q versus q curve. The leading eigenvalue of the transfer matrix is determined by:

$$\lambda^2(q) - \lambda(q)\,\mathrm{Tr}\,[\mathbf{T}] + \det\,[\mathbf{T}] = 0 \tag{A4.10}$$

and is given explicitly by:

$$\lambda(q) = \frac{\sigma_p^q(00) + \sigma_p^q(11)}{2} + \left\{ \left[\frac{\sigma_p^q(00) - \sigma_p^q(11)}{2} \right]^2 + \sigma_p^q(01)\sigma_p^q(10) \right\}^{1/2} \quad \text{(A4.11)}$$

The exponent τ is readily calculated from this result as

$$\tau = \frac{-\ln(\lambda)}{\ln(\Omega)} \quad \text{(A4.12)}$$

where in the present case Ω is equal to 2. The generalized dimensions are given by Eq. A4.2 and the multifractal spectrum may be calculated from Eq. 4.45.

Using the above results, the scaling functions can be extracted from the multifractal spectrum. At the extrema of the spectra ($f = 0$), one can assign:

$$\alpha_{min} = d_{\infty} = \frac{\ln(\sigma_p(00))}{-\ln(2)} \quad \text{(A4.13a)}$$

$$\alpha_{max} = d_{-\infty} = \frac{\ln(\sigma_p(11))}{-\ln(2)} \quad \text{(A4.13b)}$$

The conservation equations (Eq. A4.9) are then used to determine the other parameters. The third independent parameter, f_{max} (or d_0), is fixed by the value of Ω and is unity for a binary process on a linear sequence. Chhabra et al. (1989) have pointed out the problems associated with extracting information from multifractal spectrum. Typically, experimental data produce simple convex curves that are not very distinctive. These curves are readily fitted with three parameters, so often 2×2 models are completely adequate. This behavior makes it difficult to discriminate physical models solely on the basis of curve fits of the multifractal spectrum.

The connection between the sequence generating functions and the transfer matrix is well illustrated with the example of a 2×2 P model. The equivalence of the transfer matrix and generating function approaches can be demonstrated in two ways. First, using the Boltzmann weights given in Section 4.3.1, identical expressions can be derived, as with the Feigenbaum scaling functions used in this section. Alternatively, the scaling functions can be incorporated directly in the sequence generating function formalism. This second course is followed here. The connection between the transfer matrix and sequence generating function approaches can be made directly via the Feigenbaum scaling function. Using the notation given in Section 4.3.1, it can be represented as:

$$\sigma_p\,(n+1;\, i_0, \ldots, j_0) = \frac{p(n+1;\, i_0, \ldots, j_0)}{p(n;\, i_0, \ldots, j_0)}$$

$$= \frac{w(n+1;\, i_0, \ldots, j_0)/Z(n+1,\, 1)}{w(n;\, i_0, \ldots, j_0)/Z(n,\, 1)}$$

$$= \frac{w(n+1;\, i_0, \ldots, j_0)}{w(n;\, i_0, \ldots, j_0)x^*\,(1)} \tag{A4.14}$$

where the last equality in Eq. A4.14 is obtained from $Z(n, 1) = x^*\,(1)^n$. The sequence generating functions are derived by analogy with Eqs 4.19 and 4.20. First, consider a sequence containing zeros. As the sequence must be initiated with a 1, the generating function is given by:

$$U(q,x) = \frac{\sigma_p^q(01)}{x} + \frac{\sigma_p^q(00)\sigma_p^q(01)}{x^2} + \frac{\sigma_p^q(00)\sigma_p^q(00)\sigma_p^q(01)}{x^3} + \ldots \tag{A4.15a}$$

$$= \frac{\sigma_p^q(01)}{x} \sum_{i=1}^{\infty} \left(\frac{\sigma_p^q(00)}{x}\right)^i = \frac{\sigma_p^q(01)}{x - \sigma_p^q(00)} \tag{A4.15b}$$

A similar function is defined for the 1 sequence and is given by:

$$V(q,x) = \frac{\sigma_p^q(10)}{x - \sigma_p^q(11)} \tag{A4.16}$$

The eigenvalue, $\lambda(q)$, of the transfer matrix problem is now given by the largest root of Eq. A4.17 (equivalent to Eq. 4.23):

$$1 - U\,(q,x)\;V\,(q,x) = 0 \tag{A4.17}$$

and for the binary model this gives Eq. A4.11. Thus the two methods yield identical results. The sequence generating function method presented in this section is slightly different from the method in Section 4.3 where the development was based on Boltzmann weights rather than Feigenbaum scaling factors. In the notation used in Section 4.3, one has:

$$\lambda(q) = \frac{x^*(q)}{x^{*q}(1)} = (\Gamma_n(q))^{1/n} \tag{A4.18}$$

This is a result of the different definitions of the sequence generating functions (Eqs A4.13 and A4.16 versus Eqs 4.19 and 4.20) in the two sections. Usually, Eqs 4.19 and 4.20 present less cumbersome forms and facilitate calculations.

References

Bak, P., and Bruinsma, R. 1982. One-Dimensional Ising Model and the Complete Devil's Staircase. *Phys. Rev. Lett.* **49**, 249–251.

Balafas, J., and Dewey, T. G. 1995. A Multifractal Analysis of Solvent Accessibilities in Proteins. *Phys. Rev. E* **52**, 880–887.

Beck, C., and Schlögl, F. 1993. *Thermodynamics of Chaotic Systems*. Cambridge University Press, Cambridge, pp. 248–251.

Bruinsma, R., and Bak, P. 1983. Self-Similarity and Fractal Dimension of the Devil's Staircase in the One-Dimensional Ising Model. *Phys. Rev.* B **27**, 5824–5825.

Bryngelson, J. D., and Wolynes, P. G. 1987. Spin Glasses and the Statistical Mechanics of Protein Folding. *Proc. Natl. Acad. Sci. USA* **84**, 7524–7528.

Bryngelson, J. D., and Wolynes, P. G. 1989. Intermediates and Barrier Crossing in a Random Energy Model (with Applications to Protein Folding). *J. Phys. Chem.* **93**, 6902–6915.

Bunde, A., Havlin, S., and Roman, H. E. 1990. Multifractal Features of Random Walks on Random Fractals. *Phys. Rev. A* **42**, 6274–6277.

Chan, H. S., and Dill, K. A. 1991. Polymer Principles in Protein Structure and Stability. *Annu. Rev. Biophys. Biophys. Chem.* **20**, 447–490.

Chhabra, A. B., Jensen, R. V., and Sreenivasan, K. R. 1989. Extraction of Underlying Multiplicative Processes from Multifractals via the Thermodynamic Formalism. *Phys. Rev. A* **40**, 4593–4611.

Csordás, A., and Szépfalusy, P. 1989 Singularities in Rényi Information as Phase Transitions in Chaotic States. *Phys. Rev. A* **39**, 4767–4777.

Dewey, T. G. 1993. Fractal Aspects of Protein Structure and Dynamics. *Fractals* **1**, 179–189.

Dewey, T. G. 1994a. Multifractals and Order-Disorder Transitions in Biopolymers. *in Fractals in the Natural and Applied Sciences, IFIP Transactions* A-41 ed. Novak, M. M. (North-Holland) pp. 89–100.

Dewey, T. G. 1994b. Sequence Generating Functions and the Multifractal Formalism: Applications to Biopolymers. *Fractals* **3**, 9–22.

Dewey, T. G. 1994c. Order-disorder Transitions in Finite Length Biopolymers: A Sequence Generating Function Approach. *J. Chem. Phys.* **100**, 7648–7653.

Evertsz, C., and Lyklema, J. W. 1987. Multifractal Nature of Truly Kinetic Random Walks. *Phys. Rev. Lett.* **58**, 397–400.

Feder, J. 1988. *Fractals*. Plenum Press, New York, pp. 66–103.

Feigenbaum, M. J., Jensen, M. H., and Procaccia, I. 1986. Time Ordering and the Thermodynamics of Strange Sets: Theory and Experimental Tests. *Phys. Rev. Lett.* **57**, 1503–1506.

Feigenbaum, M. J., Procaccia, I., and Tel, T. 1989. Scaling Properties of Multifractals as an Eigenvalue Problem. *Phys. Rev. A* **39**, 5359–5372.

Fisher, M. E. 1966. Effect of Excluded Volume on Phase Transitions in Biopolymers. *J. Chem. Phys.* **45**, 1469–1473.

Frauenfelder, H., Sligar, S. G., and Wolynes, P. G. 1991. The Energy Landscapes and Motions of Proteins. *Science* **254**, 1598–1603.

Halsey, T. C., Jensen, M. H., Kadanoff, L. P., Procaccia, I., and Shraiman, B. I. 1986. Fractal Measures and their Singularities: The Characterization of Strange Sets. *Phys. Rev. A* **33**, 1141–1151.

Havlin, S., and Bunde, A. 1989. Probability Densities of Random Walks in Random Systems. *Physica D* **38**, 184–191.

Lifson, S. 1964. Partition Functions of Linear-Chain Molecules. *J. Chem. Phys.* **40**, 3705–3710.

Litan, A., and Lifson, S. 1965. Statistical–Mechanical Treatment of Multistranded Polynucleotide Molecules. *J. Chem. Phys.* **42**, 2528–2532.

Meisel, L. V., Johnson, M., and Cote, P. J. 1992. Box-counting Multifractal Analysis. *Phys. Rev. A* **45**, 6989–6996.

Novotný, J., Bruccoleri, R., and Karplus, M. 1984. An Analysis of Incorrectly Folded Protein Models: Implications for Structure Prediction. *J. Mol. Biol.* **177**, 787–818.

Pietronero, L., Evertsz, C., and Siebesma, A. P. 1988. Fractal and Multifractal Structures in Kinetic Critical Phenomena. In *Stochastic Processes in Physics and Engineering*, ed. S. Albeverio, P. Blanchard, M. Hazewinkel and L. Streit, Reidel, Dordreoht, pp. 253–278.

Poland, D., and Scheraga, H. A. 1966. Occurrence of Phase Transitions in Nucleic Acid Models. *J. Chem. Phys.* **45**, 1464–1469.

Poland, D., and Scheraga, H. A. 1970. *Theory of Helix–Coil Transitions in Biopolymers*. Academic Press, New York.

Redner, S. 1990. Random Multiplicative Processes: An Elementary Tutorial. *Am. J. Phys.* **58**, 267–273.

Richardson, J. S. 1981. The Anatomy and Taxonomy of Protein Structure. *Adv. Protein Chem.* **34**, 167–339.

Sarkar, S. K. 1987. Multifractal Description of Singular Measures in Dynamical Systems. *Phys. Rev. A* **36**, 4104–4106.

Stanley, H. E. 1991. Fractals and Multifractals: The Interplay of Physics and Geometry. In *Fractals and Disordered Systems*, ed. A. Bunde and S. Havlin, Springer-Verlag, Berlin, pp. 1–49.

Stanley, H. E., and Meakin, P. 1988. Multifractal Phenomena in Physics and Chemistry. *Nature* **335**, 405–409.

Strait, B. J., and Dewey, T. G. 1995. Multifractals and Decoded Walks: Applications to Protein Sequence Correlations. *Phys. Rev. E* **52**, 6588–6592.

Tél, T. 1988. Fractals, Multifractals, and Thermodynamics. *Z. Naturforsch., Teil a* **43**, 1154–1174.

Vicsek, T. 1992. *Fractal Growth Phenomena*. World Scientific, Singapore, pp. 89–97.

West, B. J., and Shlesinger, M. F. 1989. On the Ubiquity of 1/*f* Noise. *Int. J. Mod. Phys.* **3**, 795–819.

Wolynes, P. G. 1992. Spin Glass Ideas and the Protein Folding Problems. In *Spin Glasses and Biology*, ed. D. L. Stein, World Scientific, Singapore, pp. 225–259.

Zimm, B. H., and Bragg, J. K. 1959. Theory of the Phase Transition between Helix and Random Coil in Polypeptide Chains. *J. Chem. Phys.* **31**, 526–535.

5

Fractal Diffusion and Chemical Kinetics

In the preceding chapters, the utility of the fractal formalism for describing biopolymer structures was explored. We now turn to issues of the dynamics and chemical kinetics associated with biological systems. It is not difficult to find examples of nonexponential kinetics in biochemistry. Quite often this nonexponential behavior is a result of the complexity of biopolymer structure. Again, the fractal formalism can be especially useful in describing these phenomena. It provides a direct and explicit mathematical connection between structure and dynamics. This connection is achieved though a quantity known as the spectral dimension. In the present chapter, the consequence of a fractal medium for diffusion and reaction dynamics is considered.

We begin with a general discussion of diffusion in biological systems. There is a significant literature, mostly theoretical, on the influence of "reduction in dimensionality" on receptor-mediated processes. When recognition events are confined to low-dimensional substrates, such as a biological membrane or the contour of a DNA molecule, there is a considerable enhancement of the effectiveness of the search process. These dimensional considerations are extended to include bimolecular associations on fractals. As will be seen, the fractal dimensionality of a system can also strongly influence its dynamics. After a discussion of diffusion, we consider chemical kinetics on fractals. An introductory description of both transient and steady state reaction kinetics in restricted geometries and on fractals is given. These scenarios yield strange effects, such as time-dependent "rate constants" and anomalous reaction orders.

To provide a concrete kinetic application of these concepts, the kinetics of isotope exchange in proteins is treated. The rate of isotope exchange in proteins is nonexponential and can cover eight orders of magnitude in time. A model is developed that relates the anomalous kinetics of this process to a structural parameter, the fractal surface dimension of proteins. The exchange reaction is envisioned to occur in the codimensional space surrounding the fractal surface of the protein. A scaling relationship is derived to describe the nonexponential kinetics. This relationship uses experimental values for the surface dimension of a protein (as discussed in Chapter 2) as an unadjusted parameter and accurately fits a range of experimental data for the protein lysozyme.

Biomembranes also provide a milieu where anomalous diffusion occurs. Phase-separated separated domains can exist in such membranes, providing restricted geometries for diffusion and reaction. Abrupt changes in the connectivity of these domains have dramatic effects on the chemical kinetics. The transition from a collection of disconnected domains to a single, continuous one is known as percolation. At the percolation threshold, fractal structures known as percolation clusters are formed. These structures will be discussed more fully in Chapter 9. In the present chapter, diffusion and reactions within phase-separated membranes is treated in a qualitative fashion. Additionally, the problem of diffusion through connected regions is also considered. Again, this is discussed within the context of the biological membranes, and is particularly pertinent to the "crowding" problem for membrane proteins.

5.1 Diffusion and Dimensionality

The importance of dimensionality in biological systems has long been recognized (Adam and Delbrück, 1968). Changes in the dimensionality of a system can greatly alter the rates at which a small, diffusable molecule binds to a receptor site. A well-studied example of this is that of the *lac* repressor binding to DNA. The *lac* repressor is a protein that binds to a specific site on DNA called the *lac* operon (Berg et al., 1981; Winter et al., 1981; Winter and von Hippel 1981). The operon is a short specific sequence encoded within the DNA. The *lac* repressor binds to the operon with a rate constant that is larger than expected for a diffusion-limited reaction. In this case, binding rates are greatly enhanced because the repressor does not find the operon sequence in a large DNA molecule via three-dimensional diffusion. Rather, it "nonspecifically" associates with the DNA and then diffuses one-dimensionally along the backbone until it locates the DNA sequence in the operon and binds specifically to it. One-dimensional diffusion greatly enhances the rate of the specific binding process, giving bimolecular rate constants that are faster than expected for diffusion in three dimensions.

In addition to the one-dimensional diffusion of DNA binding proteins, membrane receptors will doubtless show a similar enhancement of binding rates when ligands are restricted to the two-dimensional structure of the biological membrane. A number of theoretical works have considered the problem of the diffusion of a ligand from the bulk solution to the membrane surface, followed by two-dimensional diffusion in the membrane to the receptor (for a review see DeLisi (1980)). The reduction of dimension at the membrane surface results in a rate enhancement for diffusion-limited reactions. This effect has been investigated for both the steady state (Berg and Purcell, 1977) and transient (Zwanzig and Szabo, 1991) kinetics of ligand binding. These theoretical studies suggest that the reduced dimensionality greatly enhances the kinetics of the ligand binding process. However, recent work has shown that if the receptor–ligand binding step in the reaction is "chemically" or reaction-limited, the enhancement as a result of the reduction in dimensionality is greatly reduced (Axelrod and Wang, 1994). Unfortunately, there are few experimental data that bear on this issue.

The dramatic effect of dimensionality in diffusive processes is seen by a simple scaling argument (see Pfeifer, 1987; Adam and Delbrück, 1968). The source of the effect is that the classic dependence on time, t, of the root mean squared displacement, r, for a diffusing molecule is independent of dimension. For one, two, and three dimensions, the Einstein relationship, $r^2 \sim t$, holds. If a diffusing molecule is visiting sites at a constant rate then the number of sites visited, N_{visited}, is proportional to time, $N_{\text{visited}} \sim t$. Within a time t, the diffuser will have access to all the sites within a volume, r^d, where d is the spatial dimension. The total number of sites, N_{total}, available for a visit will be equal to the density times this volume. Therefore, $N_{total} \sim r^d$. The ratio, R, of sites visited to total sites is:

$$R = \frac{N_{visited}}{N_{total}} \propto \frac{t}{(\sqrt{t})^d} \tag{5.1}$$

Now consider the behavior as $t \to \infty$. In one dimension, $R \to \infty$, indicating that all sites will be visited infinitely many times. This is sometimes referred to as a recurrent or "compact" exploration of sites. In two dimensions, on average each site is visited once, i.e., $R \sim 1$; in three dimensions, $R \to 0$ and most sites are never visited. The three-dimensional case is nonrecurrent (or noncompact), while two dimensions is the border between recurrent and nonrecurrent exploration. The mean diffusion time τ for a molecule to reach a specific target site in a space of diameter L is proportional to L^2 for one and two dimensions, and is proportional to L^3 in three dimensions. Thus, a biological system can greatly enhance the targeting of a small molecule to a receptor by reduction of dimensionality.

Pfeifer (1987) has discussed the generalization of this dimensionality effect for diffusion on fractal structures. For such processes, the mean squared displacement is now given by:

$$r^2 \sim t^{\bar{d}/D} \tag{5.2}$$

where D is the fractal dimension of the structure in which the diffuser is embedded, and $\bar{d}$ is a parameter known as the spectral dimension. Another way of viewing Eq. 5.2 is that diffusion on fractals has a time-dependent diffusion "constant." Equation 5.2 can be obtained from the solution of a classical diffusion equation that has been modified to account for the fractal geometry of the embedding space (see O'Shaughnessy and Procaccia, 1985). This equation takes the form:

$$\frac{\partial p(r,t)}{\partial t} = \frac{1}{r^{D-1}} \frac{\partial}{\partial r} \left[K(r) r^{D-1} \frac{\partial p(r,t)}{\partial r} \right] \tag{5.3}$$

where $p(r,t)$ is the probability that the diffusing particle is at position r at time t, D is the fractal dimension of the space accessible to the diffuser, and $K(r)$ is a distance-dependent diffusion constant. Although Equation 5.3 has been successfully used to model fractal diffusion, it has been considered before in other contexts (for an interesting comparison of diffusion equations with distance-dependent diffusion constants, see van Kampen, (1981)). For the fractal case, it is assumed that $K(r)$ follows a power law:

$$K(r) = Kr^{-\theta} \tag{5.4}$$

where θ is called the exponent of anomalous diffusion. The solution of Eq. 5.3 with Eq. 5.4 is:

$$p(r,t) = \frac{(2+\theta)}{D\Gamma\,[D/(2+\theta)]} \left[\frac{1}{K(2+\theta)^2\, t} \right]^{D/(2+\theta)} \exp\left[-\frac{r^{2+\theta}}{K(2+\theta)^2\, t} \right] \tag{5.5}$$

where Γ is the gamma function and the probability has been normalized over a fractal space, i.e., a volume element of $r^{D-1}\, dr$. From computer simulations of diffusion on fractal lattices, it was seen that Eq. 5.5 provides a good quantitative description of such processes (see O'Shaughnessy and Procaccia, 1985).

The "nontraditional" Einstein relationship derived from Eq. 5.5 is:

$$\langle r^2(t) \rangle = [K(2+\theta)^2 t]^{2/(2+\theta)} \frac{\Gamma[(D+2)/(2+\theta)]}{\Gamma[D/(2+\theta)]} \tag{5.6}$$

where $\langle r^2(t) \rangle$ is the mean squared displacement. Equation 5.6 shows a deviation from the "classical" diffusion equation, i.e., $\langle r^2(t) \rangle = Dt$. It is convenient to define a "walk" dimension, d_w, as:

$$\langle r^2(t) \rangle \sim t^{2/d_w} \tag{5.7}$$

In the above example, $d_w = 2 + \theta$. Using Eq. 5.2, the spectral dimension is given by:

$$\bar{d} = \frac{2D}{d_w} \tag{5.8}$$

Equation 5.8 will appear repeatedly in this chapter and in Chapter 7. It provides a convenient means of calculating the spectral dimension. The spectral dimension is an interesting quantity as it combines a structural parameter, D, and a dynamical parameter, d_w. It provides a bridge between structure and dynamics. Our goal in the next section is to relate this diffusion behavior to chemical kinetics. In particular, the scaling laws for "time-dependent" rate constants will be related to the spectral dimension.

5.2 Fractal Chemical Kinetics

In textbook descriptions of macroscopic chemical kinetics, the rate constant is treated as a time-independent entity. However, this condition only applies for homogeneous, well-stirred solutions. When the reaction occurs in a restricted geometry and depletion zones develop, concentration fluctuations can dominate the kinetics of the process. Under these conditions, the rate "constant" will have a time dependence (see Kang and Redner, 1984; Zumofen et al., 1985; Kopelman, 1988). Considering a bimolecular reaction of the type, A + B → products, one has:

$$\frac{-d[\mathrm{A}]}{dt} = k(t)[\mathrm{A}][\mathrm{B}] \tag{5.9}$$

where the time-dependent rate constant, $k(t)$, has the form:

$$k = k'\,t^{-\mathrm{h}} \tag{5.10}$$

where k' is a time-independent constant and the exponent h is restricted to values between 0 and 1.

Most theories attempt to relate h to the spectral dimension, $\bar{d}$. This relationship depends on the universality class of the reaction (Kang and Redner, 1984). For instance, the reaction A + A → products gives $h = 1 - \bar{d}/2$, while the reaction A + B → products gives $h = 1 - \bar{d}/4$. The spectral dimension, defined in Eq. 5.8, is related to the fractal dimension, D, of the supporting medium and of the walk dimension. For normal, diffusive processes in a homogeneous system $d_w = 2$, and D will merely be the Euclidean dimension. In these cases $\bar{d} = D$, and anomalous kinetics still occurs. Fractal geometry is not a prerequisite for anomalous chemical kinetics. Generally, time-dependent rate constants, as in Eq. 5.10, give non-exponential behavior. Depending on the type of reaction and concentration conditions, a variety of integrated rate expressions are possible (see Pfeifer, 1987). Typically these are power laws or stretched exponentials. A stretched exponential has the functional form $\exp(-At^{\alpha})$, where $0 < \alpha < 1$. It gets its name from the characteristic "long tail" of the decay.

5.2.1 Fluctuations and Anomalous Scaling

A theoretical development based on the diffusion equation (Eq. 5.3) can be used to derive a time-dependent rate constant as in Eq. 5.10. This can be done using the approach of Eigen (1954). Alternatively, simpler scaling arguments can be employed and this treatment is followed for heuristic reasons. For reactions of the type A + B → products, fluctuations can effect the rate law over two different time domains. For short times, local fluctuations in the density difference between

A and B, $(\rho_A - \rho_B)$, will dominate the reaction kinetics and the reactant concentration will decay as $t^{-\alpha}$. In the long-time regime, spatial fluctuations of the majority species dominate the reaction kinetics. In such cases, the reactant concentration will follow a stretched exponential, $\exp(-At^{\alpha})$. The exponent α can be related to the fractal properties of the system, as described by Kang and Redner (1984) and it is seen that $\alpha = 1 - h$. We present a simplified version of those arguments.

Consider the number of reactants B in a reaction volume with fractal dimension D. This number, N_B, is given by:

$$N_B \sim N_B(0) \pm \delta N_B \sim \rho_B(0) r^D \pm \sqrt{\rho_B(0) r^D} \tag{5.11}$$

where $N_B(0)$ is the average number of reactants in the volume and δN_B is the number fluctuation. The parameter r is the distance a molecule can diffuse in time t, and the square root term is a result of the standard relationship from statistical mechanics, $\delta N_B \sim \sqrt{N_B(0)}$. In unstirred solutions, concentration fluctuations of the majority species will dominate the reaction kinetics. Taking B as the majority species, the concentration fluctuations, $\delta\rho_B$, will scale as:

$$\delta\rho_B \sim \frac{\delta N_B}{r^D} \sim r^{-D/2} \tag{5.12}$$

Now consider the volume explored by the reactant during a time t. The generalized diffusion law found in Eq. 5.7 is used, i.e. $r \sim t^{1/d_w}$. Substituting Eq. 5.7 into Eq. 5.12, one obtains:

$$\delta\rho_B \sim t^{-D/2d_w} \sim t^{-\bar{d}/4} \tag{5.13}$$

which gives $\alpha = \bar{d}/4$ and $h = 1 - \bar{d}/4$.

5.2.2 *Scaling of Depletion Zones*

In low dimensions, as the reaction proceeds depletion zones develop. These regions segregating domains of reactants from each other will grow deeper and deeper as time goes on (Figure 5.1). The depth of these zones can be estimated, again from scaling arguments (Redner and Leyvraz, 1991). Consider the region separating a domain of reactant A from a domain containing reactant B to have thickness or gap distance ℓ_{AB}. A scaling Ansatz is made such that the gap distance grows as $\ell_{AB} \sim t^{\varsigma}$. Our goal is to relate ς to the dimensionality of the embedding space. During a time, t, reactions will occur between molecules at the boundary of the domains. At the boundary, the distance between like molecules will be the same as the distance between different ones and will be approximately ℓ_{AB}. Therefore, the number of molecules, $N_{boundary}$, at a boundary domain with diffusive length r is given by $N_{boundary} \sim (r/\ell_{AB})^{d-1}$. The number density at the boundary will then be $\rho_{boundary} \sim (r/\ell_{AB})^{d-1}/r^d \sim t^{-1/2}\, t^{\varsigma(1-d)}$, where in the second proportionality it is assumed that diffusion is classical, $r^2 \sim t$, and that the scaling

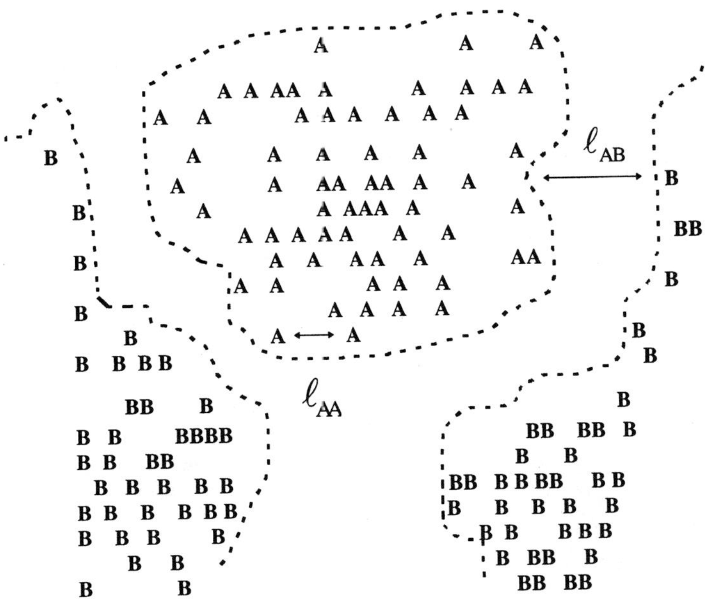

FIGURE 5.1 A schematic representation of the segregation of reactants, A and B, into domains in two dimensions. The width of the depletion zone, or gap distance, is designated as ℓ_{AB}. Note that the density within a domain decreases as the edge of the domain is approached. (Figure adapted from Render and Leyvraz, 1991.)

Ansatz is evoked. Reactions will occur on the time scale that it takes a molecule to diffuse across the gap distance. Therefore:

$$\frac{d\rho_{boundary}}{dt} \sim \frac{\rho_{boundary}(t)}{l_{AB}^2} \sim t^{-1/2} t^{\varsigma(1+d)} \tag{5.14}$$

Taking the time derivative of Eq. 5.13 and equating it to Eq. 5.14 allows the exponent, ς, to be determined. Assuming classical diffusion and Euclidean dimensions, one obtains:

$$\varsigma = \frac{2+d}{4(1+d)} \tag{5.15}$$

This gives ς values of 3/8 and 1/3 in one- and two-dimensional systems, respectively. This is in good agreement with results obtained by computer simulation (Redner and Leyvraz, 1991). As the depletion zone grows, the "local" concentration of reactants decreases, and consequently the rate of reaction declines.

5.2.3 Anomalous Reaction Order

So far, we have only considered the transient behavior of a low-dimensional, reacting system. The steady state regime also shows interesting and anomalous behavior. If a source is present, a steady state regime can be achieved. In this regime, by definition one has a time-independent rate constant, i.e., Eq. 5.10 is no longer valid. Instead anomalous reaction orders are observed (see Kopelman, 1988). For instance, in the reaction A + B → products considered above, the steady state rate law scales as:

$$-\frac{d[\mathrm{B}]}{dt} \sim \frac{\delta\rho_{\mathrm{B}}}{t} \sim \delta\rho_B^{1+4/\bar{d}} = k_{ss}[\mathrm{B}]^x \qquad (5.16)$$

where Eq. 5.13 has been used and k_{ss} is the steady state rate constant. The reaction order x is given by:

$$x = 1 + 4/\bar{d} = 1 + (1-h)^{-1} \qquad \bar{d} < 2 \qquad (5.17)$$

Thus, reaction orders are greater than the value of 1, as expected by the mass action law. In one dimension, an order of 5 will be obtained for A + B → products. Reactions of the type A + A → products will give similar, but lower, anomalous reaction orders. This case is also especially interesting because it shows segregation and self-ordering of the two reactants. It would not be surprising, given the various restricted geometries present in living cells, that such effects might play a role. For instance, enzymes such as lipases catalyze reactions involving membrane constituents and are, therefore, restricted to two dimensions. Such phenomena remain to be explored.

5.2.4 Scaling in Chemically Controlled Reactions

While the importance of diffusion in biological systems has been emphasized, paradoxically it is difficult to find biochemical reactions with bimolecular rate constants at the diffusion limit. There are only two or three enzymes that catalyze reactions at the diffusion-controlled limit. Even simple processes such as oxygen binding to hemoglobin are not diffusion controlled. The reason for the slowness of most biochemical reactions is the large number of orientation and recognition events required to accomplish the reaction. This requirement for specific ligand–polymer contacts slows the reaction and prevents most biochemical reactions from being diffusion controlled. Thus, for the bulk of biochemical reactions, especially metabolic processes, diffusion is not an important consideration. It is only in reactions that involve small numbers of reactants or restricted geometries that diffusion plays a key role. Such reactions are usually involved in sensory or regulatory processes.

When considering fractal reaction kinetics, the diffusional step influences the rate of reactions that are not strictly diffusion controlled. This is a consequence of the time dependence (Eq. 5.10) of the diffusion-controlled rate constant. This is seen by considering the following standard mechanism in solution phase kinetics:

$$A + B \underset{k_{-1}}{\overset{k_1}{\leftrightarrow}} A \bullet B \overset{k_2}{\rightarrow} C \tag{5.18}$$

where A•B is an outer sphere (or encounter) complex. Traditionally, the encounter complex is envisioned as having a specific structure. For instance, in ionic reactions it would correspond to hydrated ions in contact with each other. However, this "species" may be viewed in more general terms as two molecules merely being in the vicinity of each other. The rate constants k_1 and k_{-1} are the diffusion-controlled, forward and reverse rate constants, respectively, and k_2 is the unimolecular rate constant for the "chemical" step. From general considerations (Eigen, 1954), the equilibrium constant, K_1, for the formation of the encounter complex, $K_1 = k_1/k_{-1}$, can be determined. This will be independent of the kinetics of the process and independent of time. Thus, when the diffusion-controlled forward rate constant follows a time dependence, as in Eq. 5.10, there must be an identical time dependence for the reverse rate constant: $k_{-1}(t) = k_1(t)/K_1$. This gives a time-dependent unimolecular rate constant.

An alternate argument for comparable scaling of the reverse rate comes from the approach of Kang and Render (1984) described above. Their scaling arguments were based in part on conservation considerations. For the diffusion-controlled reaction of the type A + B → products, a constant of motion of the system is $([A(t)] - [B(t)])$. The conservation of this quantity is exploited to yield an equation of the form of Eq. 5.3. In the reversible, chemically limited mechanism, $([A(t)] - [B(t)])$ is still a constant of motion and Kang and Redner's arguments still hold. Thus, the rate constant k_1 takes the form $k_1(t) = k_1' t^{-h}$. One could argue for a comparable dependence for k_{-1}. For the reversible reaction A + B ↔ C, there are now two constants of motion: $([A(t) - B(t)])$ and $([A(t)] + [C(t)])$. The second conservation relationship gives $[C(t)] = C_{tot} - A(t) = C_{tot} - A'\ t^{h-1}$, where C_{tot} is the total amount of reactant present and A' is a constant. The form derived from this second relationship implies an identical scaling of the reverse constant as for the forward rate constant.

The effect of a fractal diffusion process on the overall rate of the reaction is most easily seen under conditions where a steady state is achieved for the encounter complex (Dewey, 1994, 1995). Under these conditions, the observed forward rate, k_f, is given by:

$$k_f = \frac{k_1 k_2}{k_{-1} + k_2} = \frac{k_1' k_2 t^{-h}}{(k_1'/K)t^{-h} + k_2} \tag{5.19}$$

where fractal rate constants have been introduced in the right-hand equality. To obtain expressions for the time course of the concentration of A or B, k_f must be integrated over time (Dewey, 1995). However, one can directly see the behavior of the system by examining the scaling for two limiting cases. For short times the t^{-h} term will dominate the denominator in Eq. 5.19 and cancel the dependence in the numerator. Consequently, k_f will be independent of time. In cases where one reactant is in excess, i.e., [A] << [B], an exponential decay in the limiting reactant will occur. This will have a reciprocal time constant of $K_1 k_2$. For the long-time behavior, k_2 will dominate the denominator and the fractal time dependence will be retained. For limiting A, the time course will be a stretched exponential of the form:

$$[A] \sim \exp(-k_1' t^{1-h}) \tag{5.20}$$

Thus, it is seen that fractal diffusion will influence the time course of a reaction even if the reaction is not diffusion limited. This is a very significant result for biological systems, where diffusion limited reactions are not common. The above analysis is based on the assumption of steady state conditions for the encounter complex. More general developments are possible and have been explored theoretically (Dewey, 1995). These more general cases show that Eq. 5.20 is retained in the long-time limit.

5.3 Isotope Exchange Kinetics in Proteins

One of the legacies of the great success of X-ray crystallography was the view of biopolymers as rigid, static structures. With the advent of a number of kinetic techniques, it was shown that contrary to the static view, biopolymers in solution are dynamic, fluctuating entities (for an interesting discussion of protein fluctuations, see Cooper (1976). One of the oldest and most successful of these kinetic methods was the observation of hydrogen exchange (Woodward et al., 1982; Englander and Kallenbach, 1984). In this approach, a protein is exposed for an extended period of time in a solution of deuterium or tritium oxide. Over a period of days, the exchangeable hydrogens in the protein will be removed and replaced with the corresponding isotope. The protein is then removed from the isotope solution and returned to a normal aqueous solvent. The rate of hydrogen isotope exchange is measured by infrared (IR) or nuclear magnetic resonance (NMR) spectroscopy for deuterium exchange or by scintillation counting of radioactivity for tritium exchange. Two-dimensional NMR techniques have now reached a level of sophistication such that for small polypeptides every exchangeable proton can be monitored (Englander and Mayne, 1992).

When total isotope exchange is observed, the exchange kinetics is non exponential and can cover a range of eight orders of magnitude in time. The exchange reaction can be either base (OH^-) or acid (H_3O^+) catalyzed by the solvent and is not a diffusion-limited reaction. For proteins, some of this broad temporal distribution could be a result of intrinsically different exchange rates for protons on different amino acids. However, in most cases the large time range reflects the different rates for exchange at different depths in the protein. Even hydrogen atoms that appear buried in X-ray structures can be exchanged, albeit at a very slow rate.

Arrhenius plots for isotope exchange show two different temperature regimes. These regimes have been attributed to different exchange mechanisms. At high temperatures, the protein unfolds and directly exposes buried sites to the solvent.

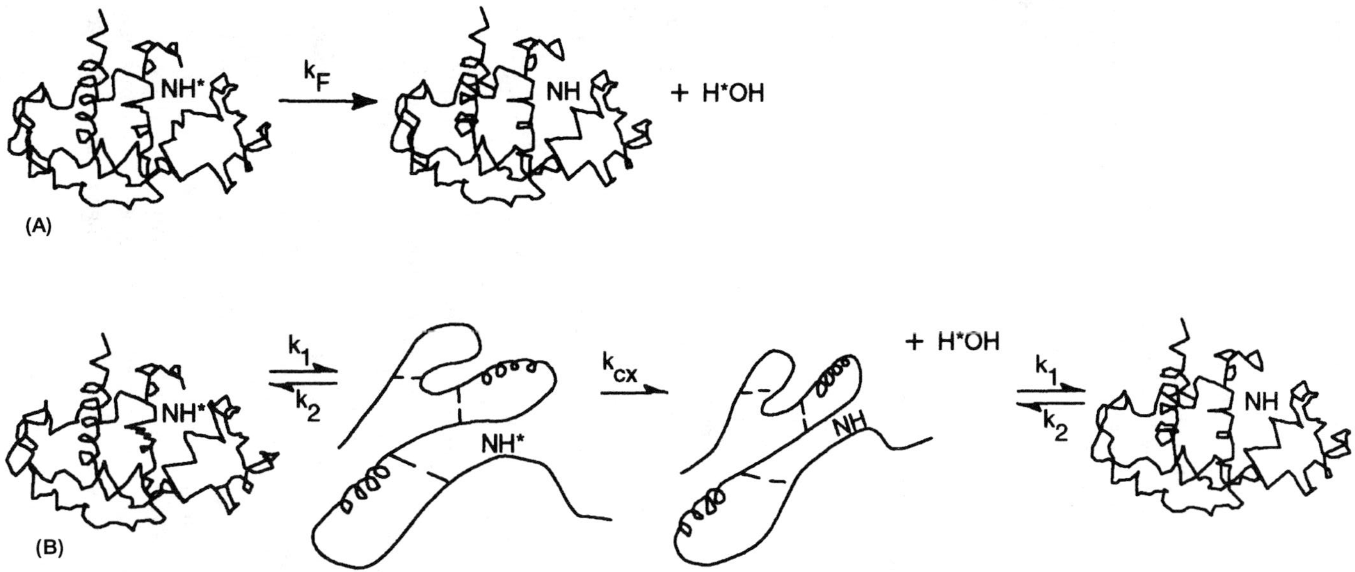

FIGURE 5.2 The two-process model for hydrogen exchange from native proteins. Exchange of isotopically labeled hydrogen, NH*, with solvent hydrogen atoms is shown for one exchanging site. (A) Exchange from the folded state. The apparent activation energy for this process varies from 20 to 40 kcal/mol. (B) Exchange via unfolding. k_1 and k_2 are the rate constants of unfolding and folding for the major, cooperative denaturation transition; k_{ex} is the chemical exchange rate constant. The apparent activation energy for the unfolding exchange process varies between 60 and 120 kcal/mol, and is related to the enthalpy of the denaturation transition. (Figure from Woodward et al., 1982. Reprinted by permission of Kluwer Academic Publishers.)

The energy requirement for unfolding of the protein structure indicates a high activation energy for this process. In the second mechanism, exchange occurs from the folded state of the protein and is a process with a low activation energy. Two general models for low energy exchange have been proposed. In one model, hydronium or hydroxyl ions must migrate through the protein to exchange sites. This mechanism is analogous to defect diffusion models in condensed phase physics and is in keeping with fluorescence data on oxygen quenching (Welch, 1986). These data show that molecular oxygen can collisionally quench the fluorescence of tryptophan residues buried deep in the protein. In the second model, protein fluctuations bring buried residues to the surface and expose them to the solvent. During these fluctuations, the structural integrity and compactness of the protein are maintained. Arrhenius plots show linear behavior at low temperatures, which is consistent with one of the two "folded mechanisms." At high temperatures a change in activation energy occurs, which is more in keeping with protein unfolding and partial denaturation. Figure 5.2 illustrates these two different mechanisms. The following discussion and analysis will be restricted to the low activation energy process.

5.3.1 Models of Exchange in Lysozyme

Figure 5.3 shows the results of an experiment that observes the time dependence of the "out-exchange" of tritium from the protein lysozyme (data from Wickett et al. (1974). Lysozyme is a small enzyme with a large cleft that contains the catalytic site. The left-hand graph in Figure 5.3 shows data for enzyme without an inhibitor bound in the cleft, and the right-hand graph shows the data when the enzymatic inhibitor is bound. Bound molecules tend to restrict the dynamics of a protein, and this is reflected in the different time scales of the two plots. In both cases, the plots of log-(tritium unexchanged) versus time shows considerable curvature. This is indicative of a stretched exponential dependence and these data points are accurately fitted by such a function. This dependence occurs over a wide temperature range, but is still below the "high temperature regime" where the protein will denature partially. On shorter time scales, hydrogen exchange is often observed to be exponential (data not shown). Thus, the dependence predicted by the steady state analysis in Section 5.2 is followed.

The major contribution to the exchange reaction is from hydrogen atoms in the amide backbone of the protein. The reaction is of the type $A + B \rightarrow C + D$, where A is the tritiated amide linkage in a protein, B is H_3O^+ (or OH^-, depending upon the pH), C is the amide linkage with a hydrogen atom replacing the tritium, and D is TH_2O^+ (or OT^-). The time course of the reaction at long times can be represented by a macroscopic rate law that has a time-dependent rate constant, as in Eq. 5.9. Since experiments are carried out in buffered solutions, [B] is essentially a constant and the reaction is pseudo-first-order in A. Assuming a scaling form for the rate constant as in Eq. 5.10, the rate expression can be integrated to give the stretched exponential form of Eq. 5.20. The value of the exponent can be predicted and used as an unadjusted parameter in fitting the experimental data. As seen in the previous section, the exponent is related to the spectral dimension, $\bar{d}$, of the system, and the

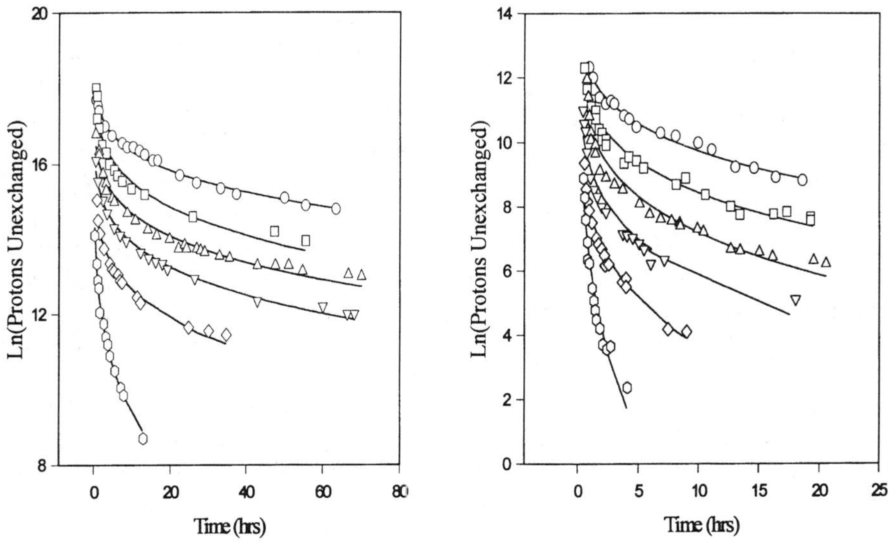

FIGURE 5.3 Plot showing the hydrogen exchange kinetics of lysozyme. Log of the concentration of unexchanged protons versus time is shown for six different temperatures: (○) 2°C, (□) 10°C, (△) 16°C, (▽) 23°C, (◇) 31°C, ( ) 41°C. Curves obtained from a two-parameter, nonlinear least squares fit to a stretched exponential (Eq. 5.26) with the parameter α held fixed at 0.21 for all temperatures. (Left) Lysozyme with an inhibitor bound; (right) without inhibitor. (Data from Wickett et al., 1974.)

functional form will depend on the universality class of the reaction. Because this is a bimolecular process, the exponent is given by:

$$1 - h = \bar{d}/4 = D/2d_w \tag{5.21}$$

To estimate $(1 - h)$ from Eq. 5.21, several assumptions must be made regarding the nature of the exchange reactions. As discussed above, there are two general views on the mechanism of exchange. The first is that exchange occurs as a result of solvent penetration through the protein matrix. This penetration occurs as a result of protein fluctuations. The second view is that fluctuations expose new regions of the protein to the surface, and thus allow exchange with the solvent. These two models are represented in Figures 5.4 and 5.5. As seen from the figures, both models represent a reaction in a low-dimensionality space because there are spatial regions that are accessible to the protein side-chains but not accessible to solvent. The reaction must occur in the space accessible to both solvent and reactants and is the intersection of the set of points characterizing the accessible regions for solvent and protein side-chains. As discussed in Chapter 1, the dimensionality of the intersection of two sets is given by:

$$D = D_{solvent} + D_{protein} - 3 \tag{5.22}$$

where $D_{solvent}$ and $D_{protein}$ are the dimensionalities of the two regions accessible to each reactant. Figure 5.4 and 5.5 provide a visualization of how the regions might appear. In Figure 5.4 the short line segments represent the space filled by

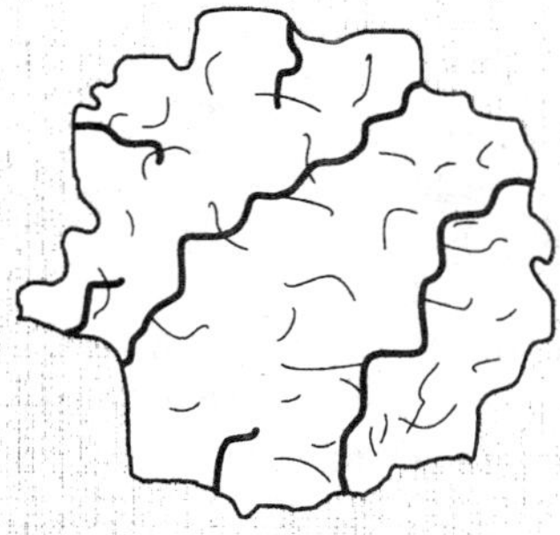

FIGURE 5.4 Schematic diagram showing the overlap in domains accessible to solvent and to protein side-chains. A cross-sectional area of a protein is depicted. Long line segments represent paths through the protein that are accessible to solvent. Short line segments are spatial regions accessible through fluctuations to protonated groups on an amino acid. The intersection of the solvent accessible and side-chain accessible region is the space in which the exchange reaction occurs. This model suggests that such a space could be a fractal "dust" with a dimension less than 1.

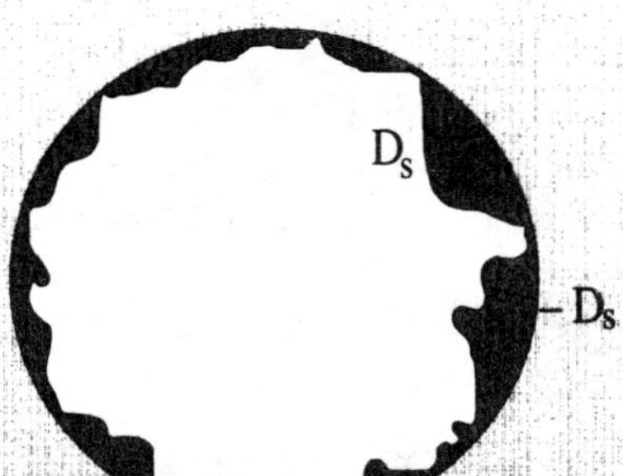

FIGURE 5.5 Schematic diagram depicting the hydrogen exchange reaction occurring at the protein surface. In this model, the solvent cannot penetrate the protein and is confined to the "codimensional" space (dark area) excluded by the surface fractal. This space has a fractal dimension of $3–D_S$, where D_S is the fractal dimension of the protein surface. The model assumes that the reaction occurs in this codimensional space. For this to happen the side-chains of the protein must be able to fill the three-dimensional space via fluctuations.

fluctuations in the exchangeable side-groups within the protein. The longer line segments represent paths of migration of the solvent as it diffuses through the protein. The intersection of these two regions represents the space in which the reaction occurs. From this figure one can readily see how this intersecting region might be a limited collection of points, i.e., $D < 1$.

While Figure 5.4 has a certain appeal, a potential problem with the model is that one might not expect D to be the same for lysozyme with and without bound inhibitor. Binding of an inhibitor to an enzyme will restrict the internal dynamics of lysozyme and cause decreased overlap of reactant regions. Also, as the temperature is changed the fluctuations should increase, and it might be anticipated that the overlap of the two regions would increase, resulting in an increase in D with temperature. This effect is not seen. A further argument against the defect diffusion model is that the charge and hydration spheres of a hydronium or hydroxyl ion would create a large activation barrier to diffusion within a protein matrix. Additionally, mixed aqueous solvent effects are observed on the reaction kinetics when it is not anticipated that the second solvent would be capable of penetrating the protein structure. Such effects are inconsistent with the penetration model (Calhoun and Englander, 1985).

Figure 5.5 represents an alternative explanation for the value of D. In this model, the protein interior fluctuates in such a manner that over a long time all the crevices within the protein are filled. Thus, the protein smoothes out its surface and achieves a dimension of 3 via such fluctuations. The solvent, on the other hand, cannot penetrate the protein, but through fluctuations still has access to the areas within the crevices. Considering the protein as a surface fractal of dimensionality, D_S, the solvent occupies the "codimensional" space, $3 - D_S$. (See Chapter 1 for a discussion of codimensional spaces.) With the application of Eq. 5.22, one now sees that:

$$D = 3 - D_S \tag{5.23}$$

In an extremely careful study on lysozyme, a surface dimension of 2.17 was determined (Pfeifer et al., 1985). This gives a value of D of 0.83, and a value of $1 - h$ of 0.21. This value was used, without adjustment, to fit the data in Figure 5.3. As can be seen, excellent agreement is obtained between the theoretical curves and the experimental data.

The experimental curves in Figure 5.3 were fitted using an equation of the form:

$$f = A \exp\left[-(t/\tau)^{0.21}\right] \tag{5.24}$$

where f is the fraction of tritium unexchanged, and A and τ, are adjustable parameters. As can be seen from Figure 5.3, accurate fits can be obtained at all temperatures with one, unadjusted value of $(1 - h)$. The parameter A represents the fraction of tritium at time $t = 0$. As can be seen from the figures, comparable values for A will be obtained at all temperatures. It is unlikely that the solution structure of the protein varies significantly over this temperature range, because structure-sensitive properties, such as enzymatic activity, are not compromised. Consequently, there are no gross morphological changes in the enzyme. Therefore, it is anticipated

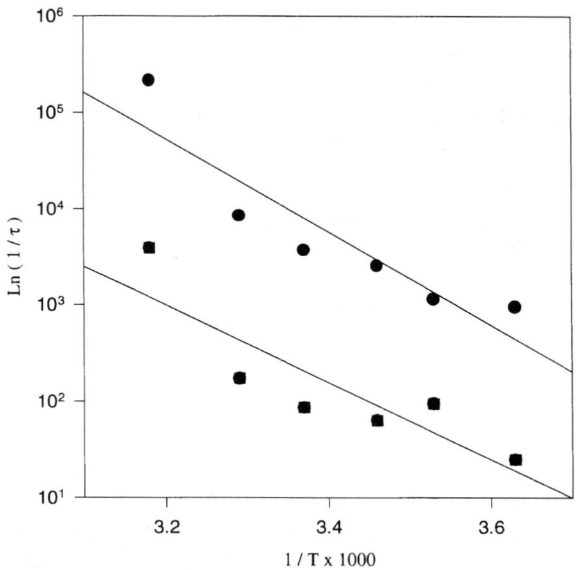

FIGURE 5.6 Arrhenius plot (log(reciprocal relaxation time) versus reciprocal temperature) for the fitted parameters obtained using Eq. 5.24 and the data in Figure 5.3. Top line (●), is lysozyme with bound inhibitor; bottom line (■), without inhibitor.

that the fractal surface dimension will be insensitive to changes in temperature and the scaling exponent will not have a large temperature dependence. Additionally, it is unlikely that the binding of an inhibitor will grossly alter the surface morphology. Consequently, both curves, i.e., the one for data obtained with inhibitor and the one for data obtained without inhibitor, are adequately fitted using the same fixed exponent. Note, however, the longer times occurring in the presence of inhibitor.

In addition to fitting the amplitudes in Eq. 5.24, the relaxation times were also obtained. Arrhenius plots derived from these fitted relaxation times are shown in Figure 5.6. From the slopes of these plots, activation energies of 18 ± 3 kcal/mol without bound inhibitor and 24 ± 4 kcal/mol with bound inhibitor were obtained. These are comparable to the value of 17.5 kcal/mol observed for hydrogen exchange in free peptides (Englander and Poulsen, 1969). For aqueous solutions under these conditions, it is anticipated that approximately 4 kcal/mol of the activation energy is due to viscosity effects and the remainder is due to the chemical barrier. In addition to accurately fitting the data in Figure 5.3, the temperature dependence of the relaxation times gives a realistic value for the activation energy. It is interesting that the addition of the inhibitor to the enzyme has only a small effect on the activation energy, while the time scale of the reaction has increased significantly. The inhibitor effect appears in the entropy of activation, rather than in the activation energy.

As discussed in Chapter 2, a large number of proteins show a similar surface dimension (approximately 2.2). This model would then predict that the parameter $(1 - h)$ would not change from protein to protein. As a check on the generality of these results, a second data set has been considered (Dewey, 1994) (data from Ellis et al. (1975). The early time regime of this data can be fitted to an exponential decay, as predicted from Eq. 5.12. The long-time regime shown is adequately fitted by a stretched exponential. In this instance, $(1-h)$ was treated as an adjustable parameter in Eq. 5.20. The fitted values for $(1-h)$ are 0.23 (20°C) and 0.16 (30°C). Because of the limited number of data points, there is significant error in the parameters of the nonlinear fit. Higher temperatures (> 40°C) do not fit the stretched exponential well. This is most likely due to the high activation energy process contributing to the kinetics. Thus, both the lysozyme and trypsin inhibitor data analysis suggest that, over a limited range of temperatures, $(1-h)$ is a constant having a value close to 0.2.

5.3.2 Reactions on Spheres and Surface Fractals

To elaborate on the surface model, a simple example is considered using the scaling arguments given in Section 5.2. First, consider the protein to be a simple sphere of fixed radius, R, containing reactants A. The A molecules are confined within the sphere, but fluctuations can bring buried molecules into a region on the surface. This sphere is embedded in a uniform medium containing the reactant B. As a result of fluctuations during a time t, B molecules will enter a shell of width ℓ, surrounding the sphere. The time, t, is chosen so that $\ell^2 \sim t$, i.e., diffusion is assumed to scale according to a classical Einstein relationship. If anomalous diffusion is to be incorporated in the model, one would have: $\ell^{d_w} \sim t$. The number of B molecules, N_B, in the shell will be $N_B \sim \rho_B(0)R^2\ell$, where $\rho_B(0)$ is the initial number density of B. Since number fluctuations will be proportional to the square root of the number of particles, the number density will vary as $\delta\rho_B \sim \sqrt{N_B}/R^2\ell \sim \ell^{-1/2}$. From the time dependence of ℓ, one now has $\delta\rho_B \sim t^{-1/4}$. Note that R has no time dependence and is merely the radius of the protein. When fluctuations dominate the reaction, the reaction has an apparent time dependent rate constant, as in Eq. 5.10. For the above scaling relationship, the exponent for the rate constant is given by: $1-h = 1/4$.

This model can be extended to consider the case where the surface of the sphere has a fractal dimension, D_s. The number of reactants B in the reaction volume with fractal dimension D is given by (see Eq. 5.11):

$$N_B \sim \rho_B(0)r^D \pm \sqrt{\rho_B(0)r^D} \tag{5.25}$$

where r is the distance a molecule can diffuse in time t, and the square root term is a result of number fluctuations, δN_B. We now make the assumption that the reaction volume consists of the boundary volume of the fractal surface. The fractal dimension of the boundary volume is given by $D = 3 - D_S$. The concentration fluctuations, $\delta\rho_B$, will scale as:

$$\delta\rho_B \sim \frac{\delta N_B}{r^{d_f}} \sim r^{-d_f/2} \sim r^{-(3-D_s)/2} \tag{5.26}$$

Again, it is assumed that at long-time scales, diffusion of the reactant into and through the reaction volume will be classical, i.e., $r^2 \sim t$. That is the walk dimension of the reactant is 2. Parenthetically, one would also anticipate that in the long-time limit the walk dimension of protein side-chains is also 2. Protein side-chains will diffuse through fluctuations in the protein structure, akin to monomer diffusion in polymer melts (deGennes, 1982). In this latter case, the short-time diffusion will be anomalous, while the long-time limit will be classical, i.e., $d_w = 2$. It is proposed that proteins will show comparable behavior. However, if the hydroxyl (or hydronium ion) is the dominant species, it will dictate the course of the reactant, and consideration of the protein dynamics is less important. With the assumptions of the model, it is seen that $\delta\rho_B \sim t^{-(3-D_S)/4}$. Thus, one has $1 - h = (3 - D_S)/4 \approx 0.21$.

5.3.3 *Reaction Widths in Proteins*

As the reaction proceeds, exchange will occur from tritiated groups deeper and deeper in the protein structure. Consequently, a domain within the protein will arise that separates tritiated residues from the solvent. Such domains can be seen by combining neutron diffraction with isotope exchange (Kossiakoff, 1982). The size of the reacted domain increases with time and this dependence can be estimated using the scaling arguments discussed in Section 5.2. Consider the region separating the tritiated portion of the protein (reactant A) from the solvent (reactant B) to have thickness, ℓ_{AB}. Again, a scaling Ansatz is made such that $\ell_{AB} \sim t^{\varsigma}$. In the simple spherical case treated above, one can make the association between ℓ and ℓ_{AB}. In this case, ς is equal to 1/2. (This is to be compared with values of 3/8 and 1/3 for simple one- and two-dimensional systems discussed previously.) For the refinement to a sphere with a fractal surface, one now has:

$$\varsigma = \frac{3 - D_s}{2} \approx 0.42 \tag{5.27}$$

It is possible to design neutron diffraction experiments to determine the rate of growth of the depletion zone. In principle, this information could be used to discriminate between isotope exchange mechanisms. If solvent penetration, as discussed above, occurs as a result of channels in the protein, the kinetics could strongly resemble a one-dimensional system. In this case, $\varsigma = 0.375$ and it would be difficult in practice to distinguish this behavior from the fractal surface model (Eq. 5.27). Thus, the length of the depletion zone is probably not a good parameter for discriminating between models.

5.3.4 *Fractal Effects on Chemically Controlled Reactions*

The activation parameters observed in Figure 5.6 and also those observed for simple peptides in solution are much higher than expected for diffusion-controlled reactions in aqueous solutions. Typically, a diffusion-controlled rate constant will have an activation energy in the range 0–6 kcal/mol. The isotope exchange energy is clearly well above this level. Yet the formalism of Eqs 5.9 and 5.10 pertains to diffusion-controlled reactions. On the time scale of the tritium exchange experiment, this reaction reaches the long-time domain discussed in Section 5.2, and becomes

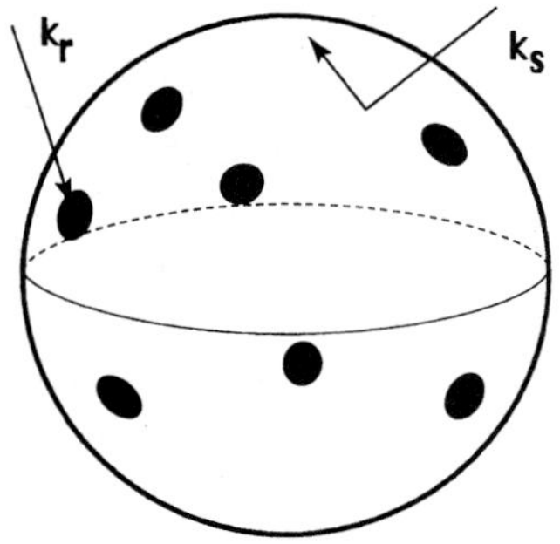

FIGURE 5.7 Schematic representation of the reaction of a ligand with a sphere covered with receptors. Two rate constants characterize this reaction. The constant k_S characterizes the collision of the ligand with the sphere. Reflective boundary conditions are assumed. The constant k_r is the rate constant associated with the binding of the ligand to a receptor. The theory of Zwanzig and Szabo (1991) accounts for the diffusive interference resulting from the multiple sinks on the sphere. In the isotope exchange problem, the disconnected receptors are replaced by the boundary volume, another disconnected space.

diffusion limited. The reaction is also unusual in that reactants diffuse to, but do not diffuse within, the "reaction volume." This is a consequence of the boundary volume having a dimensionality less than 1. To derive an expression for the diffusion steps in the above process, the formalism and results of a kinetic problem treated by Zwanzig and Szabo (1991) are examined. These workers considered the problem of ligands in the bulk solution diffusing and binding to receptors partially covering the surface of a sphere. This problem is illustrated in Figure 5.7.

This time-dependent treatment of the problem considers the diffusive interference of different receptors on the sphere. It proves convenient to define the Laplace transform of the diffusive rate constant, $\hat{k}(s)$, by:

$$\hat{k}(s) = \int_0^\infty e^{-st}\, k(t)\, dt \tag{5.28}$$

In this problem, two diffusional fluxes were considered: flux to the surface of the sphere and flux to the receptors. It was shown that the Laplace transform of the overall, diffusive rate constant could be related to the corresponding transforms of rate constants associated with these two diffusive steps. This gives:

$$\frac{1}{\hat{k}(s)} = \frac{1}{\hat{k}_{sphere}(s)} + \frac{1}{\hat{k}_{receptor}(s)} \tag{5.29}$$

where $\hat{k}_{sphere}(s)$ is the Laplace transform of the bimolecular rate constant for a ligand colliding with a sphere. It is given by:

$$\hat{k}_{sphere}(s) = \frac{4\pi R D_{diff}}{s}\left(1 + R\sqrt{\frac{s}{D_{diff}}}\right) \tag{5.30}$$

where D_{diff} is the sum of the diffusion constants of the sphere and the ligand, and R is the radius of the sphere. In the original work, the receptor was taken to be a partially absorbing disk and an appropriate expression was derived for that situation (Zwanzig and Szabo, 1991).

Equation 5.29 is actually quite general and can be used to describe rate processes that proceed through multiple domains. For our purposes, the boundary volume of the protein is considered to be the "receptor," $\hat{k}_{bv}(S) = \hat{k}_{receptor}(S)$. From the dimensional arguments presented in this section, one has:

$$\hat{k}_{bv}(S) \propto s^{-\bar{d}/4} \tag{5.31}$$

The long-time regime will correspond to the region where $s \to 0$. From Eqs 5.29–5.31, it is seen that this is the regime where $\hat{k}_{bv}(s)$ will dominate. Under these conditions, the diffusional rate constant in the forward direction will be given by $k_1(t) = k_1' t^{-h} = k_1' t^{1-d/4}$. To account for the overall time course of the reaction, one must consider the effects of the reversibility of the reaction and the influence of the chemical steps. As argued in Section 5.2, the diffusional dissociation step will have a comparable time dependence as the association step. In the long-time limit, one then has: $k_{-1}(t) = k_{-1}' t^{-h} = k_{-1}' t^{1-d/4}$. Using the long-time formalism given in Section 5.2, it is seen that the stretched exponential dependence of Eq. 5.26 is retained.

5.3.5 Summary of Hydrogen Exchange in Proteins

Hydrogen exchange in proteins has been discussed in detail because it represents an experimental setting in which a fractal structural parameter is related to the dynamics of the system. Hydrogen exchange in proteins is viewed as a solvent-catalyzed reaction between a hydronium or hydroxyl ion and the fractal surface of the protein. However, the protein is not considered as a static structure, but rather as having units that fluctuate to the surface. These fluctuations do not change the overall shape of the surface and maintain a constant surface fractal dimension. These fluctuations are considered to be similar in nature to those in polymer melts (deGennes, 1982). This is not an inappropriate analogy, considering the dense interior of most proteins. It should be emphasized that only the long-time behavior of these systems is considered, a regime where units in the fluctuating protein show normal diffusive behavior, i.e., $d_w = 2$. A number of studies have focused only on the short-time regime, and quite often these data show simple exponential behavior. In the steady state model discussed in the previous section, the short-time regime will be exponential and crossover to a stretched exponential in the long-time limit. It is particularly interesting that fractal diffusion processes will strongly affect this long-time limit, even for reactions that are not diffusion limited. This analysis also demonstrates that isotope exchange data are a potentially useful tool for exploring the fractality of protein surfaces under a variety of conditions. Clearly, at high temperature or under denaturing conditions such a model is inappropriate. However, data for acid-denatured proteins also show nonexponential behavior, and ultimately may be interpretable by a fractal model relevant to the protein structure known as the molten globule (see Shakhnovich and Finkelstein, 1989).

5.4 Diffusion and Reaction in an Archipelago

There are a number of other biological diffusion processes that have fractal aspects. Of particular recent interest are diffusion and reaction in biological membranes.

Biomembranes can be viewed as a two-dimensional fluid, and the kinetic processes in this fluid are far from simple. There are two features that create inhomogeneities in the membrane and result in restricted diffusion. First, in most native membranes, proteins occupy a significant fraction of the surface area. These are large objects compared to lipids and act as obstacles to diffusion. Additionally, most membrane proteins have a boundary layer of tightly bound lipids. This affinity also serves to obstruct lipid diffusion. The second common inhomogeneity is due to phase separation of lipids. Often, biomembranes will consist of islands of one phase surrounded by a sea or continuous region of a second phase (Thompson et al., 1992). Diffusion in either phase is restricted by the presence of the second phase. Obviously, diffusion within the island domains is very constrained, yet diffusion in the continuous phase will also show the effects of island obstacles. It is also not uncommon for membrane proteins to aggregate into domains, also forming large impenetrable islands. These sources of heterogeneity in biomembranes can result in anomalous diffusion for proteins as well as lipids. Such effects have been observed experimentally using fluorescence photobleaching techniques. Additionally, advances in fluorescence microscopy have allowed the tracking of single particles diffusing in the membrane. Often, the issue is not whether anomalous diffusion occurs in biological systems, but rather on what time scale it appears.

Factors that control membrane heterogeneity in native membranes are difficult to isolate and characterize. A large body of work exists on model systems, and it shows that domains can be manipulated by a number of physical and compositional parameters. It is possible to vary a parameter, usually temperature, so that island domains in the membrane increase in size. At some point, these domains will connect and become continuous, and what was previously the continuous region will break up into islands. This point is known as the percolation threshold, and the domain structure is a percolation cluster. Percolation clusters are fractal and have been investigated extensively. Percolation in biomembranes will be discussed in detail in Chapter 9. Presently, we consider the kinetic effects of the islands, i.e., the domain structure just below (or above, depending on the phase being considered) the percolation threshold.

It is instructive to consider the effects of island domains on a simple chemical reaction. There have been a number of computer simulations on the simple dimerization reaction $A + A \rightarrow A_2$ and the self-annihilation reaction $A + A \rightarrow O$. These are worth examining to provide a qualitative picture of how domains influence reaction equilibria and kinetics. In addition to the computer work, a number of similar, simple bimolecular reactions have been explored experimentally. These include spin exchange between fatty acid spin labels and collisional quenching of fluorescent fatty acids by spin labels. Although these reactions are different from those studied by computer simulations, it is anticipated that both sets of reactions will have similar qualitative behavior.

Thompson et al. (1992) have used numerical simulations to determine the effects of "disconnection" on dimerization. Figure 5.8 shows the results for a reversible dimerization reaction. In this case, the equilibrium concentration of dimer was determined for a connected and disconnected domain. The ratio of these concentrations is plotted versus the number of reactants A per domain. This ratio

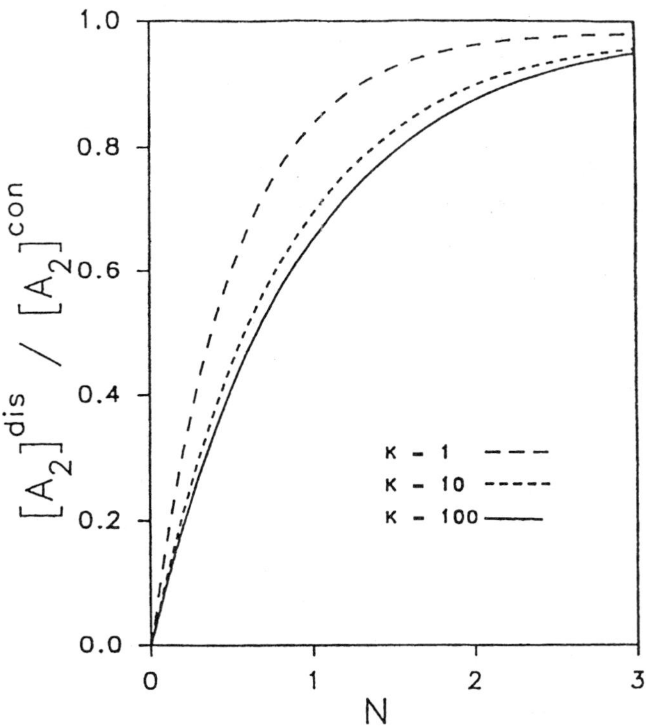

FIGURE 5.8 The ratio of the concentration of dimer in a disconnected system to that in a connected system as a function of N, the initial average number of monomers of A per domain. The system is initially all monomer and is then allowed to equilibrate. Calculations were performed for equilibrium constants of 1, 10, and 100. (From Thompson et al., 1992. Reprinted by permission of Gordon and Breach Publishers.)

shows how the domain structure alters the normal (connected) equilibrium properties. In the limit of a very small number of reactants, single particles will be trapped in separate domains, and dimerization will be impossible. Thus the concentration of dimers in disconnected domains will fall to zero at that limit. In the high-concentration limit, local interactions will dominate and the domain structure has minimal effect. Thus, the ratio of disconnected and connected dimer concentration approaches 1. Interestingly, if one were determining an "apparent" equilibrium constant from concentrations of monomer and dimer without a knowledge of the domain structure, a smaller value would be obtained than the true constant.

Reaction kinetics of a bimolecular annihilation reaction in disconnected domains has also been investigated using Monte Carlo simulations (Newhouse and Kopelman, 1986). The steady state kinetics of A + A annihilating each other on islands showed a rate law with an anomalous reaction order (see Eq. 5.16 for the A + B reaction). In this computer experiment there is a steady state production of A

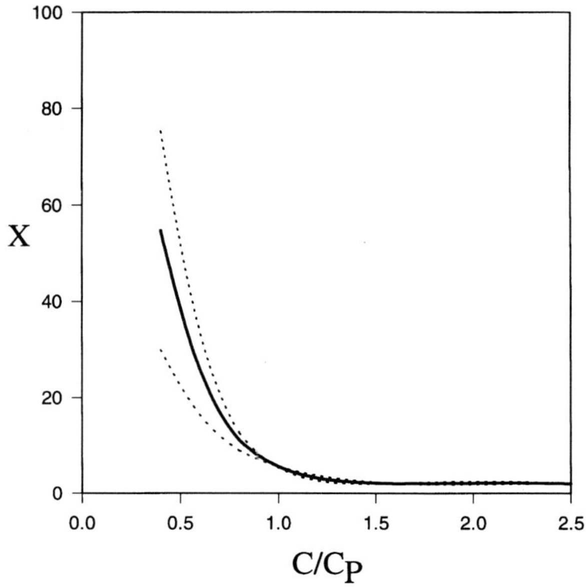

FIGURE 5.9 Results of a computer simulation of the rate of A + A → O at steady state. The anomalous reaction order, *X*, is plotted versus the reduced concentration of guests covering a two-dimensional lattice. (----) Crude error bounds. (Data taken from Newhouse and Kopelman, 1986.)

at "guest" sites on a two-dimensional lattice. These guest sites form the reaction domain. When the concentration of guest sites reaches the percolation threshold, C_P, then the reaction domain is continuous and the reaction order, X, equals 2, the classical value. Below the percolation threshold, the fractal dimension of the guest sites is less then 2. Because the reactants diffuse classically on the guest sites, the walk dimension, d_w, is 2 and the spectral dimension is equal to the fractal dimension. For this reaction, the order is given by: $X = 1 + 2/\bar{d}$. Below the percolation threshold, X will increase above 2. As the topology becomes more disjointed, i.e., smaller islands, the guest sites will form a fractal "dust" and $\bar{d} \rightarrow 0$. This results in an infinite reaction order. These effects are illustrated in Figure 5.9, were the reaction order is plotted as a function of the reduced guest concentration (C/C_p). The expected sharp rise in X is seen.

The results of these kinetic and equilibrium simulations provide an intuitive picture of the role of disconnected domains. Domains with small coverage of reactants do not promote the reaction in the sense that the "equilibrium" favors the monomeric form. The extreme limit of this case is when there is one reactant per island. On the other hand, the rates of the reactions are strongly enhanced as the domain size decreases. This is a result of the high local concentration of reactants. Again, an extreme case would be the great enhancement of two reactants on a shrinking island. Reactions on disjointed geometry can provide serious problems for the interpretation of experimental data. This is especially true in the case of equilibrium measurements of macroscopic parameters. Without prior knowledge of the domain structure or how to manipulate it, experimental equilibrium constants must be viewed as "apparent," since they will reflect not only the energetics of the reactions but also the geometric constraints.

The diffusion of particles in the continuous region surrounding island domains is also of considerable interest and has a number of biological implications. Anomalous diffusion of membrane proteins has been observed in single-particle tracking experiments (Ghosh and Webb, 1990; Ghosh, 1991). Using computer-enhanced video microscopy, a fluorescent-modified form of the low-density lipoprotein–receptor complex could be monitored. The mean square displacement for individual particles diffusing laterally in the cell membrane was measured. The time dependence of this displacement was anomalous at short times, i.e., $d_w \neq 2$ in Eq. 5.7, and a crossover to normal diffusive behavior was seen at long times. Similar anomalous diffusion has been observed for membrane proteins using the fluorescence photobleaching recovery technique (Brust-Mascher et al., 1993). Nagle (1992) has recently pointed out problems associated with the conventional analysis of photobleaching experiments when anomalous diffusion occurs. These problems result from observing diffusion over a limited time scale. Diffusion constants reported in the literature using this technique must be considered pertinent only within the time-domain of the experiment.

In a recent series of papers, Saxton (1989, 1992, 1993a, b, 1994) has considered the effects of obstacles on two-dimensional lateral diffusion using Monte Carlo simulations. If these obstacles are immobile, percolation theory can be used to understand long-range diffusional effects. Above the percolation threshold, long-range diffusion is totally hindered. However, even below the percolation threshold there will be a short-time regime in which the diffusion is anomalous. This time regime increases as the threshold is approached. As the fraction of the two-dimensional lattice covered with obstacles increases, the time regime that shows anomalous diffusion also increases.

These two different temporal regimes are a result of the characteristic lengths of the system. The obstacles formed by randomly filling a lattice are fractal over short distances and appear homogeneous at distances longer than the correlation length. These structures are discussed more fully in Chapter 9. Because of this fractal behavior, there is a "crossover length," R^*, characterizing diffusion of a tracer. Below this crossover length, the mean-square displacement of the tracer is described by:

$$\langle r^2 \rangle = D(C)t^{2/d_w} \tag{5.32}$$

where the diffusion constant is a function of C, the fraction of the lattice covered by obstacles. The walk dimension is also a function of C and this dependence, obtained from computer simulations, is shown in Figure 5.10. In this figure the dimensionless parameter C/C_p is used, where C_P is the fraction of coverage at the percolation threshold. This parameter allows for the comparison of results from different lattices. In the limit of small C (no obstacles), the walk dimension approaches 2 and normal diffusion holds. At the percolation threshold, diffusion is anomalous at all time scales, and d_w approaches 2.87, the anticipated result in two dimensions

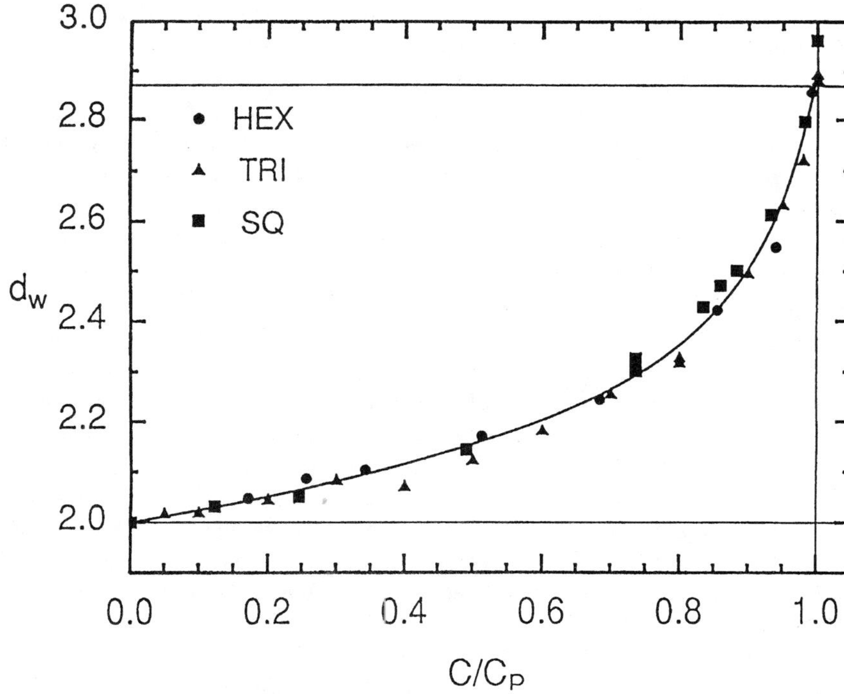

FIGURE 5.10 Walk dimension, d_w, as a function of obstacle concentration for three different geometries: point obstacles on a square lattice (SQ), point obstacles on a triangular lattice (TRI), and hexagonal obstacles of unit radius on a triangular lattice (HEX). The reduced parameter C/C_P is used so all lattices can be represented equivalently. C_P is the lattice coverage at the percolation threshold. (Figure from Saxton, 1994.)

(Havlin and Bunde, 1991). The crossover length between the anomalous and normal diffusive regime scales as:

$$R^* \sim |C - C_p|^{-\nu + \beta/2} \tag{5.33}$$

where the scaling exponents in two-dimensions are $\nu = 4/3$ and $\beta = 5/36$. Thus, it is seen that as C approaches C_P, R^* goes to infinity, indicating that anomalous diffusion will occur at all length scales.

Using the Monte Carlo simulations, it is possible to explore the experimental conditions that are likely to show anomalous diffusion and crossover to normal diffusion (Saxton, 1994). Fluorescence photobleaching experiments have a time resolution in the millisecond range, and video tracking experiments typically have a resolution of 33 ms. Fluorescence quenching or excimer formation experiments have time scales of the order of microseconds or faster. The crossover time for anomalous diffusion depends on both the unobstructed diffusion constant and on the coverage of obstacles, C. For most experimental situations, it is anticipated that photobleaching and video tracking will observe normal diffusion, while the faster time regime of fluorescent quenching and excimer experiments allows anomalous diffusion to be observed. However, proteins can have diffusion constants that are lower than anticipated from hydrodynamic arguments. For instance, when the diffusion constant is below 10^{-10} cm^2/s, anomalous diffusion will be observed even with the slower experimental techniques. With these considerations, it is not surprising that anomalous diffusion has been observed over a wide range of time scales. It is the challenge for both the experimentalist and the theorist to piece together the disparate information obtained using a variety of techniques.

5.5 Other Diffusional Processes

Another very complicated diffusion problem involves modeling reaction dynamics by diffusion over a potential surface (see Bagchi and Fleming, 1990). The biological focus of such modeling is found in the work of Frauenfelder et al. (1988), and involves photodissociation of carbon monoxide (CO) from myoglobin. When photodissociation occurs at low temperatures the dissociated CO is entrapped in the protein and either diffuses out of the protein or back to the heme pocket. The kinetics of this process is nonexponential and covers may orders of magnitude in time. Initial diffusion models provided only a semiquantitative description of the experimental data (Agmon and Hopfield, 1983a, b). Recently, a sophisticated treatment involving diffusion on a two-dimensional potential surface has been presented and used to analyze the myoglobin data in more detail (Agmon and Rabinovich, 1992). These models predict power law behavior when there is diffusional anisotropy. Under high anisotropy conditions, there are critical jumps in power law exponents (Rabinovich and Agmon, 1993). While this latter condition may not be of experimental relevance for this system, it is of great theoretical interest. Protein dynamics will be discussed more fully in Chapter 6.

5.6 Summary

When considering diffusion and reaction in inhomogeneous environments, structure and dynamics can be linked by a parameter known as the spectral dimension. The spectral dimension is related to a structural parameter, the fractal dimension of the medium, and a dynamical parameter, the walk dimension. As shown in this chapter, the spectral dimension appears in kinetic scaling laws. In transient kinetics, it appears as an exponent for a time-dependent rate constant. In steady state kinetics, the spectral dimension appears in expressions for an anomalous reaction order. Such behavior occurs when fluctuations dominate the reaction. Interestingly, we have seen that, even for chemically controlled reactions, anomalous diffusion greatly influences the reaction rate. With these considerations, a model of protein fluctuations has been described that accurately fits data on the rate of isotope exchange. This predictive model uses the fractal surface dimension of the protein as an unadjusted parameter. Anomalous diffusion appears in biomembranes as a result of inhomogeneous domains. Domain formation can strongly influence reaction and diffusion of small molecules within both the disconnected and connected domains.

In addition to chemical kinetics effects, these domains can result in departures from bulk equilibrium behavior. Some biophysical phenomena can be model as diffusion across a potential energy surface. This also can give rise to complicated rate behavior.

References

Adam, G., and Delbrück, M. 1968. Reduction of Dimensionality in Biological Diffusion Processes. In *Structural Chemistry and Molecular Biology.* ed. A. Rich and N. Davidson, Freeman, San Francisco, pp. 198–215.

Agmon, N., and Hopfield, J. J. 1983a. Transient Kinetics of Chemical Reactions with Bounded Diffusion Perpendicular to the Reaction Coordinate: Intramolecular Processes with Slow Conformational Changes. *J. Chem. Phys.* **78**, 6947–6959.

Agmon, N., and Hopfield, J. J. 1983b. CO Binding to Heme Proteins: A Model for Barrier Height Distributions and Slow Conformational Changes. *J. Chem. Phys.* **79**, 2042–2053.

Agmon, N., and Rabinovich, S. 1992. Diffusive Dynamics on Potential Energy Surfaces: Nonequilibrium CO Binding to Heme Proteins. *J. Chem. Phys.* **97**, 7270–7286.

Axelrod, D., and Wang, M. D. 1994. Reduction-of-Dimensionality Kinetics at Reaction-Limited Cell Surface Receptors. *Biophys. J.* **66**, 588–600.

Bagchi, B., and Fleming, G. 1990. Dynamics of Activationless Reactions in Solution. *J. Phys. Chem.* **94**, 9–20

Berg, H., and Purcell, E. M. 1977. Physics of Chemoreception. *Biophys. J.* **20**, 193–239.

Berg, O. G., Winter, R. B., and Von Hippel, P. H. 1981. Diffusion-Driven Mechanisms of Protein Translocation on Nucleic Acids. 1. Models and Theory. *Biochemistry* **20**, 6929–6948.

Brust-Mascher, I., Feder, T. J., Slattery, J. P., Baird, B., and Webb, W. W. 1993. FRP data on mobility of cell surface proteins reevaluated in terms of temporally constrained molecular motions. *Biophys. J.* **64**, A354.

Calhoun, D. A., and Englander, S. W. 1985. Internal Protein Motions, Concentrated Glycerol, and Hydrogen Exchange of Myoglobin. *Biochemistry* **24**, 2095–2100.

Cooper, A. 1976. Thermodynamic Fluctuations in Protein Molecules. *Proc. Natl. Acad. Sci. USA* **73**, 2740–2741.

deGennes, P. G. 1982. Kinetics of Diffusion-Controlled Processes in Dense Polymer Systems. I. Nonentangled Regimes. *J. Chem. Phys.* **76**, 3316–3321.

DeLisi, C. 1980. The Biophysics of Ligand–Receptor Interactions. *Q. Rev. Biophys.* **13**, 201–230.

Dewey, T. G. 1994. Fractal Analysis of Proton Exchange Kinetics in Lysozyme. *Proc. Natl. Acad. Sci. USA* **91**, 12 101–12 104.

Dewey, T. G. 1995. Chemically-Controlled Reaction Kinetics on Fractals: Application to Hydrogen Exchange in Lysozyme. *Fractals* **3**, 251–267.

Eigen, M. 1954. Über die Kinetik sehr schnell verlaufender Ionenreaktionen in wässeriger Lösung. *Z. Phys. Chem.* **12**, 176–183.

Ellis, L. M., Bloomfield, V., and Woodward, C. K. 1975. Hydrogen–Tritium Exchange Kinetics of Soybean Trypsin Inhibitor (Kunitz). Solvent Accessibility in the Folded Conformation. *Biochemistry* **14**, 3413–3419.

Englander, S. W., and Kallenbach, N. R. 1984. Hydrogen Exchange and Structural Dynamics of Proteins and Nucleic Acids. *Q. Rev. Biophys.* **16**, 521–655.

Englander, S. W., and Mayne, L. 1992. Protein Folding Studied Using Hydrogen-Exchange and Two-Dimensional NMR. *Ann. Rev. Biophys. Biol. Struct.* **21**, 243–265.

Englander, S. W., and Poulsen, A. 1969 Hydrogen–Tritium Exchange of Random Chain Polypeptides *Biopolymers* **7**, 329–393.

Frauenfelder, H., Sligar, S. G., and Wolynes, P. G. 1988. The Energy Landscapes and Motions of Proteins. *Science* **254**, 1598–1603.

Ghosh, R. N. 1991. Mobility and Clustering of Individual Low-Density Lipoprotein Receptor Molecules on the Surface of Human Skin Fibroblasts. Ph.D. thesis, Cornell University, Ithaca, NY.

Ghosh, R. N., and Webb, W. W. 1990. Evidence for Intra-Membrane Constraints to Cell Surface LDL Receptor Motion. *Biophys. J.* **57**, 286A.

Havlin, S., and Bunde, A. 1991. Percolation II. In *Fractals and Disordered Systems*, ed. A. Bunde and S. Havlin. Springer-Verlag, Berlin, pp. 965–149.

Kang, K., and Redner, S. 1984. Scaling Approach for the Kinetics of Recombination Processes. *Phys. Rev. Lett.* **52**, 955–958.

Kopelman, R. 1988. Fractal Reaction Kinetics. *Science* **241**, 1620–1626.

Kossiakoff, A. A. 1982. Protein Dynamics Investigated by the Neutron Diffraction–Hydrogen Exchange Technique. *Nature* **296**, 713–721.

Nagle, J. F. 1992. Long-Tail Kinetics in Biophysics? *Biophys. J.* **63**, 366–370.

Newhouse, J. S., and Kopelman, R. 1986. Reaction Kinetics on Clusters and Islands. *J. Chem. Phys.* **85**, 6804–6806.

O'Shaughnessy, B., and Procaccia, I. 1985. Analytical Solutions for Diffusion on Fractal Objects. *Phys. Rev. Lett.* **54**, 455–458.

Pfeifer, P. 1987. Characterization of Surface Irregularity. In *Preparative Chemistry Using Supportive Reagents*. ed. P. Laszlo, Academic Press, New York, pp. 13–32.

Pfeifer, P., Welz, U., and Wippermann, H. 1985. Fractal Surface Dimension of Proteins: Lysozyme. *Chem. Phys. Lett.* **113**, 535–540.

Rabinovich, S., and Agmon, N. 1993. Scaling and Critical-Like Behavior in Multidimensional Diffusion Dynamics. *Phys. Rev. E* **47**, 3717.

Redner, S., and Leyvraz, F. 1991. Spatial Organization in Two-Species Annihilation. *J. Stat. Phys.* **65**, 1043–1056.

Saxton, M. 1989. Lateral Diffusion in an Archipelago: Distance Dependence of the Diffusion Coefficient. *Biophys. J.* **56**, 615–622.

Saxton, M. 1992. Lateral Diffusion and Aggregation: A Monte Carlo Study. *Biophys. J.* **61**, 119–128.

Saxton, M. 1993a. Lateral Diffusion in an Archipelago: Dependence on Tracer Size. *Biophys. J.* **64**, 1053–1062.

Saxton, M. 1993b. Lateral Diffusion in an Archipelago: Single-Particle Diffusion. *Biophys. J.* **64**, 1766–1780.

Saxton, M. 1994. Anomalous Diffusion Due to Obstacles: A Monte Carlo Study. *Biophys. J.* **66**, 394–401.

Shakhnovich, E. I., and Finkelstein, A. V. 1989. Theory of Cooperative Transitions in Protein Molecules. I. Why Denaturation of Globular Protein is a First-Order Phase Transition. *Biopolymers* **28**, 1667–1694.

Thompson, T. E., Sankaram, M. B., and Biltonen, R. L. 1992. Biological Membrane Domains: Functional Significance. *Comments Mol. Cell. Biophys.* **8**, 1–15.

van Kampen, N. G. 1981. *Stochastic Processes in Physics and Chemistry*. North-Holland, Amsterdam, pp. 291–303.

Welch, G. R. (1986) *The Fluctuating Enzyme*. Wiley, New York.

Wickett, R. R., Ide, G. J., and Rosenberg, A. 1974. A Hydrogen-Exchange Study of

Lysozyme Conformation Changes Induced by Inhibitor Binding. *Biochemistry* **13**, 3273–3277.

Winter, R. B., and Von Hippel, P. H. 1981. Diffusion-Driven Mechanisms of Protein Translocation on Nucleic Acids. 2. The *Escherichia coli* Repressor–Operator Interaction: Equilibrium Measurements. *Biochemistry* **20**, 6948–6960.

Winter, R. B., Berg, O. G., and Von Hippel, P. H. 1981. Diffusion-Driven Mechanisms of Protein Translocation on Nucleic Acids. 3. The *Escherichia coli lac* Repressor–Operator Interaction: Kinetic Measurements and Conclusions. *Biochemistry* **20**, 6961–6977.

Woodward, C., Simon, I., and Tüchsen, E. 1982. Hydrogen exchange and the dynamic structure of proteins. *Mol. Cell. Biochem.* **48**, 135–159.

Zumofen, G., Blumen, A., and Klafter, J. 1985. Concentration Fluctuations in Reaction Kinetics. *J. Chem. Phys.* **82**, 3198–3206.

Zwanzig, R., and Szabo, A. 1991. Time Dependent Rate of Diffusion Influenced Ligand Binding to Receptors on Cell Surfaces. *Biophys. J.* **60**, 671–678.

6

Are Protein Dynamics Fractal?

A molecule in a liquid is in a densely packed but disordered setting. It is also in a very dynamic setting. Typically, it will have a handful of neighbors with which it is constantly colliding. At room temperature, approximately one collision occurs every picosecond (10^{-12} s). If we could "tag" two neighbors and follow their motion, we would find that they collide with each other about a 1000 times before they drift apart, unlikely to meet again. If we were to "tag" all the molecules within a sphere, for the first few nanoseconds (10^{-9} s) the sphere would remain intact. As time goes on, untagged molecules would enter the sphere, while tagged molecules would diffuse outward. The sphere would have appear to be growing, but the tagged molecules are being diluted.

Now consider another sphere of tagged molecules. This time all the tagged species are connected by chemical bonds. As we observe the molecules, the sphere will again grow as a result of "invasions" of the solvent. However, this time it will stop when it reaches the radius of gyration of the polymer. The collisions with solvent molecules are now doing something very different. Instead of an intricate and densely packed set of billiard balls bouncing around elastically, we now have bonds bending, twisting, and stretching. The translational energy of the solvent is transferring energy to various deformation modes of the polymer. There is a random distribution of energy entering and leaving our "sphere." The sphere contains the random walk of a polymer chain. The overall energy transfer process is again one of these problems involving a convolution of random processes – random energy input on a random lattice. As discussed in Chapter 4, this should give rise to multifractal behavior.

While it is easy to describe a picture of polymer dynamics, putting this picture into a mathematical model is not so simple. The energy transfer modes within the polymer will depend on the nature of the bonding and internal connectivity of the polymer. This makes proteins a particularly challenging problem. Most work involves sorting through the phenomenology of protein dynamics and experimentally exploring the multitude of relaxation processes that are occurring. In the next section, an introductory discussion of protein dynamics is presented. The time, length, and energy scales of an array of different processes are discussed. Considering the complexity and range of protein dynamical processes, it is not surprising that protein dynamical processes often have a nonexponential time dependence. In Section 6.2 we examine a number of physically plausible and simple models that generate nonexponential behavior. We see that these models can be classified within two general schemes: structural disorder and dynamic disorder. Within both schemes, fractal behavior is possible. In Section 6.3 the concept of fractal time is introduced, and again we see that simple models are sufficient to generate fractal time. This presentation is qualitative in nature, emphasizing why protein dynamics might be fractal. In the final section, we examine in more detail the kinetics of ion channel opening. This is a phenomenology that has been investigated extensively using the fractal formalism.

6.1 Introduction to Protein Dynamics

There is an enormous and sometimes confusing literature on protein dynamics. This is a result of the diversity of techniques and time scales used to probe proteins. Virtually every available window for observing proteins yields a different dynamical picture. Each dynamical regime is, in some way, associated with a specific structural domain. The coupling between regimes occurs in a complicated fashion, and it is not clear how relaxation processes at one time scale effect similar processes in another regime. Indeed, one of the unresolved problems of molecular biophysics is to understand the relationship and coupling between this multitude of protein relaxation processes.

The motional properties of proteins are a direct consequence of their structural properties (for reviews see Brooks et al., 1988; Nall and Dill, 1991; Creighton, 1992). The rigid, covalently bonded polypeptide chain provides a dynamic frame for the protein. At a given instance, a protein might undergo a wide variety of motions as a result of thermal motions. There is a constant background of collisions with solvent molecules as well as collisions between the closely packed amino acid side-chains within the protein. The bonding forces along the chain constrain these dynamic responses in certain ways. The internal structural constraints can channel thermal fluctuations into larger collective modes. Current thinking is that structure and fluctuations are intimately entwined.

Each peptide linkage has two flexible internal rotations, the angles ϕ and ψ, and a rigid torsional angle ω. In addition to these backbone rotations, the amino acid side-chains have some degree of flexibility. These side-chains usually have bonds that are flexible enough for internal rotation. As an added complication to the bonding constraints of the polypeptide backbone, disulfide linkages provide additional bonding interactions that bridge the polymer structure. Such interactions and the weaker "salt links" that result from opposing charge pairs provide additional structural features that constrain a protein's dynamics.

The density of amino acids within a protein is essentially the same as that in an amino acid crystal. Because of the close packing and covalent constraints, motions within a protein behave much as rigid bodies within a fluid. Internal rotations of side-groups, such as the phenyl groups of tyrosine or the isopropyl groups of valine or leucine, can undertake seemingly chaotic librations as a result of structural constraints. These groups are limited to relatively small internal motions because of the high energy of deformation of bond lengths and angles. While many of these motions are of no consequence for the function of the protein, there may be crucial sets of functionally important motions (FIMs) (see Frauenfelder et al. 1988). These are concerted fluctuations that drive the protein into a new and functionally important conformational state. FIMs provide a path for the protein to enter entirely new regions of conformational space. Most likely, these critical, functional motions of a protein involve displacements of groups associated with torsional oscillations about rotationally permissive single bonds. In this view, the majority of the internal dynamics will not lead to any new conformational territory. For instance, high-frequency vibrations that occur within the local groups are thought not to play a role in the concerted, conformational motions of a protein.

The existence of fluctuations has been established using a number of experimental techniques, including hydrogen isotope exchange, as discussed in Chapter 5. However, our understanding of the biological role of fluctuations remains incomplete. Fluctuations show local and global character and both are expected to be important. In a protein, as in other rigid condensed systems, structural changes arise from correlated fluctuations. Perturbations, such as ligand binding, can produce conformational changes by introducing a structural bias that directs fluctuations along a reaction path. This allows transitions from one structure to another to occur with a minimal driving force. Fluctuations can be regarded as searching out the path or paths that lead to conformational transitions.

In considering the internal motions of proteins, there is an interesting interplay between dynamics and thermodynamics. Thermodynamics emphasizes the equilibrium conformations, while dynamical studies have focused on large displacements from equilibrium. In certain cases, some features of the dynamics may be unimportant because the time scale is faster than the phenomenon of interest. An example might be the fast local relaxation of atoms that are part of a much slower hinge-bending motion or "breathing" mode of an enzyme. In this case, only the time scale of the large-scale mode would be expected to be involved in determining the rate of the catalytic process. In other situations, the detailed aspects of the atomic fluctuations are a significant factor. This may be the case in the oxygen storage protein myoglobin, where local side-chain motions appear to be essential for the entrance and exit of ligands.

The range of dynamical processes of proteins and their functional role is summarized in Table 6.1 (Brooks et al., 1988). A spectrum of internal motions has been identified. These motions show displacements of 0.01–100 Å, energies of

TABLE 6.1 Summary of internal motions of proteins (Brooks et al., 1988)

I. *Local Motions (0.01–5 Å, 10^{-15}–10^{-1} s)*

(a) Atomic Fluctuations
1. Small displacements required for substrate binding (many enzymes)
2. Flexibility required for "rigid-body" motion (lysozyme, liver alcohol dehydrogenase, l-arabinose binding protein)
3. Energy "source" for barrier crossing and other activated processes
4. Entropy "source" for ligand binding and structural changes

(b) Side-Chain Motions
1. Opening pathways for ligand to enter and exit (myoglobin)
2. Closing active site (carboxypeptidase)

(c) Loop Motions
1. Disorder-to-order transition covering active site (triose phosphate isomerase, penicillopepsin)
2. Rearrangement as part of rigid-body motion (liver alcohol dehydrogenase)
3. Disorder-to-order transition as part of enzyme activation (trypsinogen–trypsin)
4. Disorder-to-order transition as part of virus formation (tobacco mosaic virus, tomato bush stunt virus)

(d) Terminal Arm Motion
1. Specificity of binding (λ-repressor–operator interaction)

II. *Rigid-Body Motions (1–10 Å, 10^{-9}–1 s)*

(a) Helix Motions
1. Induction of larger scale structural change (insulin)
2. Transitions between substates (myoglobin)

(b) Domains (Hinge-Bending) Motions
1. Opening and closing of active-site region (hexokinase, liver alcohol dehydrogenase, l-arabinose binding protein)
2. Increasing binding range of antigens (antibodies)

(c) Subunit Motions
1. Allosteric transitions that control binding and activity (hemoglobin, aspartate transcarbamoylase)

III. *Larger Scale Motions (> 5 Å, 10^{-7}–10^{4} s)*

(a) Helix–Coil Transition
1. Activation of hormones (glucagon)
2. Protein folding transition

(b) Dissociation/Association and Coupled Structural Changes
1. Formation of viruses (tomato bushy stunt virus, tobacco mosaic virus)
2. Activation of cell fusion protein (hemagglutinin)

(c) Opening and Distortional Fluctuations
1. Binding and activity (calcium-binding proteins)

(d) Folding and Unfolding Transition
1. Synthesis and degradation of proteins

0.1–100 kcal/mol, and time scales of 10^{-15}–10^{3} s. These physical scales are usually correlated. For instance, an increase in the amplitude of the fluctuations will also correspond to an increase in the energy and in a longer time scale. However, some motions can be slowed by their setting. For instance, an aromatic ring flip occurs on the time scale of seconds in a protein, while for a small molecule in solution it can occur on a picosecond time scale. The difference is that localized events in proteins can often be slowed by high activation barriers. These barriers result from the larger scale, concerted motion required to create a ring flip within a protein.

6.2 The Origins of Nonexponential Behavior

6.2.1 Random Multiplicative Processes and the Log–Normal Distribution

Protein dynamics provide a rich range of interconnected events occurring on many time scales. Given this, it is not surprising that protein systems can exhibit nonexponential behavior. To see, in general terms, how this nonexponential behavior may arise, we consider a complicated process such as a conformational change in a protein. We return to the random multiplicative processes discussed in Chapter 4. For a complicated process to occur, a number of independent lower order processes must be successfully completed. The completion of our complicated task of changing the conformational state of a protein is assumed to happen during a characteristic time period, τ. The probability, P, of the conformational change occurring during this time is given by:

$$P(\tau) = p_1 p_2 \dots p_n \tag{6.1}$$

where p_j is the probability of the lower order process occurring. For protein dynamics, this lower order process may be librational modes, bond rotations of side-chains, concerted motions, or collective modes. One can well imagine that a number of these events must occur, either simultaneously or in sequence, for a conformational change to be triggered. If each of these subprocesses has a characteristic time scale, τ_j, then the probability of a process being completed during time, τ, is given by $p_j = \tau_j/\tau$. Because of the multiplicative nature of these probabilities, it is sometimes simpler to consider:

$$\log P(t) = \log p_1 + \log p_2 + \dots + \log p_n \tag{6.2}$$

Now, if the p_j is an independent random variable so is $\log p_j$. If the appropriate moments exist for this new distribution and n is sufficiently large, then the central limit theorem can be applied (West and Shlesinger, 1989). This implies that $\log P$ follows a normal distribution and P is log–normal:

$$P(\tau) = \frac{1}{\sqrt{2\pi\tau\sigma^2}} \exp\left\{ -\frac{(\log \tau - \log \bar{\tau})^2}{2\sigma^2} \right\} \tag{6.3}$$

where $\bar{\tau}$ and σ^2 are the mean and variance of the τ_i terms. The prefactor to the exponential is required to normalize the probability function. The above arguments are completely general and can be applied to any complicated task that requires a collection of independent subtasks to be completed before the task itself is accomplished. For a colorful discussion of log–normal distributions, see West and Shlesinger (1989).

Equation 6.3 has an interesting time dependence. At short times, it follows a power law dependence, $P(\tau) \propto \tau^{-1/2}$, and at longer times it shifts to the more complicated exponential law. Although such a simple model may not be realistic, it shows a number of features that appear in more sophisticated treatments. First, it demonstrates how easily one can get away from the "exponential regime." The present case shows a transition from one rate law (power law) to another (log–normal). Neither of these are exponential. This model also shows how "power law" behavior can be almost as pervasive as exponential. In the ion channel opening experiments discussed later in this chapter, the probability of ion channel opening or closing is give by $P(\tau) \propto \tau^{-\alpha}$ with $0 < \alpha < 1$. This is quite similar to the short-time dependence predicted by the multiplicative model. Similar rate laws have been derived by Millhauser using a reptation model adapted from polymer physics (Millhauser, 1990; Oswald et al., 1991). While at first appearance the reptation model is very different from the multiplicative process, the two models are actually mathematically very similar.

6.2.2 Static and Dynamic Disorder Models

While protein dynamics are doubtless quite complex, with processes extending over many orders of magnitude in time, this does not necessarily imply that they are fractal. As will be discussed, complex dynamics may be treated with other nonfractal models. Indeed the log–normal distribution just discussed has no intrinsic fractal feature to it. Consequently, it is important to examine the assumptions of those models that generate fractal dynamics to assess whether they can be realistically applied to proteins. In addition to fractal models, there are more general situations that predict nonexponential relaxation behavior in complex systems. These are divided into two categories (Zwanzig, 1990; Dewey and Spencer, 1991; Dewey, 1992): the static disorder model and the dynamical disorder model.

From a variety of theoretical (Brooks et al., 1988) and experimental results (Frauenfelder et al., 1988, 1991), it is seen that the potential energy surface of a protein is an extremely rugged landscape. Transitions across such a landscape can result in nonexponential behavior. When considering such transitions, there are three different temperature regimes that yield distinctly different behavior. These regimes are static disorder, dynamic disorder, and rapid equilibration, and are illustrated in Figure 6.1. The static disorder regime occurs at low temperature, where each molecule is locked into a local well in the potential energy surface. The transition from one conformation to another results from transitions directly out of the local wells. Since each protein will have been locked into a different local minimum, each one will have a slightly different activation energy for a given conformational process. As the temperature is increased, the protein is no longer locked into a well and transitions occur between local minima. At these intermediate temperatures, the local transitions can be on a comparable time scale as the larger conformational

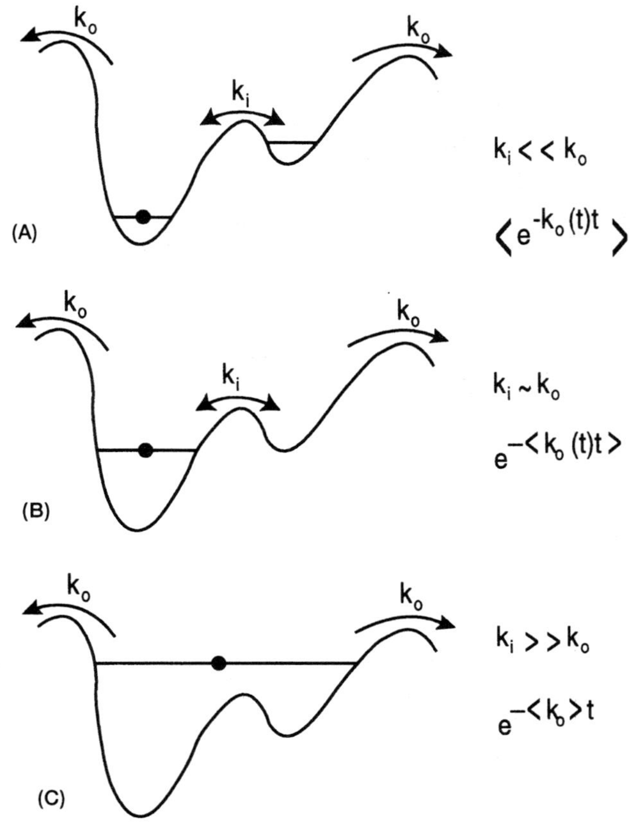

FIGURE 6.1 Illustration of kinetic regimes for a complicated potential energy surface. (a) At low temperature each species is locked into a substate, and a distribution of activation energies occurs as a result of having many substates. This condition is called static disorder. (b) At intermediate temperature the rate of exchange between substates is comparable to the rate out of the energy well. The internal dynamics influences the dynamics of removal from the well. This is dynamic disorder. (c) At high temperature fast equilibration between substates occurs. An average rate of disappearance from the well occurs. This rapid equilibration limit gives exponential behavior.

transition. In this case, the protein does not have time to equilibrate, yet it "samples" other local states. The internal dynamics will then influence the overall transition rate. This condition is the dynamical disorder regime. At high temperatures, local transitions occur rapidly and the internal states are essentially equilibrated. The high-temperature regime results in conventional chemical kinetics, i.e., exponential processes.

Experimental evidence for the static disorder model exists from the extensive work of Frauenfelder and coworkers on the photodissociation of carbon monoxide bound to myoglobin. In these studies, evidence from Raman scattering suggests that conformational substates exist in a protein. The details of this evidence have been described in a number of review articles (Frauenfelder et al., 1988, 1991). Here, we focus on the kinetic model, as opposed to spectroscopic evidence. At the very low temperatures at which these experiments were performed, individual proteins are locked into specific conformational substates. Because experimental observables are based on an ensemble of molecules, each protein will see a slightly different activation barrier. Thus, an average must be made over all possible activation energies. Assuming a unimolecular, conformational change for the protein, (see Figure 6.1a), one has a microscopic rate constant, $k(E)$, associated with each conformational substate and its separate activation energy E. The microscopic rate law for a protein in conformational state A undergoing a change to a state B is given by:

$$\frac{-d[\mathrm{A}]}{dt} = k(E)[\mathrm{A}] \tag{6.4}$$

[A] is the concentration of the conformational state and $k(E)$ is a rate constant that follows a normal Arrhenius law:

$$k(E) = k_0\, e^{-E/RT} \tag{6.5}$$

where k_0 is the pre-exponential frequency factor. Equation 6.4 can be solved for each state, and the resulting time dependence is averaged over the multitude of conformational states. To perform this average the probability of finding a barrier of a given height must be assumed. The rate law for the average decay, $\langle A(t)\rangle$, is then given by:

$$\langle A(t)\rangle = A(0) \int dE\, \rho(E)\, e^{-k(E)t} \tag{6.6}$$

where $A(0)$ is the initial amount of A present and $\rho(E)$ is the density of states with barrier E. Note that E and $k(E)$ are not functions of time. Thus there is no dynamic interchange between substates, and Eq. (6.6) represents a static disorder problem.

There are two ways of handling data analysis based on Eq. 6.6. First, one can assume a distribution for the density of states, solve Eq. 6.6, and see how the resulting function fits the experimental data. This is the approach taken by Ansari et al. (1985) and is discussed shortly. Alternatively, one can determine the functional form of $\langle A(t)\rangle$ using experimental data. Subsequently, a Laplace transform of Eq. 6.6 and integration provides a direct determination of $\rho(E)$. Liebovitch and Tóth (1991) explored this inversion method for a variety of cases, including when $\langle A(t)\rangle$ is a stretched exponential.

In previous applications to protein dynamics (Ansari et al., 1985), a uniform box distribution was assumed for $\rho(E)$:

$$\begin{aligned} \rho(E) &= (E_{mas} - E_{min})^{-1} = \Delta^{-1} && E_{min} \le E \le E_{mas} \\ \rho(E) &= 0 && \text{otherwise} \end{aligned} \tag{6.7}$$

In these studies, Eq. 6.6 was numerically integrated using Eq. 6.7. (Note, however, that the integral is exactly solvable (Dewey, 1992).) This static disordered model could adequately fit experimental data on reassociation of carbon monoxide (CO) to myoglobin at low temperatures (Ansari et al., 1985). It was later shown that the effects observed in these data were due to kinetic hole-burning, and other analyses are more appropriate (see Šrajer et al., 1986; Šrajer and Champion, 1991; Dewey and Bann, 1992). Nevertheless, there is additional experimental evidence (for a review, see Frauenfelder et al. (1988, 1991) that adds support for the existence of conformational substates in a protein. It is perhaps not surprising that at the low temperatures studied (60–160 K) a complex structure such as a protein should have a number of substate conformations available to it. At these low temperatures the small barriers between substates become significant and have a profound effect on the dynamics of the system.

As the temperature increases there can be a natural progression from static disorder to dynamic disorder. This progression was investigated theoretically by Agmon and Hopfield (1983a,b), who treated the high temperature regime as a bound diffusion problem. In this dynamical model, the "reactant" CO molecule diffuses over a potential energy surface. This theory provided qualitative agreement with the photodissociation data on myoglobin (see also Agmon and Rabinovich, 1992). For such a dynamical disorder model, the rate constant changes with time as a result of temporal fluctuations in some physical parameter, $B(t)$. When $B(t)$ is a random function in time and fluctuations occur on the same time scale as the conformational transition then dynamic disorder occurs. The rate law in this case is

$$\frac{-d[A(t)]}{dt} = k(B(t))\,[A(t)] \tag{6.8}$$

and the integrated form is

$$\langle A(t)\rangle = A(0)\exp\left\{-\int_0^t k(B(s))\,ds\right\} \tag{6.9}$$

In very simplistic terms, it is seen that static disorder is an average over exponentials (Eq. 6.6), while dynamic disorder is the exponential of an average (Eq. 6.9). If B fluctuates very rapidly, then $k(B(s))$ can be replaced by its time average, $\langle k\rangle$. An exponential decay is then recovered:

$$\langle A(t)\rangle = \mathrm{A}(0)\exp(-\langle kt\rangle) \tag{6.10}$$

This is the high-temperature limit given in Figure 6.1.

Specific models of dynamical orders fall into two categories: discrete and continuous (Zwanzig, 1990). In the discrete model, the value of $B(t)$ fluctuates between a fixed set of values (see van Kampen, 1981). In continuous models, $B(t)$ is determined from the solution of a Langevin equation. Both approaches usually require numerical solutions to the problem. The previously mentioned applications to CO recombination in myoglobin used the Langevin approach (Agmon and Hopfield, 1983a,b; Agmon and Rabinovich, 1992). This model was considered applicable for the high-temperature, low-viscosity limit of these experiments. It gives "non-Kramers kinetics" and provides suitably nonexponential behavior. However, a quantitative analysis of the data was not achieved. Also, when the CO photodissociation is performed close to physiological temperatures, the CO can actually leave the myoglobin. This results in a normal, bimolecular diffusional process that shows simple exponential kinetics. It is difficult to test the "bound diffusion" model stringently, because of the limited temperature regime to which it has access. At the very low temperatures used in Frauenfelder studies, structural disorder will occur. At high temperatures, bimolecular dissociation complicates the issue.

In principle, one might expect a system to proceed from static disorder to dynamic disorder to conventional kinetics as the temperature is increased. To date, no systems have been observed that convincingly follow these transitions. Such observations are problematic, for a number of reasons. First, while it is easy to discriminate between exponential and nonexponential behavior, it can be extremely difficult to distinguish between nonexponential behavior of static versus dynamic disorder. Basically, most nonexponential models provide very good fitting functions. One must either go to elaborate statistical analyses to discriminate between models (Korn and Horn, 1988; Marshall, 1989; Sansom et al., 1989) or, ideally, to develop new experiments that allow a more direct testing of them (Dewey, 1992). Additionally, there may be multiple fluctuation variables, $B(t)$. As the temperature is changed, the system may move out of a dynamic regime for one variable and into a regime for a second variable. Thus, a single equation of the form and parameters of Eq. 6.9 may not be adequate. Despite these problems, nonexponential behavior has been investigated extensively, and successful models have been constructed (Hamill and Funabashi, 1977; Bagchi and Fleming, 1990), often relying on the physical appeal of the model.

Where do fractal dynamic models fall within the context of structural and dynamic disorder? It will be seen that both classes of disorder model support fractal behavior. Fractal models can be considered as a subset of these more general models. Two microscopic models, the hierarchical and defect-diffusion model, provide physically plausible descriptions of protein dynamics and show fractal behavior. Concepts such as the hierarchy of states and FIMs arise naturally in these models. Such concepts have been supported by other argument and have physical appeal. In the next section, fractal kinetic models, or "fractal time," are introduced. The conceptual basis for applying such models to protein dynamics is then considered.

6.3 Fractal Time

6.3.1 Long Tail Distributions and Fractal Time

Time-dependent scaling laws involving noninteger exponents appear pervasively in the kinetics of complex physical systems. Frequently these systems are frozen or glassy materials, although simple heterogeneous systems show similar behavior (see Chapter 5). Many of these diverse processes have been described using a concept known as fractal time (Shlesinger, 1988). Fractal time describes highly intermittent, self-similar behavior that does not possess a characteristic time scale. If the average time for an event is finite, then it has a characteristic time scale. Fractal time, on the other hand, does not have such a finite average time; rather, the average time is infinite. Examples of phenomena that show this lack of scale are electron–hole recombination in amorphous material, particle dynamics in turbulent flow, and phase rotation in a Josephson junction. While these examples appear far removed from protein dynamics, the conceptual basis for them is actually quite close to the concepts that exist in the biophysical literature. While the first modern application of fractal time was in 1963 with Berger and Mandelbrot's description of transmission errors in telephone networks, the basic concepts can be traced back to the "Petersburg Game" of Bernoulli's (see Shlesinger, 1988 for a colorful historical background). This "game" captures the fundamental features of fractal time.

In our present treatment, we illustrate this example of fractal time by putting it into a biological setting. Consider a cell membrane that has a population of ion channels. During a given long time period, a channel will flicker into an open conformation for a short time period, τ, with a probability of 1/2. When the channel is open an ion diffuses down a transmembrane concentration gradient and passes through the channel. This results in a change in electrical potential across the membrane, a quantity that can be easily and accurately measured. Let the probability of the channel opening for twice the length of time be 1/4. This longer open period results in the transport of two ions across the membrane. Following this scenario, the probability of an opening of duration $n\tau$ is $(1/2)^n$, with a concomitant transport of 2^n ions. Now, an order of magnitude more ions are transported with a likelihood that is an order of magnitude less than for the previous time period. The average number of ions transported is infinite, as determined by:

$$1 \text{ ion} \times \frac{1}{2} + 2 \text{ ions} \times \left(\frac{1}{2}\right)^2 + \ldots + 2^n \text{ ions} \times \left(\frac{1}{2}\right)^{n+1} + \ldots = \infty \qquad (6.11)$$

The paradox of the situation is that the median number of ions transported per channel is 1. The ion channel has a long-tail distribution, i.e., the mean and the median are very different. Also, the average transport for an ensemble of channels is infinite. The infinite behavior of an average of a random variable is a hallmark of fractals. While this is a fanciful example, long tail distributions with infinite moments do arise in a number of physical situations and are no longer considered to be paradoxical.

For a physical process in which the time between events can be considered as a random variable, the average time between events $\langle t \rangle$ is a natural parameter to consider. If $\langle t \rangle$ is finite than the system has a natural scale or characteristic time. Events in such a system will occur at an average rate proportional to $1/\langle t \rangle$. If, on the other hand, $\langle t \rangle$ should be infinite (or much longer than the measurement time) than fractal time occurs. In this case, the sequence of event times looks as dissimilar to a constant rate process as possible. Points in this sequence appear self-similar, clustered in intervals as one might find in a random Cantor set. When fractal time occurs the time correlation functions are characterized by the "stretched" exponential. The corresponding behavior in the frequency domain is a broad (non-Debye) amplitude dispersion curve.

6.3.2 A Chemical Kinetic Model with Fractal Time

To put this discussion on a more concrete footing, a specific model is considered that shows fractal time. Consider a conformational transition between two states of a protein that fits into the static disorder scheme. For a protein with a rugged conformational surface, there will be many different substates from which the transitions can occur. For a given substate with barrier height E, the probability of a transition occurring between some time t and $t + dt$ is given by: $k(E) \exp [- k(E)t]\, dt$. For a system where many of the substates are populated, the probability of a transition must be weighted by the distribution of $k(E)$. The overall transition probability, $\psi(t)$, then becomes:

$$\psi(t) = \int k(E) \exp [- k(E)t] \rho(k)\, dk \tag{6.12}$$

where $\rho(k)$ describes the distribution of rate constants arising from the various substates. It is often more natural to average over barrier heights, E, rather then rate constants. As a result of the assumed Arrhenius behavior (Eq. 6.5), Eq. 6.12 can result in complicated integrals involving exponentials of exponentials.

If the barrier heights are assumed to be distributed in a Poisson fashion, then the distribution function of these heights, $f(E)$, is given by:

$$f(E) = (kT_0)^{-1} \exp (- E/kT_0) \tag{6.13}$$

where kT_0 is the mean barrier height. To change variables in the integral in Eq. 6.12, one uses $f(E)\, dE = \rho(k)\, dk$, and finds

$$\psi(t) = \int k(E) \exp [- k(E)t] f(E)\, \mathrm{d}E \tag{6.14}$$

The asymptotic behavior of this integral can be determined using Eq. 6.5 (Shlesinger, 1988), and this is given by

$$\psi(t) \sim t^{-1-\beta} \tag{6.15}$$

where $\beta = T/T_0$. The function $\psi(t)$ can be used to calculate $\langle t \rangle$ from its first moment. This model gives an interesting temperature dependence. At low temperatures, $T < T_0$, $\langle t \rangle$ is infinite, while with $T > T_0$, $\langle t \rangle$ becomes finite. Thus a transition from fractal time to conventional kinetics is made as the temperature increases.

6.3.3 Hierarchical and Defect-Diffusion Models

One could debate whether the physical basis of this model (Eq. 6.13) is appropriate for proteins. However, there are other fractal models that have more physical appeal. They are the hierarchical model and the defect-diffusion model. Palmer and Stein developed a model for spin relaxations in glasses and amorphous materials that demonstrate kinetic behavior which is fractal in time (Palmer et al., 1984; Stein, 1985). In the original model, there is a hierarchy of relaxation events. In this hierarchy, the fast relaxations must occur before a slower set of events occur. Relaxations within a tier are considered to be uncorrelated. However, relaxation times from one tier to the next are related by an empirical law, and this establishes

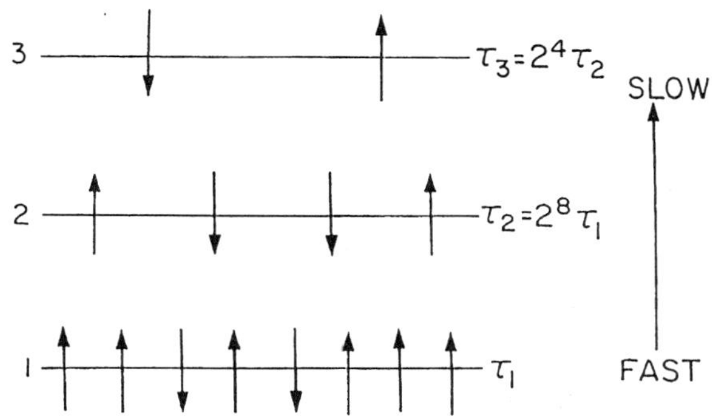

FIGURE 6.2 Schematic representation of a hierarchical relaxation model. Spins (or rotamers) must exist in a specific configuration before relaxation can proceed at the higher level. (Adapted from Klafter and Drake, 1992.)

the final form of the function describing the overall relaxation. This model is illustrated in Figure 6.2.

In initial work (Palmer et al., 1984), the following simple relationship was used to relate the relaxation time on one tier, τ_n, to that on the next highest tier, τ_{n+1}:

$$\tau_{n+1} = 2^{\mu_n}\tau_n \tag{6.16}$$

where μ_n is the number of spins (or, in our case, rotamers) in a given level. In this model the spins, or rotamers, in the next state are only free to relax after the μ_n spins of the previous state have reached one specific configuration (see Figure 6.2). The relaxation function, $\psi(t)$, for the hierarchy of processes is then given by

$$\psi(t) = \sum_n \Omega_n \exp(-t/\tau_n) = A \exp[-(t/\tau)^\beta] \tag{6.17}$$

where the sum is over all tiers, ω_n is a weighting factor for the population of each state, and A and β are the amplitude and fractional exponents of the "stretched" exponential.

This hierarchical model may be an appropriate description of protein conformational changes. In these models, a set of relaxation processes in one subset of conformational states must occur before a second set proceeds. One way of visualizing this process is that most of the bond rotations in a protein are physically constrained. Those rotations that are not constrained must reach, in concert, a specific configuration in order to free a second set of rotations. The barriers for the second tier of processes are now low enough to allow these slower relaxations to occur. Once the second tier rotations reach the appropriate configuration, a third tier of processes is now accessible. The process then continues from tier to tier and to progressively longer time scales. This model has conceptual, but not quantitative, similarities with the model proposed by Frauenfelder and coworkers in analyzing the dynamics of myoglobin after photodissociation of CO. From the temperature dependence of this process, Frauenfelder proposed that four tiers occur in a

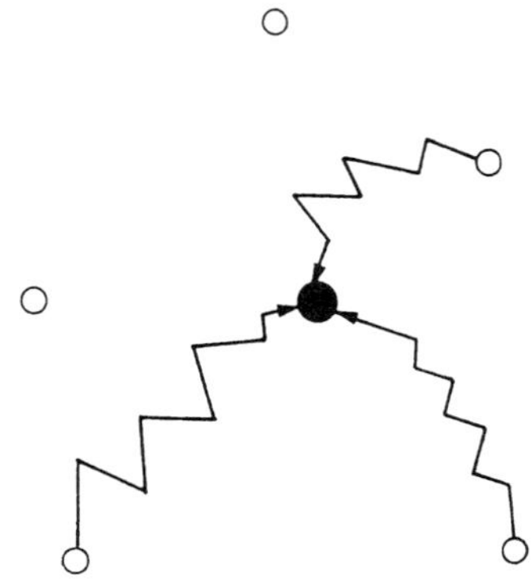

FIGURE 6.3 Schematic representation of a defect-diffusion model. Defects (○) must diffuse through a complex medium and combine with a dipole (or rotamer) before relaxation can occur. Parallel processes from multiple defects result in a distribution of relaxation times.

hierarchical, glass-like transition (Ansari et al., 1985). The main feature determining the dynamics of this model is the conformational substates within a given tier. The Frauenfelder model represents a hierarchical model in which individual tiers follow a nonfractal, static disorder model. Palmer's model has very similar physical motivations and yet gives fractal time behavior. In a similar vein, the Markov chain model, in which the rate constants are related by a geometric progression, will also give similar behavior (Liebovitch, 1989). The key ingredient in all these models is a hierarchical relationship in rate phenomena.

A second model, the defect-diffusion model, may also provide a reasonable starting point for exploring protein dynamics. The defect-diffusion model has been used to describe dielectric relaxation in amorphous material (Klafter and Blumen, 1985; Klafter and Shlesinger, 1986). In this model, a dipole is frozen in place until a lattice defect diffuses into it. Upon fusion with the defect, the dipole abruptly relaxes to a new orientation. Many defects are moving through the medium and each one of these will have a different probability of combining with the dipole. The relaxation of a given dipole reflects a parallel sequence of events. This resulting parallel distribution of "waiting" times confers the stretched exponential behavior to the relaxation process. The defect diffusion model is depicted in Figure 6.3.

These ideas can be adapted to proteins by adopting the concept of FIMs. As mentioned previously, Frauenfelder suggested that certain, highly specific motions within a protein are key to functionality. If this is so, then diffusion of defects through the protein to the specific regions associated with FIMs could trigger a conformational change. Presumably a number of defect diffusion pathways exist within the protein and lead to this region. Upon combination of the defect with the sensitive region, an FIM is generated. The FIMs then trigger a conformational change of functional significance. Because of the parallel defect-diffusion paths, the conformational transition will show fractal time behavior. Läuger (1988) has proposed a defect-diffusion model for ion channel kinetics, and a one-dimensional version of this has been solved analytically (Condat and Jäckle, 1989). These models show power law scaling and can adequately fit some of the data on channel kinetics.

The defect-diffusion and hierarchical models provide two very different microscopic descriptions of protein dynamics. Despite this, they generate the same "stretched" exponential behavior for conformational relaxation. These models from condensed matter physics are not implausible models for protein dynamics. Ultimately, however, plausibility is not an issue. The crux of the problem is devising experimental methods that discriminate between models on grounds other than statistical fits to kinetic data.

6.4 Ion Channel Kinetics

To date, the most comprehensive effort to establish fractality in a system has been the work done by Liebovitch on ion channel dynamics (see Liebovitch and Tóth, 1990a,b). Consequently the focus of this section will be on ion channels. Ion channels appear in a wide variety of cellular settings. In nerve cells they generate and propagate the action potential, in muscle they control calcium levels that initiate contraction, in sensory cells they trigger electrical changes in signal transduction, and in epithelial cells they are part of the pathway of net ionic movement across cell layers. Such channels are membrane-bound proteins and are typically made up of several subunits. Channels span the membrane and can permit transport of ions from one side to another. This is the result of open conformations that allow free access of the ions. Typically, the thermodynamic driving force for ion transport is the concentration and electrical gradients that exist across the membrane. The channels offer a means for ions to diffuse down their concentration gradients. Channels will fluctuate between open and closed conformations, and thus the resulting ionic current also fluctuates.

The spontaneous conformational fluctuations of individual channel molecules can be measured because of the associated ionic conductivity changes. A cell membrane will have a very low conductivity as a result of the hydrophobic nature of the lipid bilayer. This creates a large activation energy for ion transport. However, when a channel opens, a hydrophilic path becomes available through which ions can diffuse. The great sensitivity of conductivity measurements can be used to observe the opening and closing of individual channels. A technique known as the patch clamp technique, developed by Neher and coworkers, allows these measurements to be made (Sakmann and Neher, 1983). This technique is illustrated in Figure 6.4. In this method, a conductivity electrode is made from a micropipette the tip size of which is approximately 1 micron. The cell membrane is impaled with this tip and a small piece adheres to the opening of the micropipette. Because the cell membrane forms a high electrical resistance seal across the tip, the current measured across the patch is due to ions moving exclusively through the channels. Recordings can be made with the patch attached to the cell, or the micropipette can be removed and exposed to the surrounding solution. Depending upon the technique used, the membrane can be oriented either right side or inside out with respect to the surrounding solution.

The beauty of the patch clamp technique is that extremely small currents can be measured. This is achieved using electronic techniques where the voltage between

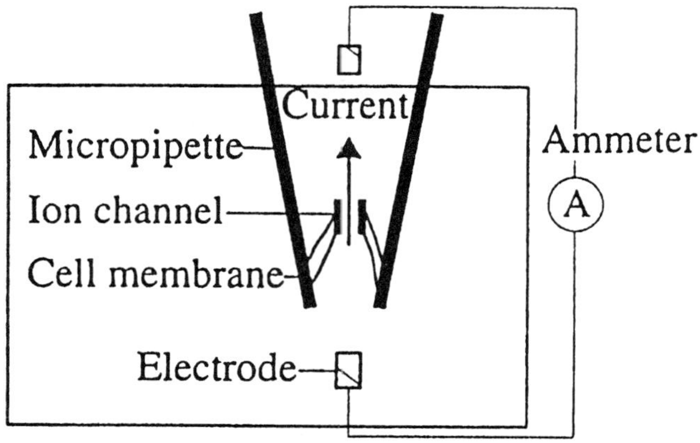

FIGURE 6.4 Schematic representation of a patch clamp experiment. A micron sized patch of membrane is sealed in a glass micropipette. The current measured across the patch shows the sequence of open and closed conformations of the channel. Individual proteins can be monitored using this technique. (From Liebovitch and Tóth, 1990a. Reprinted by permission of Biomedical Engineering Society.)

the electrode and a ground in the external buffer solution are held constant ("clamped") and small changes in current through the membrane patch are converted to voltages, which are then amplified. Channels will open for periods of milliseconds. During this time, of the order of 10,000 ions will diffuse through the channel. This charge displacement gives picoamp changes in the conductivity, an amount that is easily measured. The surface density of channels in most biological systems is relatively low, and one can have a single channel in a one square micron area on the tip of the micropipette. This means that the dynamics of a single protein can be observed. There is no other technique of comparable accuracy that can be used to make such an observation.

An example of the experimentally observed opening and closing of ion channels is shown in Figure 6.5, where the current is plotted as a function of time. The current follows the fluctuation of the channel between open and closed states. As can be seen, there is significant variation in the length of time for which a given open or closed state lasts. Such patch clamp traces are analyzed using dwell-time histograms. To construct such a histogram, one considers a time interval, Δt, and counts how many open (or closed) events had durations less than or equal to Δt. Next one determines how many events had durations between Δt and $2\,\Delta t$. The entire histogram is constructed by determining the number of events, $N(i\,\Delta t)$, in each of i intervals. The probability density, $f(t)$, is the probability of an event having a duration within a certain interval. Thus $f(i\,\Delta t) = N(i\,\Delta t)/\Delta t\, N_{tot}$, where N_{tot} is the total number of events. The cumulative probability, $P(t)$, is the probability that a channel is open for a duration greater than t and is equal to the integral of $f(t)$ over time. In chemical kinetic parlance, $f(t)$ is equal to the $\psi(t)$ in Eq. 6.12, and $P(t) = -\,d\Psi/dt$.

The information concerning protein dynamics is contained in the behavior of the functions, $f(t)$ and $P(t)$. At this point it appears that no "universal" or generic

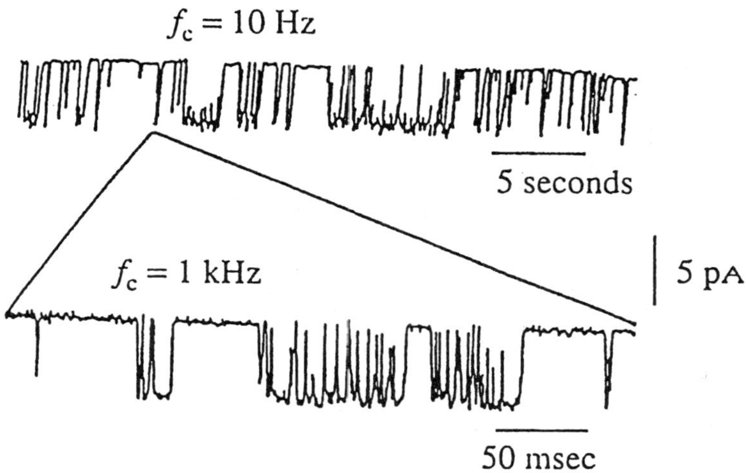

FIGURE 6.5 Trace showing the current through a membrane patch versus time. The opening and closing of an adenosine triphosphate (ATP) sensitive potassium channel is observed. Expansion of the time scale shows similar statistical sequences of openings and closings. The trace visually illustrates the fractal nature of the phenomenon. (From Liebovitch and Tóth, 1990b. Reprinted by permission of the *Annals of the New York Academy of Sciences.*)

behavior is seen. Some channels give simple exponential behavior for $f(t)$. This is referred to as Markov behavior. However, other channels appear to show fractal time behavior. Inspection of the recording shown in Figure 6.5 shows the self-similar characteristic of a given channel system. To the eye, a change in time scale does not show appreciably different fluctuation behavior. A plot of $P(t)$ versus t for a channel that shows fractal behavior is shown in Figure 6.6. As can be seen from the logarithmic scale of the plot, a simple exponential will not describe this behavior. Liebovitch et al. (1987) found that for a number of different channels accurate fits could be obtained with $f(t)$ and $P(t)$ of the form

$$f(t) = At^{1-D} \exp\left[-\left(\frac{t}{\tau}\right)^{2-D}\right] \tag{6.18}$$

$$P(t) = \exp\left[-\left(\frac{t}{\tau}\right)^{2-D}\right] \tag{6.19}$$

Such a fit is shown in Figure 6.6. Some categories of channels are more conveniently fitted by forms that appear to be asymptotic limits of Eqs 6.18 and 6.19. Thus good fits can be observed in certain cases to forms such as $P(t) \propto t^{-\alpha}$ and $f(t) \propto t^{-\alpha-1}$. In such cases, the best fit often gives values for α of 1/2. Referring to our example in Eq. 6.15, this gives infinite $\langle t \rangle$, i.e., fractal time.

While data from some channels clearly show fractal time, this phenomenon is not universal. This dependence of the dynamics on the specific system under

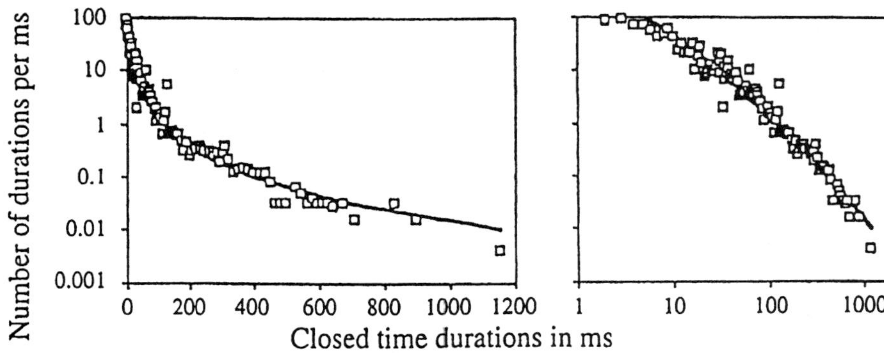

FIGURE 6.6 Plots of the number of closed time events versus the duration for a potassium channel in the apical membrane of a corneal endothelial cell. The left-hand plot is semilogarithmic plot, while right-hand plot is a log–log plot. The experimental data were fitted using the function in Eq. 6.18 and are characteristic of fractal time. (From Liebovitch and Tóth, 1990a. Reprinted by permission of Elsevier Publishing Co., Inc.)

consideration has created controversy in the field. Additionally, more complicated phenomenology has been observed that requires even more sophisticated models than the simple fractal ones (Dewey and Bann, 1992). There will probably be no single model that will encompass this diverse range of behaviors of ion channels. In order to sort through this phenomenology, microscopic physical models of channel gating will ultimately be required. This will allow independent testing of models that goes beyond statistical analysis of dwell-time histograms. A number of attempts have already been made to construct physical models to explain the complicated dynamics (Lüger, 1988; Condat and Jäckle, 1989; Liebovitch et al., 1992; Oswald et al., 1991). These models are very much in the spirit of the dynamic disorder models discussed earlier.

6.5 Summary

In this chapter we have explored the complexity of protein dynamics and looked for models that can incorporate the multitude of relaxation events into a single formalism. Complicated physical processes that require the completion of several independent tasks have been described as random multiplicative processes. Such processes result in log–normal distributions and show nonexponential behavior. Indeed, a number of plausible chemical kinetic models can yield complicated scaling laws and stretched exponentials. These models fall into two main conceptual categories: static disorder and structural disorder. In both cases, processes occur that are said to be fractal in time. Fractal time is characterized by a finite mean event time and an infinite average event time. They result in long-tail time distributions. Using concepts derived from experimental results in the biophysical literature, two physically plausible models of protein dynamics are discussed. These are a hierarchical model adapted from spin–glass relaxation models and a defect-diffusion model. Both cases yield stretched exponential kinetics. The dynamics of opening and closing of ion channels is discussed. Experimental data from a number of different channels show open and closed probability distributions that are consistent with a fractal time dependence.

References

Agmon, N., and Hopfield, J. J. 1983a. Transient Kinetics of Chemical Reactions with Bounded Diffusion Perpendicular to the Reaction Coordinate: Intramolecular Processes with Slow Conformational Changes. *J. Chem. Phys.* **78**, 6947–6959.

Agmon, N., and Hopfield, J. J. 1983b. CO Binding to Heme Proteins: A Model for Barrier Height Distributions and Slow Conformational Changes. *J. Chem. Phys.* **79**, 2042–2053.

Agmon, N., and Rabinovich, S. 1992. Diffusive Dynamics on Potential Energy Surfaces: Nonequilibrium CO Binding to Heme Proteins. *J. Chem. Phys.* **97**, 7270–7286.

Ansari, A., Berendzen, J., Bowne, S. F., Frauenfelder, H., Iben, I. E. T., Sauke, T. B., Shyamsunder, E., and Young, R. D. 1985. Protein states and protein quakes. *Proc. Natl. Acad. Sci. USA* **82**, 5000–5004.

Bagchi, B., and Fleming, G. R. 1990. Dynamics of Activationless Reactions in Solution. *J. Phys. Chem.* **94**, 9–20.

Brooks III, C. L., Karplus, M., and Pettitt, B. M. 1988. Proteins: A Theoretical Perspective of Dynamics, Structure and Thermodynamics. *Adv. Chem. Phys.* **LXXI**. 7–21.

Condat, C. A., and Jäckle, J. 1989. Closed-Time Distribution of Ionic Channels. *Biophys. J.* **55**, 915–925.

Creighton, T. E. 1992. *Protein Folding*. W. H. Freeman, New York.

Dewey, T. G. 1992. Probing Dynamical Disorder Kinetics with Chemical Relaxation Techniques. *Chem. Phys.* **161**, 339–350.

Dewey, T. G., and Bann, J. 1992. Protein Dynamics and $1/f$ Noise. *Biophys. J.* **63**, 594–598.

Dewey, T. G., and Spencer, D. 1991. Are Protein Dynamics Fractal? *Comments Mol. Cell. Biophys.* **7**, 155–171.

Frauenfelder, H., Parak, F., and Young, R. D. 1988. Conformational Substates in Proteins. *Ann. Rev. Biophys. Biophys. Chem.* **17**, 451–480.

Frauenfelder, H., Sligar, S. G., and Wolynes, P. G. 1991. The Energy Landscapes and Motions of Proteins. *Science* **254**, 1598–1603.

Hamill, W. H., and Funabashi, K. 1977. Kinetics of Electron Trapping Reactions in Amorphous Solids; A Non-Gaussian Diffusion Model. *Phys. Rev. B.* **16**, 5523–5527.

Klafter, J., and Blumen, A. 1985. Models for Dynamically Controlled Reactions. *Chem. Phys. Lett.* **119**, 377–382.

Klafter, J., and Drake, J. M. 1992. Relaxation in Spatially-Restricted Structures. In *Dynamics and Mechanisms of Photoinduced Transfer and Related Phenomena*, ed. N. Mataga, T. Okada, and H. Masuhara. Elsevier Science/North Holland, Amsterdon, pp. 345–361.

Klafter, J., and Shlesinger, M. F. 1986. On the Relationship among Three Theories of Relaxation in Disordered Systems. *Proc. Natl. Acad. Sci. USA* **83**, 848–851.

Korn, S. J. and Horn, R. 1988. Statistical Discrimination of Fractal and Markov Models of Single-Channel Gating. *Biophys. J.* **54**, 871–877.

Läuger, P. 1988. Internal Motions in Proteins and Gating Kinetics of Ionic Channels. *Biophys. J.* **53**, 877–884.

Liebovitch, L. 1989. Analysis of Fractal Ion Channel Gating Kinetics: Kinetic Rates, Energy Levels, and Activation Energies. *Math. Biosci.* **93**, 97–115.

Liebovitch, L. S., Fischbarg, J., and Koniarek, J. P. 1987. Ion Channel Kinetics: A Model Based on Fractal Scaling rather than Multistate Markov Processes. *Math. Biosci.* **84**, 37–68.

Liebovitch, L. S., Selector, L. Y., and Kline, R. P. 1992. Statistical Properties Predicted by the Ball and Chain Model of Channel Inactivation. *Biophys. J.* **63**, 1579–1585.

Liebovitch, L. S., and Töth, T. I. 1990a. Using Fractals to Understand the Opening and Closing of Ion Channels. *Ann. Biomed. Eng.* **18**, 177–194.

Liebovitch, L. S., and Töth, T. I. 1990b. Fractal Activity in Cell Membrane Ion Channels. *Ann. N Y Acad. Sci.* **591**, 375–391.

Liebovitch, L. S., and Töth, T. I. 1991. Distribution of Activation Energy Barriers that Produce Stretched Exponential Probability Distributions for the Time Spent in Each State for the Two State Reaction A↔B. *Bull. Math. Biol.* **53**, 443–455.

Marshall, D. B. 1989. Statistical Considerations in the Analysis of Dispersive Kinetics Data as Discrete or Continuous Distributions of Rate Constants. *Anal. Chem.* **61**, 660–665.

Millhauser, G. L. 1990. Reptation Theory of Ion Channel Gating. *Biophys. J.* **57**, 857–864.

Nall, B. T., and Dill, K. A. 1991. *Conformations and Forces in Protein Folding*. American Association for the Advancement of Science, Washington, DC.

Oswald, R. E., Millhauser, G. L., and Carter, A. A. 1991. Diffusion Model in Ion Channel Gating. *Biophys. J.* **59**, 1136–1142.

Palmer, R. G., Stein, D. L., Abrahams, E., and Anderson, P. W. 1984. Models of Hierarchically Constrained Dynamics for Glassy Relaxation. *Phys. Rev. Lett.* **53**, 958–961.

Sakmann, B., and Neher, E. 1983. *Single-Channel Recording*. Plenum Press, New York.

Sansom, M. S. P., Ball, F. G., Kerry, C. J., McGee, R., Ramsey, R. L., and Usherwood, P. N. R. 1989. Markov, Fractal, Diffusion, and Related Models of Ion Channel Gating. *Biophys. J.* **56**, 1229–1243.

Shlesinger, M. F. 1988. Fractal Time in Condensed Matter. *Annu. Rev. Phys. Chem.* **39**, 269–290.

Šrajer, V., and Champion, P. M. 1991. Investigations of Optical Line Shapes and Kinetic Hole Burning in Myoglobin. *Biochemistry* **30**, 7390–7402.

Šrajer, V., Schomacker, K. T., and Champion, P. M. 1986. Spectral Broadening of Biomolecules. *Phys. Rev. Lett.* **57**, 1267–1270.

Stein, D. L. 1985. A Model of Protein Conformational Substates. *Proc. Natl. Acad. Sci. USA* **82**, 3670–3672.

van Kampen, N. G. 1981. *Stochastic Processes in Physics and Chemistry.* North-Holland, Amsterdam, pp. 203–207.

West, B. J., and Shlesinger, M. F. 1989. On the Ubiquity of 1/*f* Noise. *Int. J. Mod. Phys. B* **3**, 795–819.

Zwanzig, R. 1990. Rate Processes with Dynamical Disorder. *Acc. Chem. Res.* **23**, 148–151.

7

Fractons and Vibrational Relaxation in Proteins

In his book on random walks, Howard Berg makes the memorable statement: "Biology is wet and dynamic" (Berg, 1983). Indeed, biochemical processes are inextricably linked to the solution chemistry of the aqueous milieu. One might anticipate a continuum description, as given in Chapter 5, as being the most appropriate for the diffusive processes of biochemistry. There is, nevertheless, considerable interest in discrete or "jump" transport processes. Such processes often provide a convenient mathematical limit for studying continuum events. However, discrete models are more than mathematical devices. The starting point for the formulation of vibrational relaxation in complex media are differential-difference equations. These equations have a rich phenomenology and can give rise to complex oscillatory and chaotic processes. The focus of the present chapter will be on vibrational processes in proteins. There are a number of other biological settings that are appropriately described by jump processes. Examples as diverse as the motion of flagellated bacteria and exciton transport in photosystems fall in this realm. In the companion chapter that follows (Chapter 8), we consider discrete processes resulting from the decoding of sequence information in DNA and proteins.

To begin, the dynamics of simple one-dimensional jump processes are considered (Section 7.1). Solutions to such problems provide a connection with the results from the continuum problems considered in Chapter 5. In this context, the model of Alexander and Orbach (1982) that describes "fracton dynamics" is discussed. Jump processes with random coupling constants result in anomalous scaling of the density of states. The scaling of the density of states can be determined from experimental parameters using the temperature dependence of these parameters. Quantities such as the heat capacity, the inelastic neutron scattering intensity, and the Raman relaxation time, directly involve the density of states and will have characteristic exponents. These are discussed in Section 7.2, along with specific results for biopolymers. With such techniques anomalous temperature dependencies, i.e., a fractional scaling exponent, are often observed. The frequency dependence of the density of states can then be determined from these exponents using simple scaling arguments. Early on, Stapleton and coworkers tried to interpret the anomalous temperature dependence of the Raman relaxation rate of iron centers in proteins. This was one of the first attempts to ascribe a fractal structure to anomalous scaling. It remains controversial for both theoretical and experimental reasons. Yet Stapleton's work has stimulated a wealth of activity on vibrational relaxation in complex media. We examine the main results of this work with particular attention to effects due to geometry and those due to connectivity (Section 7.3). A number of simple polymer models will be presented that show the interplay of these two effects. The predicted scaling laws are compared with the observed results from electron spin resonance studies of iron-containing proteins. In the final section, the simple problem of a linear chain with multiple connectivities is treated using a generating function approach. From the analytic results given in this section, it is seen that complicated connectivity on its own does not yield anomalous scaling. Rather, anomalous scaling is a result of randomness and disorder in the system. The multiple connectivity model provides a starting point for future work on more complicated polymer networks.

7.1 Lattice Diffusion and Fracton Dynamics

7.1.1 Lattice Dynamics and Scaling Arguments

The dynamics of lattice walks have been studied extensively (see Montroll and Shlesinger, 1984), and even one-dimensional systems can be surprisingly complex (Lieb and Mattis, 1966; Alexander et al., 1981, Nakayama et al., 1994). Our goal in this section is to consider some simple examples, as well as to present some results for more complicated cases. This is not intended to be a rigorous presentation, but rather provides the context for discussing later results. Initially, we consider a simple random walk on a one-dimensional lattice. The "particle" on the lattice can hop one unit at a time at unit frequency. Thus, the probability of finding a particle at the ith site on the lattice will depend on the transition rate from the $i + 1$ or $i - 1$ sites. One must also account for the loss of the particle to either of these two sites, and the following dynamical equation is obtained:

$$\frac{dP_i}{dt} = P_{i+1} + P_{i-1} - 2P_i \tag{7.1}$$

The above differential-difference equation can be solved elegantly by using a generating function approach (van Kampen, 1981) or directly from combinatorial arguments. The generating function is used in Section 7.4 to solve problems involving multiply connected lattices.

For the case of the particle or "walker" starting at the origin at time $t = 0$, and ending at site n at time t, the solution is

$$P_n(t) = (1/\sqrt{4\pi t})\exp\left(-\frac{n^2}{4t}\right) \tag{7.2}$$

Equation 7.2 can be converted to the probability of finding the particle at a distance r from the origin by making the transformation $r = na$, where a is the distance between lattice points (and the distance traveled per unit time). The mean squared distance, $r^2(t)$, is determined by:

$$r^2(t) = a^2 \sum_{-\infty}^{\infty} n^2 p_n(t) \tag{7.3}$$

The sum in Eq. 7.3 can be calculated explicitly and follows the normal Einstein relationship $r^2 \sim t$.

The vibrational dynamics of a linear chain presents a problem that is virtually identical to the one-dimensional random walk. Consider a chain of point masses that are connected by massless springs. Each point or bead on the chain will experience a force as a result of displacements with respect to its neighbors. For a harmonic potential between neighbors, the equation of motion for a single point (or bead) on the chain is:

$$m\frac{\partial^2 \xi_i}{\partial t^2} = f(\xi_{i+1} - \xi_i) + f(\xi_{i-1} - \xi_i) \tag{7.4}$$

where ξ_i represents the displacement from the equilibrium position of the ith point, m is the mass of the point, and f is the force constant of the spring. Using the transformation

$$\xi_i(t) = e^{i\omega t} u_i \tag{7.5}$$

Eq. 7.4 becomes

$$(2 - \Omega)u_i = u_{i+1} + u_{i-1} \tag{7.6}$$

where Ω is a frequency variable and is equal to $m\omega^2/f$.

Although Eq. 7.6 can be solved by conventional techniques, it is a convenient starting point for a discussion of scaling arguments used in more complicated problems. Such scaling arguments can be developed using a decimation procedure (Stinchcoombe, 1988), similar to the lattice renormalization technique discussed in Chapter 3. In the decimation procedure, one derives an expression analogous to Eq. 7.6 for the decimated lattice, i.e., a lattice with sites removed. By removing every other site, a new lattice is created with a spacing a', that is twice that of the original lattice constant a. However, the form of the dynamical equation (Eq. 7.6) is taken to be invariant upon decimation (this is where the self-similarity enters). The decimated lattice will also follows an equation of this form:

$$(2 - \Omega')u_i = u_{i+2} + u_{i-2} \tag{7.7}$$

This equation can be derived from Eq. 7.6 by adding and subtracting the appropriate equations with u_{i+1} and u_{i-1} on the left-hand side. This substitution provides a relationship between the frequency variable, Ω', in the scaled expression (Eq. 7.7) and its counterpart, Ω, in the original lattice (Eq. 7.6). For the linear chain, this relationship is:

$$\Omega' = 4\Omega - \Omega^2 \tag{7.8}$$

The frequency variable, Ω', is appropriate for a lattice constant $a' = 2a$.

Equation 7.8 is a recursion relationship for the decimation process. Recursion relationships of this quadratic form have been studied extensively in one-dimensional problems that show chaotic behavior. In general, vibrational problems associated with repetitive structures will give relationships of the form (Stinchcoombe, 1988):

$$\Omega' = \lambda\Omega - \Omega^2 \tag{7.9}$$

where λ is a constant. Typically, the scaling behavior of Ω is examined about the fixed point, giving a functional equation of a form that was discussed in Chapter 1 (Eq. 1.7):

$$\Omega'(2a) = \lambda\Omega(a) \tag{7.10}$$

where the dependence on the lattice spacing is specifically shown. The solution of this functional equation provides a scaling law between the frequency parameter and the lattice spacing. It is:

$$\Omega \sim a^z \tag{7.11}$$

where

$$z = \frac{\ln \lambda}{\ln 2} \tag{7.12}$$

With these results and dynamical scaling arguments (Lewis and Stinchcoombe, 1984), the scaling of the vibrational density of states, $\rho(\omega)$, with frequency can be obtained. The density with respect to the parameter, Ω, is given by: $\rho(\Omega) = dN/d\Omega$, where N is the number of units in the polymer. For a fractal structure one has $N \sim a^{d_f}$. Using this relationship and Eq. 7.11, $\Omega \sim a^z$, one obtains $\rho(\Omega) \sim a^{d_f - z} \sim \Omega^{d_f/z - 1}$. To determine the frequency dependence of the density, one uses $\rho(\omega) = \rho(\Omega)(d\Omega/d\omega)$. Recall that $\Omega = \omega^2$. The final result is:

$$\rho(\omega) \sim \omega^{2d_f/z - 1} \sim \omega^{\bar{d} - 1} \tag{7.13}$$

where $\bar{d}$ is the spectral dimension introduced in Chapter 5. The solution of the decimation problem (Eq. 7.9) allows the scaling of the density of states to be determined by Eq. 7.13. The parameter z plays the role of the walk dimension, d_w, discussed in Chapter 5.

7.1.2 Fractons

There are a wide variety of problems in which the low-frequency modes have a density of states as described in Eq. 7.13. Such problems have been investigated extensively by Orbach, Alexander, and coworkers (Alexander et al., 1981). These problems are a modification of Eq. 7.1, such that:

$$C_i \frac{dP_i}{dt} = W_{i,i+1}(P_{i+1} - P_i) + W_{i,i-1}(P_{i-1} - P_i) \tag{7.14}$$

where now the transition probabilities, $W_{i,i+1}$, and the constant, C_i, can be treated as random variables that reflect the disorder in the system. Problems such as the harmonic chain in which dP_i/dt is replaced by $d^2 P_i/dt^2$ can be treated similarly. In these cases the force constants between units in the chain are treated as random variables, again as a result of disorder in the system. Alexander and Orbach (1982) were able to map the quantum vibrational problem into a classical diffusion problem. They were able to determine the density of states from the Greens function for the diffusional problem.

In many of the diffusion problems on disordered system, it is more convenient to work with the Laplace transform of Eq. 7.14. The transform of the probability function, $P_n(\omega)$, is directly related to the density of states. Using a method originally due to Dyson, averages can be performed over the random variables $W_{i,i+1}$ or C_i (Alexander et al., 1981) to determine $\langle P_n(\omega)\rangle$. This general approach can be used to handle a variety of problems. Using the initial site correlation function, the density of states is determined by:

$$\rho(\omega) = -\frac{1}{\pi}\langle \tilde{P}_0(-\omega + i0^+)\rangle \tag{7.15}$$

In their seminal work, Alexander and Orbach (1982) introduced the concept of the "fracton." At very low frequencies, complex networks will have vibrational modes that are delocalized and act as normal phonon modes. These phonon modes span the lattice structure. Above a cutoff frequency the vibrational modes become localized and have a density of states that follows Eq. 7.13. These localized modes will "diffuse" through the lattice much as a classical particle would, and are referred to as fractons. There are a number of different structures and mechanisms that can produce fractons. Fractons occur in disordered media and are particularly striking in percolation clusters (see Havlin and Bunde, 1991). Percolation clusters are discussed in Chapter 9.

7.2 Experimental Techniques for Determining the Density of States

7.2.1 Heat Capacity Measurements on Biological Systems

Once the relationship in Eq. 7.13 has been determined, what do we do with it? The scaling of the density of states relates structure (via d) to dynamics. The utility of such a scaling law arises in the derivation of relationships for experimental parameters. With standard statistical mechanical results, anomalous scaling, usually with respect to temperature, can be determined. For instance, the vibrational contribution to the constant-volume heat capacity, C_V, is given by:

$$C_v = k \int_0^\infty d\omega \frac{\rho(\omega)(\hbar\omega/kT)^2 e^{-\hbar\omega/kT}}{(1 - e^{-\hbar\omega/kT})^2} \tag{7.16}$$

where k is a constant, and $\hbar$ is Planck's constant divided by 2π. The anomalous temperature scaling of a fractal object is seen by inserting Eq. 7.13 into 7.16 and using a change of variables such that $x = \hbar\omega/kT$. This gives:

$$C_v = k' T^{\bar{d}} \int_0^\infty dx \frac{x^{\bar{d}-1} x^2 e^{-x}}{(1 - e^{-x})^2} \tag{7.17}$$

where k' is a new numerical constant. The integral over x is independent of temperature and can be taken as a constant. Consequently, $C_v \sim T^{\bar{d}}$. For a normal, three-dimensional crystal, the familiar T^3 power law is recovered. This result shows that from a relatively easy measurement, one has a direct determination of the spectral dimension. In practice, things are not that simple, as there are often multiple scaling regimes that are not well separated.

Extreme care must be used in interpreting anomalous scaling as a fractal phenomenon. Anisotropic materials will typically give rise to unusual scaling laws, and these can give the appearance of being fractal (Krumhansl, 1986). For instance, in the case of a crystal consisting of interacting polymer chains, the heat capacity is given by:

$$C_V = 3k \left[\frac{T}{\theta_1}\right] \int_{\theta_3/T}^{\theta_1/T} dx \frac{x^2 e^x}{(e^x - 1)^2} + \frac{9kT^3}{\theta_1 \theta_3} \int_0^{\theta_1/T} dx \frac{x^4 e^x}{(e^x - 1)^2} \tag{7.18}$$

where θ_1 and θ_3 are characteristic temperatures. At low temperatures this system shows a classic T^3 dependence, and at temperatures greater than θ_3 crosses over to a linear dependence on T, as found for polymers. Initially, the system behaves as a three-dimensional crystalline material, and the polymeric nature of the units is not observed. At higher temperatures, the regime switches and reflects the one-dimensional topology of the linear polymer.

For anisotropic layered systems, one similarly has a dependence that looks three-dimensional at low temperature and crosses over to a two-dimensional system at high temperature. Again, the high-temperature regime reflects the dimensionality of the sheet or layered structure of the material. These scaling laws are not fanciful theoretical constructs, but rather such transitions have been experimentally observed with the low-temperature heat capacity of the polyamino acid, poly(L-alanine) (Finegold and Cude, 1972). This polymer could be crystallized into two different forms, alpha helix or beta sheet, depending on the experimental conditions. The alpha helix data are best fitted by Eq. 7.18, the interacting linear chain model. The beta sheets form two-dimensional arrays and the anisotropic layered model fits this data better. Other more complicated dependences that depend on the structure of the material (Krumhansl, 1986) are possible. In principle, one can use this crossover phenomenon to distinguish between fractal and conventional models. However, in practice, scaling regimes can be limited because of the finite size of the system. Additionally, fractal systems will also show crossover regimes from localized to fracton modes, giving the appearance of an anistropic material. Clearly, one needs complementary techniques to establish fractal behavior for any given situation.

In addition to the work on polyamino acids, there have been few measurements of the low-temperature heat capacities of proteins (Edelman, 1992). Anhydrous and partially hydrated powders of proteins were studied and the heat capacities scaled as $C_v \sim T^{\bar{d}}$ for $T < 30\ K$. Although this work did not employ a fractal model, the values of $\bar{d}$ was determined directly from the data. For the proteins insulin and chymotrypsin, values of $\bar{d}$ ranged from 1.6 to 1.8 in both amorphous and hydrated samples. The effect of limited hydration was a slight increase in this exponent. Interestingly, the dipeptide glycylgycine gave an exponent of 2.3–2.5, implying that the protein structure rather than the peptide linkage causes the low values of $\bar{d}$.

7.2.2 Electron Spin Relaxation and Proteins with Iron Centers

Another experimental parameter that can give an anomalous temperature dependence for fractal structures is the Raman relaxation time. This parameter is most commonly measured using electron spin resonance (ESR). In this technique, the relaxation to equilibrium of an excited electron spin state is measured. Three different relaxation mechanisms have been identified, each with a characteristic temperature dependence. These processes are referred to as: direct relaxation, Orbach or resonance Raman relaxation, and Raman relaxation. The direct mechanism involves the absorption or emission of a phonon at the characteristic Zeeman splitting of the system. The reciprocal relaxation time, $1/T_{1D}$, for the direct mechanism has a linear dependence on temperature, T:

$$\frac{1}{T_{1\mathrm{D}}} \propto \coth\left(\frac{g\beta H}{2kT}\right) \sim T \tag{7.19}$$

where $g\beta H$ is the Zeeman splitting. Often, the direct transfer mechanism can be so efficient that the spin system can transfer Larmor energy into the vibrational system faster than the vibrational system can equilibrate. In such systems, the direct process becomes phonon-limited or bottlenecked. The relaxation rate then follows:

$$\frac{1}{T_{1\mathrm{B}}} \propto \frac{\coth\,(g\beta H/2kT)}{1 + \beta'\,\tanh\,(g\beta H/2kT)} \sim T^2 \tag{7.20}$$

where β' is the bottleneck factor. The squared dependence on temperature occurs when the bottleneck is extreme ($\beta' >> 1$).

A resonance Raman or Orbach relaxation process results in very efficient energy transfer. This process can mask direct processes. The relaxation time for the Orbach process has an exponential temperature dependence given by:

$$\frac{1}{T_{1O}} \propto \exp\left(\frac{-g\beta H}{kT}\right) \tag{7.21}$$

This resonance process occurs when the energy splitting between the ground and first excited state of the paramagnetic ion is equal to the Zeeman energy. Thus, both the direct and Orbach mechanisms involve phonon coupling at a single frequency.

In contrast to the previous two processes, the Raman relaxation involves all the high-frequency vibrational modes of the system. Typically, these can be 1–3 orders of magnitude greater than the Larmor frequency. The Raman relaxation time, T_R, for an unpaired electron system is given by (see Drews et al., 1990):

$$\frac{1}{T_R} = k \int_0^{\omega_{max}} d\omega \frac{\rho^2(\omega)\omega^4 e^{\hbar\omega/kT}}{(1 - e^{\hbar\omega/kT})^2} \tag{7.22}$$

where k is a constant. Assuming a vibrational density of states that scales as $\rho(\omega) \sim \omega^{\bar{d}}$, a substitution of variables is made and the integral in Eq. 7.22 is reduced to a quantity that is independent of temperature. The integral is then merely a proportionality constant. The temperature dependence is seen to be:

$$\frac{1}{T_{1R}} \sim T^{3 + 2\bar{d}} \tag{7.23}$$

Equation 7.23 provides the experimental means for determining the spectral dimension using the Raman relaxation time.

The Raman relaxation rate for simple Fe^{3+} salts at liquid helium temperatures shows a T^9 dependence, as would be expected for a three-dimensional solid. Iron centers in proteins, on the other hand, show a more complicated dependence. Typically, below 3 K a T or T^2 dependence occurs. Above this temperature a crossover to an anomalous temperature scaling occurs. This dependence typically ranges from $T^{5.7}$ to $T^{6.3}$ (Stapleton, 1986). However, exponents as low as 4.8 have been observed (Drews et al., 1990). This anomalous behavior was one of the first phenomena to be interpreted using a fractal model. It gives spectral dimensions between 1 and 2 for protein vibrational systems. The bulk of the data on heme proteins give $\bar{d}$ values of 1.7 in agreement with the calorimetric results mentioned above. Historically, this work greatly stimulated interest in this area. For this reason, we discuss it in some detail. In the next section, the possible spectral dimensions that can arise from polymeric structures are considered.

Raman relaxation in proteins remains extremely controversial, for a number of reasons. First, it appears that the experimental results are extremely sensitive to small changes in sample condition that should not alter the protein structure (Drews et al., 1990). This casts doubt on whether the anomalous effects are actually associated with global modes in the protein. Secondly, there has been an increased awareness that anomalous scaling can come from other sources and that fractals need not be evoked to account for noninteger power laws (see Krumhansl, 1986; Cusack and Doster, 1990).

7.2.3 Neutron Scattering from Proteins

The vibrational density of states can be determined directly using neutron scattering measurements. In frequency regimes where the inelastic scattering dominates, the contributions from elastic and quasielastic scattering are either small or easily accounted for. In these regimes, the experimentally determined, incoherent structure factor, $S(\boldsymbol{q},\omega,T)$, due to the inelastic contribution is given by (Cusack and Doster, 1990):

$$S(\boldsymbol{q},\omega,T) = \frac{\hbar \boldsymbol{q}^2}{6\omega(e^{\hbar\omega/kT} - 1)} e^{-q^2 U(\omega)} \rho(\omega) \tag{7.24}$$

where $\boldsymbol{q}$ is the wave vector, ω is the frequency of scatter radiation, and U is the Debye–Waller exponent. Frequently, the density of states is taken as being temperature independent. This allows the Debye–Waller exponent to be determined from the temperature dependence of the structure factor. Once this exponent has been obtained, the density of states, $\rho(\omega)$, can be determined directly from the experimental data. Neutron scattering has considerable advantage over the previous two techniques, since the entire functional form of $\rho(\omega)$, can be determined rather than the scaling relationship.

Neutron diffraction studies on hydrated myoglobin powders have been performed; the frequency dependence of the density of states is shown in Figure 7.1 (Cusack and Doster, 1990). These data show a characteristic crossover frequency of 25 cm^{-1}. Below this frequency the density of states scales as ω^2, which is indicative of a Debye-like three-dimensional crystal lattice. As is seen in the figure, anomalous scaling occurs above the crossover frequency. No attempt was made to interpret this region or to determine the scaling law. The temperature dependence of the Raman relaxation rate was also determined from the scattering data. As with the ESR results, a crossover regime was seen at low temperature. Below 5 K a Debye dependence of T^9 was seen, while above 5 K a significantly weaker dependence was observed. Again, no scaling law was determined in the high-temperature regime, but it has approximately the form expected from the ESR experiments. These results differ from ESR results in the low-temperature regime, where a T or T^2 dependence is typically seen. This may be due in part to the large difference in the sample conditions (lyophilized powder vs frozen glass). Additionally, the neutron scattering was measured at temperatures above 180 K, so the density of states would have to be extremely insensitive to temperature to extrapolate to the liquid helium temperatures used in the ESR experiments. It should be noted that a dynamical transition is seen in the scattering data at about 180 K, and this effect may interfere with the extrapolation to low temperature. In this transition, the scattering contribution at the characteristic frequency is drastically reduced. The characteristic

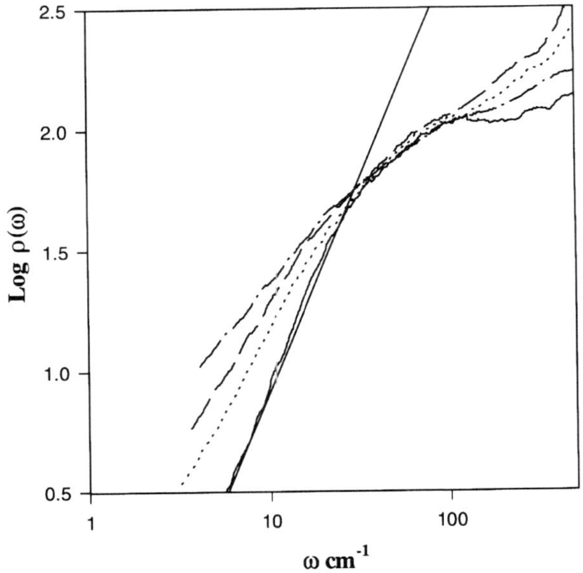

FIGURE 7.1 Log–log plot of the vibrational density of states versus frequency for myoglobin at different temperatures: (——) 180 K, (· · · ·) 250 K, (– – –) 300 K, (·–·–·) 350 K. The straight line has a slope of 2 as a three-dimensional Debye model. (Data adapted from Cusack and Doster, 1990.)

frequency observed in the neutron scattering experiment may provide a key to why the determination of the density of states is so problematic for the ESR experiments. For acoustic modes it is argued that the crossover frequency corresponds to a length scale of 40 Å. With the length of an amino acid being approximately 4 Å, there is a limited spatial regime for establishing a scaling law. Regardless of the experimental problems with ESR, the results from the limited neutron scattering work on proteins indicate that anomalous scaling of the vibrational density of states occurs. While the origin of these effects may be a result of a fractal phenomenon, this has yet to be established convincingly.

7.3 Density of States of Simple Polymer Models

In this section, scaling arguments are used to determine the spectral dimension for several, simple polymer models. In principle, these features could be incorporated in global descriptions of protein relaxation modes. Such global models are appealing in that they are usually more tractable than those involving specific, local effects. However, models based exclusively on global properties may not always be appropriate. When the dynamics associated with the specific function of a protein (enzyme catalysis, for example) are considered, local modes are normally invoked. In the case of proteins with large prosthetic groups, such as the heme proteins, local behavior will clearly dominate. Experimental and theoretical results suggest (for a review see Miller (1991)) that vibrational relaxation of a heme in a protein is much like that of a heme in solution. First, the heme relaxes intramolecularly, and this is followed by intermolecular relaxation to the surrounding protein. The protein itself subsequently transports energy to the solvent. Thus, only the second step in this complicated process would involve global modes. Relaxation of the protein could proceed through the backbone and through the multiple connections resulting from side-chain interactions. An argument against long-range transport of energy through the backbone is that multiple contacts at the surface should result in loss of energy to the solvent. A full description of vibrational relaxation in heme proteins will be quite complicated, and relaxation of the heme may or may not be strongly coupled to global relaxation modes.

Here, we focus on statistical models that provide information on the vibrational density of states of a polymer. This information is the first step in the development of dynamical models of protein relaxation that involve complete energy randomization, i.e., models in the spirit of the Rice–Ramsperger–Kassel–Marcus (RRKM) theory of chemical kinetics. Scaling arguments used to describe the density of states for polymers of different connectivity are considered. These models were originally developed to explain the Raman spin-lattice relaxation data described in the previous section. As will be seen, one such model correctly predicts the scaling exponents for a significant amount of the ESR data (Dewey, 1993). However, as discussed previously, this area remains controversial.

The density of states for three different models of polymer connectivity is now considered (Cates, 1984; Helman et al., 1984; Stapleton, 1984). These examples are illustrated in Figure 7.2. The cases are:

A. a protein that has no connectivity except along the backbone of the chain;
B. a protein that has a high density of both local and nonlocal connections, the density of these connections being as high as the connections along the chain; and
C. a protein with a low density of local connections.

All three cases can be handled by relatively simple scaling arguments. It is presumed that the polymer is a fractal structure with dimension d_f. Scaling arguments are than used to determine z, the walk dimension. This allows the spectral dimension to be calculated using Eq. 7.12. The temperature dependence of the Raman relaxation time is likewise determined from Eq. 7.23, and a comparison with the ESR results can be made. In Section 7.4, a polymer with regular, multiple connectivities is considered and is solved exactly. This example represents a starting point for developing more complicated models with random interactions. As shown in Section 7.4, when the connectivity is regular, e.g., every fourth unit has an off-backbone connection, as one might find in an α helix, the walk dimension, z, is 2.

For case A, the fractons diffuse exclusively down the length of the chain. As mentioned in Chapter 4, the distance traveled along the chain is given the unfortunate name, the "chemical" distance. It is proportional to N, the number of units covered. For simple diffusion, $N^2 \sim t$. We seek to relate this to the distance, r, traveled in Euclidean space. For a mass fractal (see Chapter 1), one has

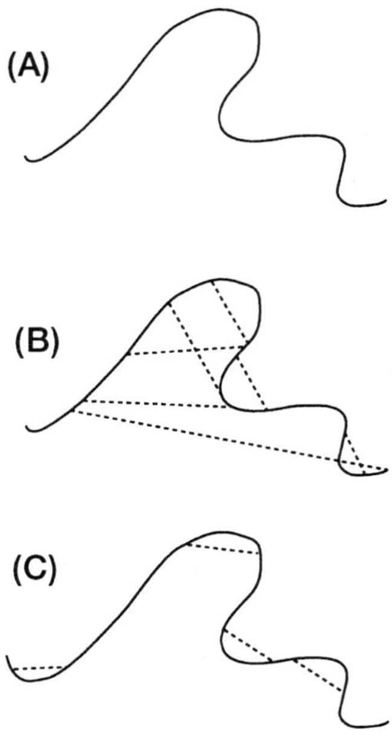

FIGURE 7.2 A schematic representation of a polymer with three different levels of chain connectivity: (A) the polymer has no off-chain connectivity; (B) the polymer has a high density of connections, with connection between distant units; (C) the polymer has only local off-chain connections.

$r \sim N^{\nu} \sim N^{1/d_f}$. The Einstein relationship becomes $r^{2d_f} \sim t$, which gives a walk dimension $z = 2d_f$. The spectral dimension in this case is $\bar{d} = 2d_f/z = 1$. With no off-chain connections, the vibrational problem is still essentially a one-dimensional problem, regardless of the fractal dimension of the structure.

Case B represents the other extreme. The connections are so plentiful that the diffuser is not confined to a path along the chain. Consequently, it is diffusing as if it where in a Euclidean space and will have a normal diffusion law, $r^2 \sim t$. This gives $z = 2$ and now the spectral dimension is equal to the fractal dimension, i.e., $\bar{d} = d_f$. Case C is a more complicated situation, and model calculations are used to argue for the scaling law (Cates, 1984) This model is one in which all the connectivities are "local," or short range. In this case, d_w is dependent on the spatial dimension in which the fractal is embedded. From model calculations involving the electrical conductivity of networks of multiply connected resistors, it is seen that $z = d_f + \tilde{z}$, where $\tilde{z}$ is 0.92 in two dimensions and 1.46 in three dimensions. This third model results in a spectral dimension that is intermediate between the two previous models. The spectral dimension is now given by:

$$\bar{d} = \frac{2d_f}{d_f + \bar{z}} \tag{7.25}$$

The results from these three cases can be compared with the experimental results obtained from spin-lattice relaxation measurements of paramagnetic centers in proteins. Values of the spectral dimension determined from the temperature dependence of $1/T_R$ range from 1.1 to 1.7, and there can be variability depending on sample preparation (Drews et al., 1990). However, in general, the experimental values fall into two categories. Hemeproteins give values of 1.67, while iron–sulfur proteins have values of 1.34. Thus, the first model (no connections) cannot account for the data, because $\bar{d} = 1$. Recall from Chapter 2 that the fractal dimension of globular proteins is approximately 3. The second model, many connections, is also ruled out, because $\bar{d}$ will be 3. The local bridge model gives $\bar{d} = 1.35$ for polymer collapse conditions, which is in excellent agreement with experimental results for iron–sulfur proteins. None of the three models fit the data for hemeproteins ($\bar{d} = 1.67$).

Is it fortuitous that the iron–sulfur proteins fit this model while the heme proteins do not? The X-ray structures of heme and Iron–sulfur proteins show very different situations for the paramagnetic center. In the heme proteins, the iron is embedded in the large prosthetic group, the porphyrin ring. In iron–sulfur proteins, such as ferredoxin, the iron is bound directly to an amino acid side-chain, a cysteine. Significant local effects due to the large porphyrin ring are to be expected with the heme proteins. The paramagnetic center in iron–sulfur proteins would doubtless be more sensitive to global relaxation modes as a consequence of its more direct connection to the protein backbone. While the agreement of the local connectivity model with the iron–sulfur data is encouraging, clearly there is a need for alternative, less ambiguous, experimental techniques that provide additional measurements of vibrational relaxation.

Apart from these ESR studies, applications of magnetic resonance techniques to fractal problems have been limited. Recently, nuclear magnetic resonance (NMR) results on the relaxation of water protons in muscle were analyzed as a fractal problem (Giugliarelli and Cannistrato, 1990). From the frequency dependence of the T_1 relaxation time of water protons, a spectral dimension of 1.11 was determined. While this was attributed to biopolymer backbone structure, the interpretation of such a complicated system is not easy or obvious.

7.4 Random Walks on Multiply Connected Linear Arrays

There has been discussion in the literature concerning the role of connectivity, as opposed to geometry, in producing anomalous scaling in vibrational systems (Elber and Karplus, 1986; Cusack and Doster, 1990). In a sense this is something of an arbitrary distinction, because the connectivity and the structure of biological macromolecule are intricately entwined. As we will see in this section, connectivity alone will not produce anomalous scaling in a one-dimensional system. Rather, anomalous scaling arises from connectivities with a distribution of force constants.

We start with the problem of a random walk on a linear chain. Initially, we treat the simple case of a chain in which only the nearest neighbors are connected. This is followed by the considerations where connections exist between remote units on the chain. Such a system could be used to model vibrational relaxation in an alpha helix, which has hydrogen bonds between every fourth residue (subunit) of the chain. The carbonyl group of the peptide linkage forms hydrogen bonds to the amide of a peptide linkage that is four amino acids away. These bonds are approximately parallel to the helix axis. The following diagram shows the connectivity of the alpha helix:

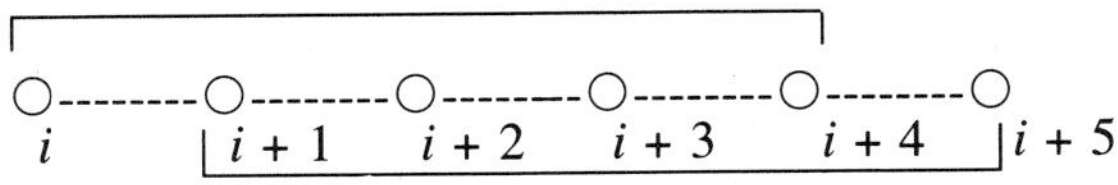

The hydrogen bonds are represented by solid lines and the covalent connectivity is given by dotted lines. Vibrations along the hydrogen bonds will doubtless have a different force constant and complicate the model. Proteins can be considered to be multiply connected polymers with random connectivity and random strength of interactions. To create a model of relaxation in proteins, one might start with a highly ordered, regular model and build in the random features. This section provides the first step by considering the ordered model.

Problems of multiple connectivity are conveniently handled using a generating function approach. The simple chain problem is considered first. Returning to Eq. 7.1, one has:

$$\frac{dP_i}{dt} = W_1(P_{i+1} + P_{i-1} - 2P_i) \tag{7.26}$$

where the coupling constant has now been introduced explicitly. The constant W_1 designates the appropriate transition rate for connections that are 1 unit apart. Following van Kampen (1981), a probability generating function, $F(z,t)$, is defined as:

$$F(z,t) = \sum_i z^i P_i(t) \tag{7.27}$$

By summing Eq. 7.26 over all i sites and using Eq. 7.27, a differential equation for $F(z,t)$ is obtained:

$$\frac{\partial F(z,t)}{\partial t} = W_1 \left(z + \frac{1}{z} - 2 \right) F(z,t) \tag{7.28}$$

Equation 7.28 is integrated with respect to time to give:

$$F(z,t) = \Omega(z) \exp \left[W_1 t \left(z + \frac{1}{z} - 2 \right) \right] \tag{7.29}$$

where $\Omega(z)$ is a constant from the integration. Choosing the boundary condition that the walker is at site n at time $t = 0$, gives $\Omega(z) = 1$. A standard relationship for the modified Bessel function, $I_k(z)$ can now be utilized:

$$\exp \left[t(z + 1/z) \right] = \sum_{k=-\infty}^{\infty} z^k I_k(z) \tag{7.30}$$

Using Eq. 7.30 in Eq. 7.29 and referring to the original series (Eq. 7.27), the probability function is seen to be:

$$p_n(t) = e^{-2W_1 t} I_{|k|} (2W_1 t) \tag{7.31}$$

The asymptotic form of the Bessel function such that $t \rightarrow \infty$ and $n \rightarrow \infty$, while $n^2 t$ remains fixed gives the familiar form for the probability function:

$$p(n,t) = \frac{1}{\sqrt{4\pi W_1 t}} \exp\left(-\frac{n^2}{4W_1 t}\right) \tag{7.32}$$

The generating function approach provides a number of mathematical conveniences. The average displacement, $< r(t)>$, and mean squared displacement, $< r^2(t)>$, can be calculated directly from it rather than from Eqs 7.32 and 7.3. This is done with:

$$\langle r(t)\rangle = \left[\frac{\partial \log F(z,t)}{\partial z}\right]_{z=1} \tag{7.33a}$$

$$\langle r^2(t)\rangle - \langle r(t)\rangle^2 - \langle r(t)\rangle = \left[\frac{\partial^2 \log F(z,t)}{\partial z^2}\right] \tag{7.33b}$$

The generating function may not always be in a form that allows for an easy or useful extraction of the probability function. In such cases one can determine the probability function from a contour integral similar to that for used to evaluate the canonical partition function from the grand canonical function (see Chapter 3). This is give by:

$$p_n(t) = \frac{1}{2\pi i} \oint z^{-n-1} F(z,t)\, dz \tag{7.34}$$

The integral in Eq. 7.34 is often evaluated using the method of steepest descents.

The case of the multiply connected linear array is now considered. The kinetic equation governing this example is:

$$\frac{dp_n(t)}{dt} = \sum_{i=1}^{\infty} W_i(p_{n+i} + p_{n-i} - 2p_n) \tag{7.35}$$

where the walker can jump to any unit that is i spaces away with a transition probability of W_i. The generating function for this case becomes:

$$F(z,t) = \exp\left(-2t \sum_{i=1}^{\infty} W_i\right) \prod_{i=1}^{\infty} \exp\,[W_i t(z^i + z^{-i})] \tag{7.36}$$

For the alpha helix problem, one has only nearest neighbor connections and connections between every fourth site. For this special case, expanding Eq. 7.36 in a series yields the probability function:

$$p_n(t) = e^{-2W_1 t}\, e^{-2W_2 t} \sum_{k=1}^{n} I_{n-4k}(2W_1 t) I_k(2W_4 t) \tag{7.37}$$

The more general form, m connections, can also be obtained and this is given by:

$$p_n(t) = \exp\left(-2t \sum_{i=1}^{m} W_i\right) \sum_{k^{(2)}=1}^{n} \sum_{k^{(3)}=1}^{n} \cdots \sum_{k^{(m)}=1}^{n}$$

$$\left[I_{n-2k^{(2)}-3k^{(3)} \ldots mk^{(m)}} (2W_1 t) \prod_{i=2}^{m} I_{k^{(i)}}(2W_i t) \right] \tag{7.38}$$

Although Eq. 7.38 provides a solution to the problem, it is not in a form that is especially easy to handle. On the other hand, the generating function (Eq. 7.36) allows a convenient determination of a number of properties of interest. For instance, using Eqs. 7.33a and 7.33b, one sees that the average distance, r, is equal to zero. The root mean squared distance is given by:

$$r^2(t) = 2t\left[W_1 + \sum_{i=2}^{m} i(i+1)W_i \right] \tag{7.39}$$

Thus, the Einstein relationship for a multiply connected linear array shows a normal one-dimensional time dependence. In other words, the walk dimension for this problem, d_w, is equal to 2. Consequently, this problem will not show anomalous behavior with regard to the density of states. The multiply connected system will appear as a normal harmonic chain. Using Eq. 7.34, the probability function can be evaluated with the method of steepest descents. Particularly convenient forms are determined when all the coupling constants are the same, $W_i = W_j$. This case gives a normal Gaussian function such as in Eq. 7.32. In the limit of the number of connections m approaching n, the function decays as an exponential.

These results indicate that a complicated linear structure cannot, by itself, yield anomalous scaling of the density of states. To achieve such scaling requires that the system show disorder. This disorder could be either dynamic (a time-dependent W_i) or structural (W_i is a random variable). Proteins clearly will have structural disorder as a result of their heterogeneous composition. The density of states for specific protein structures has been explored computationally (Elber and Karplus, 1986). In this work, protein X-ray structures were examined and the vibrational modes of these structures were determined. However, a simple connectivity was employed. Each amino acid was taken to have four connections, two along the chain and two distant ones. Distant connections had to be greater than six residues away, thus eliminating alpha helical hydrogen bonding as a connection. The normal mode equations for the protein vibrations were solved using an effective-medium approximation. The effective medium approximation is a way of incorporating the distribution of coupling constants into a single frequency-dependent variable. Using this approach, the scaling of the density of states was calculated. An "effective" fractal dimension was determined from this scaling. These give the following values: myoglobin, 1.6; cytochrome c_{551}, 1.4; and ferredoxin, 1.4 These numbers are essentially in agreement with the ESR work. Again, this emphasizes that connectivity per se does not result in anomalous scaling, but rather it is the distribution of coupling constants that give such effects.

7.5 Summary

The scaling properties of differential-difference equations used to describe vibrational relaxation were examined. Once again, the spectral dimension appears. This time it is an exponent in the scaling of the frequency dependence of the vibrational density of states. This scaling occurs in a regime where "fractons" exist. These are localized vibrations that "diffuse" through the media as a classical particle. The spectral dimension can be determined from the temperature dependence of several different experimental parameters. Specific results obtained from heat capacity, ESR, and neutron scattering measurements show that anomalous scaling occurs in a number of biological settings. However, interpretation of these exponents has been problematic. The spectral dimension for simple polymer models with different degrees of connectivity was derived and compared with ESR results. Models of vibrational relaxation in multiply connected polymers show that anomalous scaling does not result from the connectivity properties of a system. Rather it is more closely tied to the randomness of the coupling factors and the disorder in the system.

References

Alexander, S., and Orbach, R. 1982. Density of States on Fractals: Fractons. *J. Phys. (Paris) Lett.* **43**, L625–L631.

Alexander, S., Bernasconi, J, Schneider, W. R., and Orbach, R. 1981. Excitation dynamics in random one-dimensional systems. *Rev. Mod. Phys.* **53**, 176–198.

Berg, H. C. 1983. *Random Walks in Biology.* Princeton University Press, Princeton, NJ, p. 3.

Cates, M. E. 1984. Fractals with Local Bridges. *Phys. Rev. Lett.* **54**, 1733.

Cusack, S., and Doster, W. 1990. Temperature Dependence of the Low Frequency Dynamics of Myoglobin. *Biophys. J.* **58**, 243–251.

Dewey, T. G. 1993. Protein Structure and Polymer Collapse. *J. Chem. Phys.* **98**, 2250–2257.

Drews, A. R., Thayer, B. D., Stapleton, H. J., Wagner, G. C. Giugliarelli, G., and Cannistraro, S. 1990. Electron Spin Relaxation Measurements on the Blue-Copper Protein Plastocyanin. *Biophys. J.* **57**, 157–162.

Elber, R., and Karplus, M. 1986. Low-Frequency Modes in Proteins: Use of the Effective-Medium Approximation to Interpret the Fractal Dimension Observed in Electron-Spin Relaxation Measurements. *Phys. Rev. Lett.* **56**, 394–397.

Edelman, J. 1992. The Low-Temperature Heat Capacity of Solid Proteins. *Biopolymers* **32**, 209–218.

Finegold, L., and Cude, J. L. 1972. One and Two-Dimensional Structure of Alpha-Helix and Beta-Sheet Forms of Poly (L-Alanine) shown by Specific Heat Measurements at Low Temperatures (1.5–20 K). *Nature* **238**, 38–40.

Giugliarelli, G., and Cannistrato, S. 1990. Nuclear Magnetic Relaxation and Dynamics of Biopolymers. *Appl. Magn. Reson.* **1**, 85–91.

Havlin, S., and Bunde, A. 1991. Percolation II. In *Fractals and Disordered Systems*, ed. A. Bunde and S. Havlin. Springer-Verlag, Berlin, pp. 97–150.

Helman, J. S., Coniglio, A. and Tsallis, C. 1984. Fractons and the Fractal Structure of Proteins. *Phys. Rev. Lett.* **53**, 1195–1197.

Krumhansl, J. A. 1986. Vibrational Anomalies are not Generally due to Fractal Geometry: Comments on Proteins. *Phys. Rev. Lett.* **56**, 2696–2699.

Lewis, S. J., and Stinchcoombe, R. B. 1984. Density of States Exponent for Two-Dimensional Spin-Wave Dynamics at the Percolation Threshold. *Phys. Rev. Lett.* **52**, 1021–1024.

Lieb, E. H., and Mattis, D. C. 1966. *Mathematical Physics in One Dimension.* Academic Press, New York.

Miller, D. 1991. Vibrational Energy Relaxation and Structural Dynamics of Heme Proteins. *Annu. Rev. Phys. Chem.* **42**, 581–614.

Montroll, E. W., and Shlesinger, M. F. 1984. On the Wonderful World of Random Walks. In *Nonequilibrium Phenomena II. From Stochastics to Hydrodynamics*, ed. J. L. Lebowitz and E. W. Montroll. North-Holland, Amsterdam, pp. 1–122.

Nakayama, T., Yakubo, K., and Orbach, R. L. 1994. Dynamical Properties of Fractal Networks: Scaling, Numerical Simulations, and Physical Realizations. *Rev. Mod. Phys.* **66**, 381–443.

Stapleton, H. J. 1984. Comment on Fractons and the Fractal Structure of Proteins. *Phys. Rev. Lett.* **54**, 1734.

Stapleton, H. J. 1986. A Fractal Model of Electron Spin Relaxation in Proteins. *Comments Mol. Cell. Biophys.* **3**, 321–346.

Stinchcoombe, R. B. 1988. Phase Transitions. In *Order and Chaos in Nonlinear Physical Systems*, ed. S. Lundqvist, N. H. March and M. P. Tosi. Plenum Press, New York, pp. 295–340.

van Kampen, N. G. 1981. *Stochastic Processes in Physics and Chemistry.* North-Holland, Amsterdam, pp. 142–144.

8

Encoded Walks and Correlations in Sequence Data

The statistical analysis of sequence data has generated ongoing interest. The statistical properties of nucleic acid and protein sequences (see Doolittle, 1990; Volkenstein, 1994) provides important information on both the evolution and thermodynamic stability of biomacromolecules. In addition to conventional statistical approaches (for reviews, see, Karlin et al. (1991) and White (1994), fractal analyses of DNA and protein sequences have given new insight into sequence correlations (Peng et al., 1992; Voss, 1992; Buldyrev et al., 1993; Dewey, 1993; Pande et al., 1994; Balafas and Dewey, 1995). In this chapter, we consider such analyses. They represent a problem in discrete dynamics very different from those discussed in the previous chapter. Lattice walks can be constructed from sequence information in a variety of ways. These encoded walks result from assigning a specific numerical value and spatial direction to the members in the sequence. For instance, in DNA problems it is common to give purines a + 1 step on a one-dimensional lattice and pyrimidines a – 1 step (Peng et al., 1992; Voss, 1992; Buldyrev et al., 1993). Similar walks have been studied in protein sequences and have been based on a specific chemical or physical property of the monomeric unit (Pande et al., 1994). The resulting trajectories of these encoded walks can be analyzed as diffusion problems. Deviation of the encoded walk from random behavior provides evidence for long-range correlations.

In the following section, the construction and analysis of the scaling behavior of encoded walks are examined. Section 8.2 describes the results and interpretation of DNA walks. Significant controversy has surrounded encoded walks for DNA sequences, particularly with regard to correlations in noncoding or "junk" DNA (see Stanley et al., 1994). We review these points of controversy. Using a variation of the DNA walk known as the "Brownian bridge," correlations in protein sequences have been examined and these are discussed in Section 8.3. The encoded walks for proteins are based on physical parameters rather than on composition. Correlations in these parameters have been attributed to evolutionary driving forces that are shaped by physical aspects of protein structure. Using the protein sequence data, the connection between encoded walks and the multifractal formalism discussed in Chapter 4 is made. The advantage of the multifractal approach is that it allows correlations within individual protein sequences to be determined accurately. This is in contrast to encoded walks where the trajectory of an individual protein is too "noisy" to obtain good statistics. Consequently, this latter approach has been restricted to the study of "ensembles" of many protein sequences. Using the multifractal approach, the ergodicity of protein sequences is examined. For a comprehensive, nonredundant data set of protein sequences, it is seen that sequence correlations within individual proteins are the same as correlations within an ensemble of sequences. By establishing the ergodicity (in the information theory sense), the information content of a protein sequence can be statistically related to the number of probable sequences. From an analyses of protein sequences, the information content of an amino acid in a protein sequence can be determined (Section 8.5). Implications of these results for information theoretical considerations of protein sequences are discussed.

8.1 Encoded Walks and Scaling Laws

An encoded walk is generated from a DNA or protein sequence using a numerical correspondence between each subunit (amino acid or nucleic acid base) and a physical or chemical property associated with it. This correspondence generates a sequence of numbers, $\{\xi_1, \xi_2, \ldots, \xi_L\}$, where ξ_i is a numerical value associated with the amino acid in the ith position along the sequence and L is the length of the sequence. These numerical associations provide a set of directions for the walker to follow. There are many different sets of instructions that one can devise for generating a DNA walk. The simplest and perhaps most useful walk is the purine–pyrimidine rule (see Stanley et al., 1994). In this case, a one-dimensional walk is created from the sequence by assigning a step direction to each nucleic acid base in the sequence. The walker begins at the origin and steps one unit up (+ 1) when it comes to a purine (adenine (A) or guanine (G)). If the next base in the sequence is a pyrimidine (cytosine (C) or a thymine (T)), the walker steps one unit in the negative direction (– 1). The walk is generated by proceeding along the DNA sequence. The position in the sequence takes the place of the time variable in a normal one-dimensional walk problem. The generation of a trajectory from a DNA sequence using the purine–pyrimidine rule is illustrated in Figure 8.1. Of course, other rules can readily be generated for creating walks. For instance, higher dimensional walks can also be devised. The four bases can be assigned a separate direction in a two-dimensional space, i.e., step jumps of (0,1), (0, – 1), (1,0), (– 1,0) in the xy plane.

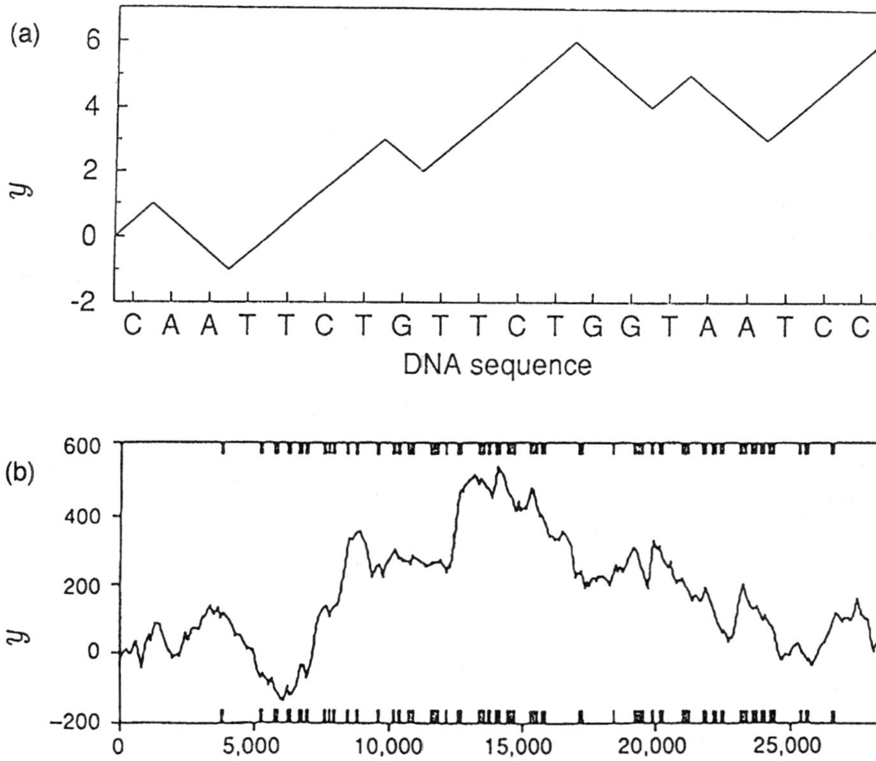

FIGURE 8.1 (a) Schematic illustration of the definition of a "DNA walk." The walker steps up when a pyrimidine (C or T) occurs in the DNA sequence, and steps down when a purine (A or G) occurs. (b) DNA walk for a DNA sequence consisting of 25,000 base pairs. (Figure adapted from Stanley et al., 1992.)

While such higher dimensional walks have been explored (Berthelsen et al., 1992; Korolev et al., 1994), the one-dimensional walk has received the most attention. This is largely due to the greater ease of handling the statistical analysis for the one-dimensional problem. Consequently, such walks will be the focus of our attention.

What is the significance of these walks created from the genetic text? Do they reveal anything about the text or are they merely a mathematical curiosity? A number of approaches can be used to examine walk trajectories to determine if there are correlations within the data set. There are three basic types of behavior that might be expected, and these have close analogies with fractional Brownian motion (see Chapter 1). The trajectories can be random, persistent or antipersistent. Random walks retain no memory of the previous excursions. A step in one direction is equally as probable as a step in the opposite direction. In a persistent walk, movement in one direction increases the probability of further movement in that direction. However, once an opposite step occurs, a trend will be favored in the new direction. Such walks show large, relatively noise-free excursions from one extreme to another. Antipersistence occurs when a step in one direction increases the

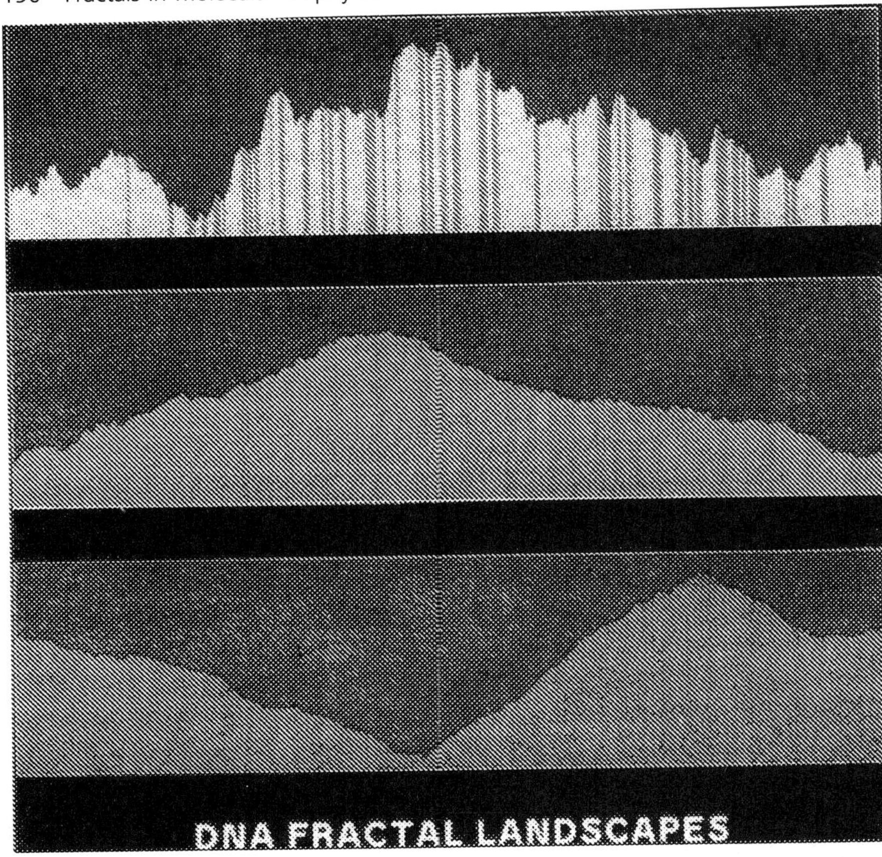

FIGURE 8.2 Fractal landscapes of DNA walks. (Top) DNA walk for the human β-cardiac myosin heavy-chain gene sequence. (Middle) DNA walk for coding regions of the gene treated as a continuous unit. (Bottom) DNA walk for bacteriophage lambda; this gene contains only coding regions. The top curve shows long-range correlations, while the other two curves are random. The top curve shows a much larger number of nucleotides. (Figure adapted from Stanley et al., 1992.)

probability of a step in the opposite direction. Antipersistent walks will appear more "noisy" than random walks. Figure 8.2 shows DNA walks for both coding and noncoding DNA sequences. Persistent behavior is seen in walks for noncoding regions, while the coding regions are thought to be random. The noncoding regions do not contain the sequence that encodes a gene product such as a protein or an RNA. The observed persistence is indicative of long-range correlations in the DNA sequence.

There are a number of mathematical devices for determining correlations within a walk. Typically, these methods determine a scaling exponent. Results from different statistical analyses can be compared using relationships between exponents. In the original description of DNA walks, the root mean squared fluctuation after l steps, $F(l)$, was determined. With the purine–pyrimidine rule, the individual displacement for the ith nucleotide, ξ_i, is + 1 for A or G and – 1 for C or

T. The total displacement after l steps is $y(l) = \Sigma_{i=1}^{i} \xi_i$. The mean squared fluctuation is given by:

$$F^2(l) = \overline{[\Delta y(l)]^2} - \overline{\Delta y(l)}^2 \tag{8.1}$$

where $\Delta y(l) = y(l_0 + l) - y(l_0)$, and l_0 is a position in the sequence. The bars indicate averages over all positions in the sequence. This is given explicitly by:

$$\overline{\Delta y(l)} = \frac{1}{N} \sum_{l_0 = 1}^{N} [y(l_0 + l) - y(l_0)] \tag{8.2}$$

where N is the number of bases in the sequence. The sum in Eq. 8.2 is the average fluctuation. The autocorrelation function, $C(l)$, is another parameter that is often used to examine fluctuations. It is given by:

$$C(l) = \overline{\xi_{l_0} \xi_{l_0 + l}} - \overline{\xi_{l_0}}^2 \tag{8.3}$$

where the bars are defined in a similar fashion as in Eq. 8.2, except now the individual displacements, ξ_i, are used. The mean squared fluctuations are related to the autocorrelation function through:

$$F^2(l) = \sum_{i=1}^{l} \sum_{j=1}^{l} C(j - i) \tag{8.4}$$

Another function that is commonly used in noise analysis is the spectral density, $S(f)$. This is the cosine transform of the autocorrelation function:

$$S(f) \propto \int C(l) \cos(2\pi f l)\, dl \tag{8.5}$$

The scaling of $F(l)$ with l provides information on the correlations within the sequence. If the sequence is a random arrangement of purines and pyrimidines, then the correlation function, $C(l)$ will be zero for all values except for $C(0)$ for which it is 1. The double sum in Eq. 8.4 is equal to l, giving $F(l) = l^{1/2}$. This is equivalent to a Hurst exponent, H, of 1/2 for fractional Brownian motion (see Chapter 1). When there are short-range correlations, the autocorrelation function decays exponentially, i.e., $C(l) = \exp(-l/R)$, where R is a characteristic length for the system. For large l, the double sum in Eq. 8.4 can be replaced by integrals and transformed into polar coordinates. This gives $F^2(l) = \int r \exp(-r/R) dr$. Solving the integral, it is seen that in this case one also has $F(l) \sim l^{1/2}$. This demonstrates that a process with short-range correlation shows the same scaling as a random process.

There can also be a situation where there is no characteristic length to the system, i.e., the system has a "fractal landscape." In such a case power law scaling for the autocorrelation function will occur, giving $C(r) \sim r^{\gamma}$, where γ is an exponent that cannot equal 1. Again converting the double sum in Eq. 8.4 into an integral, one has:

$$F(l) \sim \left\{ \int_0^l r^{1-\gamma}\, dr \right\}^{1/2} \sim l^{(2-\gamma)/2} \tag{8.6}$$

Defining a scaling exponent for $F(l)$ as:

$$F(l) \sim l^{\alpha} \tag{8.7}$$

where $\alpha \neq 0.5$, one has the relationship between exponents, $\alpha = (2 - \gamma)/2$. These exponents can also be related to those for the spectral density. Assuming a scaling of $S(f) \propto f^{-\beta}$ and using the cosine transform in Eq. 8.5, $\beta = 1 - \gamma$. This gives $\alpha = (1 + \beta)/2$. Most of the computational work on DNA has determined α from the log–log plots involving Eq. 8.1. However, Voss (1992) has used the spectral density in his work. The relationships between exponents allow comparison of the results from different methods.

8.2 Correlations in DNA Sequences

When Stanley and coworkers examined DNA sequences with the algorithm based on Eqs 8.1 and 8.4, they made two controversial findings. First, they showed that for regions of DNA that coded gene products, $\alpha = 0.5$. Thus, coding regions show no long-range correlations. When noncoding regions or intron-containing genes are analyzed, α values greater than 0.5 were observed. These values are typically in the range 0.5–0.7 (Peng et al., 1992). This result is somewhat surprising because the coding region contains the information to make proteins and transfer RNAs (tRNAs). This information is apparently conveyed in a statistically random sequence of nucleotides. The noncoding regions do not have a gene product. The physiological role of such "junk DNA" remains a mystery. Yet it is these filler, or junk, regions that show the long-range correlations. Interestingly, the percentage of junk DNA increases as one moves up the phylogenetic tree. Most bacteria, for instance, have 100% coding DNA. Invertebrates typically have about 30% coding regions, and mammals have only 5% of their DNA coding for a gene product.

There are other reasons why the DNA walk results are considered controversial. Some workers suggest that $\alpha > 0.5$ for coding regions as well as the noncoding region (Voss, 1992). Other studies argue that for all regions, both coding and noncoding, $\alpha = 0.5$. A second controversial result is that for the noncoding regions α increases with evolutionary time. For a variety of gene families and species the following values of α where obtained: bacteria 0.50 ± 0.014, invertebrates 0.525 ± 0.006, and vertebrates, 0.574 ± 0.006 (Buldyrev et al., 1993). Thus, higher order organisms have more complex fractal scaling properties in the noncoding region. Again, conflicting results appear in the literature.

To resolve these controversies, two issues must be addressed: (1) the quality of the algorithm used to analyze the data, and (2) the choice of data sets. It has been argued that the mean squared displacement algorithm used with a purine–pyrimidine DNA walk provides the most "robust" indicator of long-range correlations (Stanley et al., 1992, 1994). Additional modifications, known as detrended fluctuation analyses, have been used to improve the analysis of smaller data sets, such as those found in coding regions (Peng et al., 1994). The purpose of these algorithms is to correct the random walk for overall compositional biases, thereby removing any drift term from the fluctuations. These new analyses confirm the original results. Extensive numerical simulations also show the algorithm reliably to reflect truly random behavior, i.e., α is determined to be 0.5 in random data sets. Despite some doubts (Tsonis et al., 1993), it appears that the algorithm is accurately calculating α. The second issue regarding choice of data sets is equally tricky. Because of their short length, coding sequences have been examined as pooled data. When taken from different gene families this potentially creates problems. Additionally, strand bias can occur in these genes, and this bias is often not corrected. To circumvent such problems, recent analyses have focused on a single gene family, the myosin heavy chain (Buldyrev et al., 1993). These efforts mark the most consistent and extensive exploration of evolutionary trends in long-range correlations.

Presuming that inconsistencies in data and algorithms can be resolved, it is tempting to speculate on the causes of these observed effects. The apparent randomness of the coding region is perhaps easier to explain than are the evolutionary effects. Recent work on protein folding (Shakhnovich and Gutin, 1990) suggests that virtually all possible protein sequences could fold into a thermodynamically dominant state. There may be kinetic problems with the folding of random sequences (see Karplus and Shakhnovich, 1992). Thus, there is no requirement for correlation within a protein sequence in order to create a protein with a well-defined, native structure. An inspection of a large number of protein sequences shows that approximately 50% of all amino acids in naturally occurring proteins are polar and 50% are nonpolar (Klapper, 1977). This is again suggestive of a random arrangement of amino acids in a protein. As will be seen in the next section, proteins are not truly random, but should be considered to be "slightly edited" by evolution. In considering the DNA walks, one must bear in mind that purines and pyrimidines by themselves are not the genetic code, rather they are letters in the code. One could make the analogy between the coding region and a page of prose. The letters in the page are like the purines and pyrimidines. Words are created by combination of letters. In the case of the genetic code, all words are three letters long. The question then is: If the words are random are the letters also random? But, the translation of this code is not random because there is a nonuniform degeneracy in it. For example, leucine has six different "triplets" that code for it, while tryptophan has only one. Despite such biases, there is still a strong enough "mixing" of purine and pyrimidines in the code that virtually randomize their content. Actually, one of the "fail-safe" mechanisms built into the code is resistance to mutations that greatly alter the nature of the encoded amino acid. The third position in the triplet code is particularly robust, since virtually any mutations in that position result in the same or a similar amino acid. Such effects serve to randomize the letters within the words. Given all the above considerations, it is not surprising that coding regions in DNA have random purine–pyrimidine walks.

Correlations in noncoding DNA regions require a more elaborate explanation. An "insertion–deletion" model has been proposed to describe the evolutionary changes in these correlations (Buldyrev et al., 1993). This model mimics the conversion of a random coding DNA into an intron containing gene. Initially, a purine-rich, random array of nucleotides is taken as the coding DNA sequence. Insertions/deletions are made in this sequence by randomly choosing a location along the chain and forming a loop. As discussed extensively in Chapter 3, the probability of forming a loop of length l is taken to obey a scaling law such that $P(l) \propto l^{-c}$, where c is the scaling exponent. In these simulations, c was assumed to be 2 (see Chapter 3 for more details on loop formation). Following a model of DNA recombination, the loop is then cleaved (or annealed), and this results in the deletion (or insertion) of a given sequence. These random events are allowed to continue along the length of the DNA. When loops are inserted they are given an equal probability of maintaining the original strand or flipping to a pyrimidine-rich strand. Additionally, retroviral insertion is included in the model. The retrovirus is given a sequence that contains equal amounts of purines and pyrimidine. This recombination event again has a length dependence derived from loop formation. The retroviral insertion is important because, without it, at long "evolutionary" time the sequence returns to an α value of 0.5. Using this computer model and analyzing the sequences for correlation, one obtains increases in α with the number of insertion/deletion events, i.e., with evolutionary time. Features of this model will, doubtless, be debated, and other models that give comparable behavior will be generated. However, the important point is that one does not need a very sophisticated model to generate long-range correlations.

Grosberg et al. (1993) have proposed a very different interpretation of long-range correlations in DNA. This model is based on the physical constraints required to package DNA into higher order structures such as chromosomes. A single strand of eukaryotic DNA can be extremely long, yet is packed into a very small region. For instance, a 1 m long strand of DNA is contained in the cell nucleus, which is approximately 1 μm in size (Grosberg et al., 1993). On this large scale, the DNA must exist as a globule or collapsed polymer. Two different globule structures were discussed in Chapters 2 and 3. They are globules consisting of Gaussian coils that form dense structures by interpenetration, and the "crumpled globule," or globules made up of small, locally dense globules that close pack but do not interpenetrate. Globally, both types of globule scale as $R_g \sim N^{\nu}$, with R_g being the radius of gyration and N is the number of subunits in the polymer. The exponent, ν, is 1/3 in both cases. Locally, a similar scaling holds, except that the Gaussian coil has $\nu = 1/2$ and the crumpled globules will have $\nu = 1/3$. From a biological perspective, one of the main problems with interpenetrating globules is that they have many configurations that form knots. At a global level, knots would create difficulties for the biological functions such as transcription, translation, and the packaging of DNA. Considering these arguments, it is highly likely that DNA is not an interpenetrating globule, but rather an unknotted, crumpled globule.

To relate the sequence properties to the physical properties of DNA packaging, it was assumed that surface interactions play a key role in the association of DNA globules. The sequence length, l, is proportional to the number of units, N, so the radius of the globule scales as $R \sim l^{1/3}$. The surface area, R^2, of the globules will scale as $l^{2/3}$. If this surface requires a specific physical or chemical property for efficient chromosomal packing, then one would anticipate correlations over regions in the sequence that scale as $l^{2/3}$. Such correlations would require $\Delta y \sim l^{2/3}$, and this gives an α value of 2/3. This is in excellent agreement with the value of 0.67 ± 0.01 obtained for the human β-cardiac myosin heavy-chain gene. Other genes with a high degree of introns show similar values. This model provides a physical picture of long-range correlation, as opposed to the processing picture of the previous computer model. Since bacteria do not have higher order organization such as chromosomes, there is no need for such a physical scaling law. This is consistent with the lack of correlations in DNA sequences in prokaryotes. In principle, one could test this picture by replacing the noncoding DNA in an organism with coding (but not expressed) DNA and observe whether DNA packaging is compromised. In practice, this would be a formidable technical feat.

8.3 The Brownian Bridge and Protein Sequences

Using the methods described for DNA, sequence walks for proteins can also be created. If one were to encode protein sequences based on composition, a 20-dimensional walk (one dimension for each amino acid) would be needed, and the analysis would be intractable. Alternatively, the sequence can be encoded using a property of the amino acids. For instance, if an amino acid is hydrophobic, the walker takes a step up. If it is polar, the walker steps down. Three different encoding schemes were used to study the randomness of protein sequences (Pande et al., 1994): hydrophobic, hydrogen bonding, and ionic charge. Three different random walk trajectories can now be obtained for each protein. The randomness of such trajectories provides information about the distribution of the respective interactions along the protein sequence.

In the protein walk study, a device known as the "Brownian bridge" was developed to account for the compositional bias or "drift" term in the walk. The Brownian bridge is a walk with equal number of step-up events as step-down events. In such a walk, the net displacement is zero. To achieve this, one defines the sum of the individual displacements, $y(l)$, as in the previous section. To correct for the drift, a new variable, $Y(l)$, is defined as:

$$Y(l) = y(l) - \left(\frac{l}{L}\right) y(L) \tag{8.8}$$

where L is the total length of the sequence. Using the transformation in Eq. 8.8, the walk will always return to the origin. At the midpoint of the walk, $l = L/2$, the root mean squared displacement of $Y(l)$ will be at a maximum and will be proportional to $\sqrt{L}$ for a random walk. Plotting the root mean squared displacement, r, versus the reduced length, λ ($\lambda = l/L$), a "bridge" structure is seen, as shown in Figure 8.3.

Persistent and antipersistent behavior will show deviations from the bridge of the purely random walker. For extremely strong persistence, the walk will proceed as far as possible in one direction before turning around and returning to the origin. A "tent" is created instead of a bridge (see Figure 8.3). The maximum displacement of the tent will be $L/2$. For extreme antipersistence, the length of individual trends will be minimal and a jagged line along the origin will occur. Thus deviations above the random bridge indicate persistence, and deviations below the bridge result from antipersistence.

One difficulty in applying this technique to proteins is that the sequence of a single protein is too short to generate a smooth bridge. To reduce the noise in the trajectories, a composite of many proteins must be considered. Thus, the trajectories

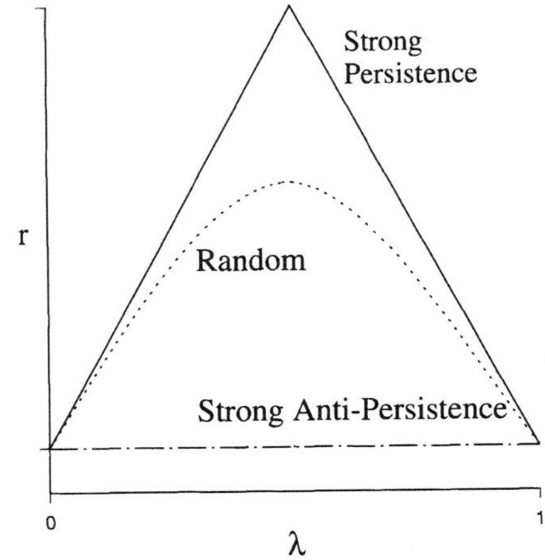

FIGURE 8.3 Root mean squared displacement, r, versus reduced distance, λ, for the Brownian bridge representation of a random walk. The random walk is the smooth arc. Persistent walks lay above the random curve, and a strongly persistent walk will appear as a "tent" (top curve). Anti-persistent walks are below the random curve. Strongly antipersistent walks will track along the origin of the line.

that are calculated represent an ensemble average over different proteins. This creates some problems in terms of weighting sequences of different length, and a correction for this was developed (Pande et al., 1994). An ensemble averaged squared displacement $\langle z^2(l)\rangle$ is defined as:

$$\langle z^2(l)\rangle = \left\langle \frac{y^2(l)}{L\,\overline{(\xi - \bar{\xi})^2}} \right\rangle \tag{8.9}$$

where the angle brackets represent averages over a large data set of different proteins and the bars represent an average within a protein sequence. For example, $\bar{\xi} = (1/L)\,\Sigma_{i=1}^{L}\,\xi_i$. The term $L\,\overline{(\xi - \bar{\xi})^2}$ eliminates the L dependence and corrects for different lengths and variances between proteins. The mean squared trajectory follows a scaling law:

$$\langle z^2(l)\rangle \sim l^{2\alpha} \tag{8.10}$$

where the exponent α will, again, equal 0.5 for a random walk. When α is greater than 0.5, the walk demonstrates persistence and an α value of less than 0.5 indicates antipersistence.

The results of an analysis of an ensemble of protein sequences are displayed in Figure 8.4. As can be seen, the three different encodings of the sequences all show nonrandom behavior. Using an empirical function, the exponent, α, associated with the walks was determined. A value for α of 0.520 was determined for walks based on hydrophilic or hydrogen bonding, and an α value of 0.470 was obtained for walks based on static charge distributions. Errors in α are typically a few percent. These results indicate that hydrophobic and hydrogen bonding walks show persistence, while Coulomb walks show antipersistence. This is not surprising from general considerations of protein structure. Hydrophobic regions tend to exist in clusters, packed in the interior of a protein. Hydrogen bonding residues are often on loop structures exposed to the solvent, and consequently will also appear as clusters.

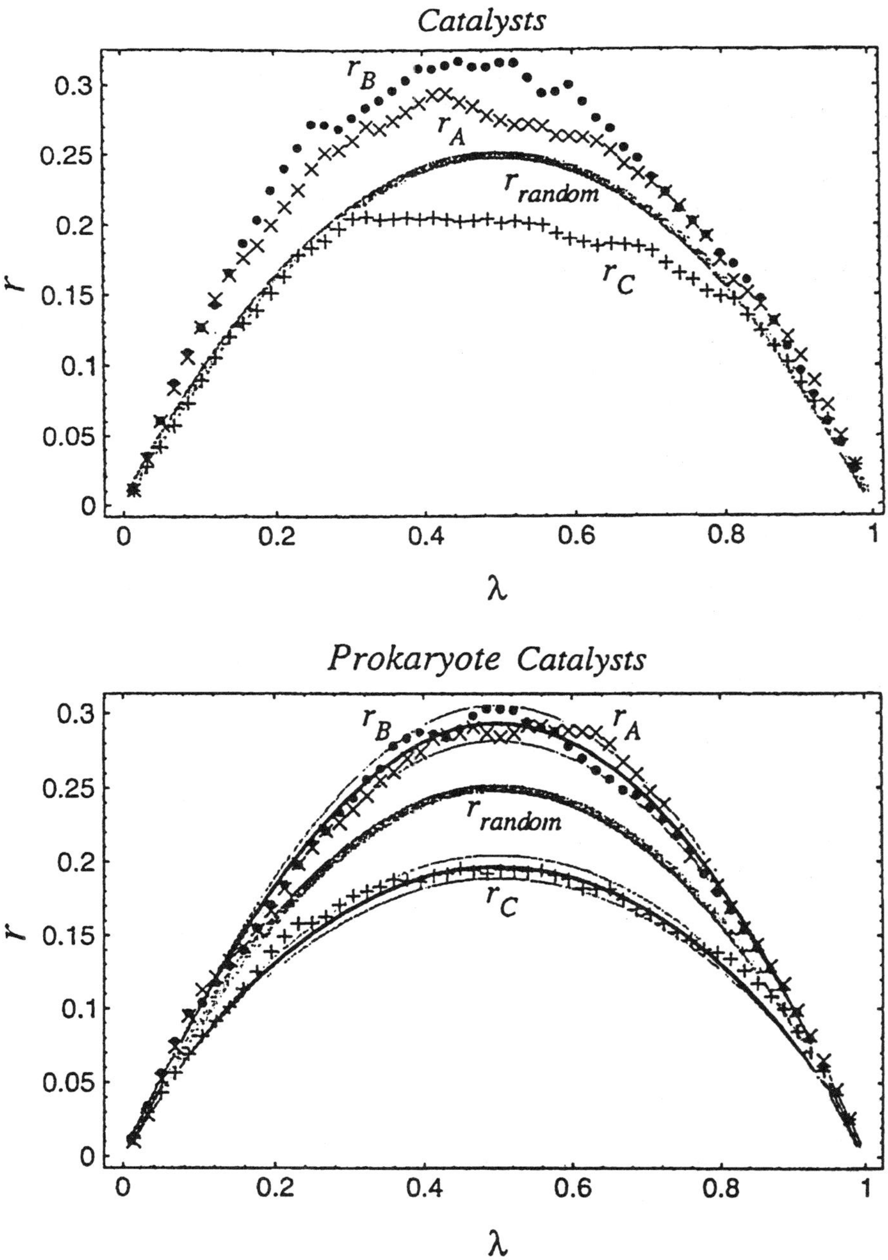

FIGURE 8.4 Brownian walks for hydrophilic (×), hydrogen bonding (●), and Coulomb (×) mappings of sequences of proteins. (Top) Plot for composite walks of globular proteins with catalytic activity. (Bottom) Plot for composites of prokaryotic enzymes. Deviations above the random curve indicate persistence (hydrophilic and hydrogen bonding) and excursions below the curve indicate antipersistence (Coulomb). (Figure from Pande et al., 1994.)

Such clusters will give correlations that appear persistent. With Coulomb interactions the protein wants to avoid the possibility of a string of unfavorable electrostatic interactions and, therefore, disperses the charges. This results in antipersistence.

What are the implications of these results for the way in which a protein evolves? If a prebiotic environment created amino acid heteropolymers at random, the number of possible protein sequences is astronomical. For instance, for a relatively small protein of 100 units, there are 20^{100} possible sequences. This number is larger than the number of all the atoms in the universe. One can speak of a "sequence space" as a multidimensional space in which all the possible protein sequences occur as vertices of a hypercube (Eigen, 1992). Such a sequence space is vast, especially with respect to the number of proteins that could have existed throughout evolution. Estimates can vary widely, but from a crude calculation the upper limit for the number of possible proteins that could have existed is 10^{48} (see Kauffman, 1993). Thus, the number of proteins that have existed is a minute fraction of the number of possible protein sequences. How could enzymes of well-defined three-dimensional structure and exquisite selectivity evolve from such a myriad of choices? There is evidence that the requirement for proper protein folding is not such a stringent one, and that proteins are very close to being random. Computer experiments show that randomly generated polymers with polar and hydrophobic subunits will fold into a well-defined, thermodynamically stable conformation. Consequently, the concept of proteins as "slightly edited random polymers" (Volkenstein, 1994) has gained wider acceptance. In evolutionary terms, proteins can be considered as arising from a dispersion of polymers in sequence space. Each of these polymers begins to search its own local region of sequence space by an "editing" process. This search leads to slight nonrandomness and to biological function. The results from the encoded walks suggests that this "editing" process is driven by physical properties, i.e., the selection process has physical criteria. These criteria result in the correlations in hydrophobicity, hydrogen bonding, and charge distributions that are observed in the Brownian walks. There will also be genetic constraints to evolution that could result in sequence correlations. It is unclear how such genetic effects would influence the clustering of physical properties.

8.4 The Connection with Multifractals

The multifractal approach described in Chapter 4 provides another method for detecting correlations in protein sequences. It has been used to explore correlations in the solvent accessibility profile and the hydrophilicity profile of a number of proteins (Balafas and Dewey, 1995; Dewey and Strait, 1995; Strait and Dewey, 1995). In this analysis, one again starts with an encoding of the sequence as in the walk problem. However, ξ_i, do not take on values of ± 1, rather they are assigned the continuous values associated either with the fractional solvent accessibility or with the hydrophilicity index. As discussed in Chapter 4, the multifractal algorithm covers the sequential array by boxes of length, l, and the trajectory within each box is calculated. The procedure is repeated with increasing box sizes and the dependence of the trajectory on box size is found. In this approach, one is not concerned with the displacement, $y(l)$, alone, but with the sum of displacements of all the boxes. Also, one is not restricted to the first moment. Rather, all moments of the displacement, $y^q(l)$ or $z^q(l)$, are considered. These moments are used to determine the "generalized" dimensions associated with the shape of the data profile. These generalized dimensions provide information on the hierarchical nature of the data set.

A "partition function," $Z_q(l)$, is defined to examine the qth moments of the sequence, and is given by:

$$Z_q(l) = \sum_{j=1}^{L/l} y_j^q(l) \tag{8.11}$$

where j labels individual boxes or sequences of length l within the complete protein sequence of length L. There will be a total of L/l of these boxes. Using a scaling Ansatz, $Z_q(l) \sim l^{-\tau(q)}$, where $\tau(q)$ is a generalized exponent. The properties of this exponent are described in Chapter 4, where it is seen how a multifractal spectrum is obtained from $\tau(q)$.

At first, it may appear as if the multifractal approach is quite different from the Brownian bridge. However, they are actually closely related. The main difference in the two approaches is that multifractals are concerned with the moment distribution of many short trajectories that make up a protein sequence, while the encoded walks focus on the root mean displacement of a single trajectory from an ensemble of protein sequences. Because the encoded walks rely on a single trajectory, an average over many proteins must be considered to generate good statistics. However, given a protein of a long enough sequence, one could, in principle, obtain a scaling law (Eq. 8.10) for a single protein rather than for an ensemble average. In the multifractal approach, the sum of trajectories generated from a given sequence replaces the ensemble average of the encoded walk. If protein sequences are ergodic in the information theory sense, then the statistics within an individual protein sequence will be the same as those within the ensemble. In such cases, a correspondence between the two approaches can be made.

Because of the finite length of a protein sequence, the number of trajectories in the multifractal approach decreases with trajectory size as L/l. Consequently, to relate the multifractal sum to the mean displacement of the encoded walks, one must normalize the sum. The scaling of $\langle z^2 \rangle$ is related to the second moment of $Z_q(l)$ by:

$$\langle z^2 \rangle \sim Z_2(l)/(L/l)^2 \sim l^{2-\tau(2)} \tag{8.12}$$

where a fractal dimension of 1 for the support, i.e., the linear sequence, is implicitly assumed. Using Eqs 8.10 and 8.12, one obtains a relationship between the walk exponent, α, and one of the multifractal exponents, $\tau(2)$:

$$\alpha = 1 - \frac{\tau(2)}{2} \tag{8.13}$$

Thus, the result of the encoded walk trajectory is related to a single point in the multifractal spectra.

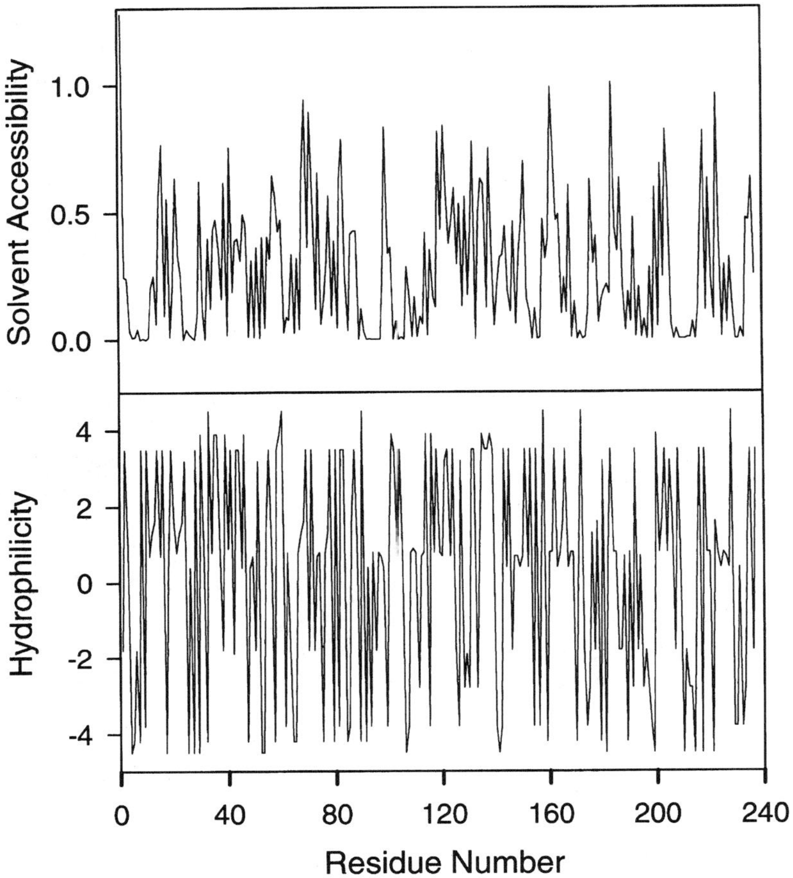

FIGURE 8.5 Hydrophilicity (top) and solvent accessibility (bottom) profiles for the protein concanavalin A. The correlations within the two profiles are statistically very different. (Data from Dewey and Strait, 1995.)

One could, of course, examine higher order moments of the average displacement of the trajectory and establish a relationship such as $\langle z^q(l)\rangle \sim l^{q\alpha(q)}$. These higher moments are related to the appropriate multifractal exponent by (Dewey and Strait, 1995):

$$\alpha(q) = 1 - \frac{\tau(q)}{q} \tag{8.14}$$

From this, it is seen that the multifractal approach has a direct correspondence with the encoded walks, and that, except for the negative moments, it provides the same information. The multifractal approach is the method of choice because it can generate scaling exponents for single protein sequences.

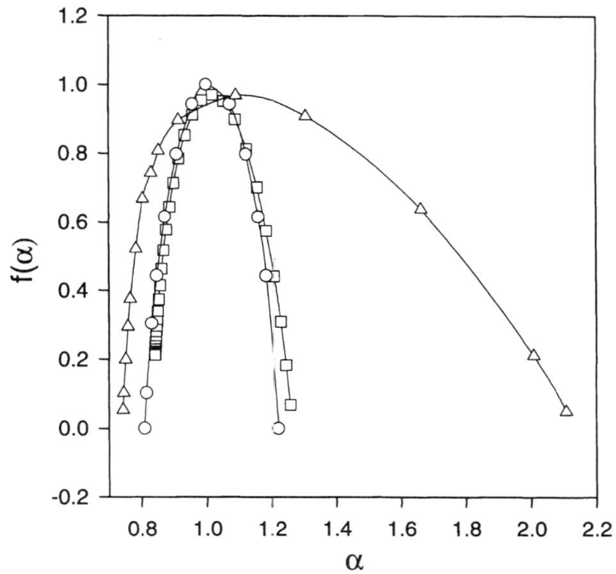

FIGURE 8.6 Multifractal spectra of concanavalin A: (□) spectrum determined from the hydrophilicity profile; (○) theoretical curve for a multiplicative binomial process used to fit the hydrophilicity curve; (Δ) spectrum determined from the solvent accessibility profile. The broader spectra indicate stronger deviations from random behavior.

The multifractal spectrum of the hydrophilicity profile (for a sample profile, see Figure 8.5) of a large data set of proteins could be accurately fit using a multiplicative binomial process (Strait and Dewey, 1995). This model provides the following relationship for $\tau(q)$ (see Feder, 1988):

$$[p^q + (1-p)^q] = 2^{-\tau(q)} \tag{8.15}$$

where p is the probability of finding a hydrophilic residue, and the factor of 2 results from treating the problem as a binary process. Figure 8.6 shows a plot of f versus α (see Chapter 4) for the protein concanavalin A. The spectrum determined from the profile shown in Figure 8.5 is shown along with the theoretical curve obtained from Eq. 8.15. Using the value of p from the fitted curve, the walk exponent could be calculated using Eqs 8.15 and 8.13.

The bottom profile in Figure 8.5 shows the solvent accessibility for concanavalin A. The multifractal spectrum corresponding to this profile is shown in Figure 8.6. Although, to the eye, the profiles in Figure 8.5 may appear similar, multifractal analysis reveals that they are correlated in very different ways. The spectrum for the solvent accessibilities is seen to be much broader than the one for hydrophilicity. This difference is particularly interesting because solvent accessibility is a parameter that reflects the folded structure of the protein, while hydrophilicity is a parameter dependent on the amino acid composition of the protein sequence. To fit the multifractal spectrum for the accessibility, one must use a more complicated model than the binomial multiplicative one. A multiplicative model with one-step memory can adequately reproduce the spectrum (see Balafas and Dewey, 1995). This model is discussed in Chapter 4.

The data sets show persistence when $\tau(2) < 1$ and antipersistence when $\tau(2) > 1$. Binomial multiplicative processes will always show persistence as $\tau(2)$ must always be positive (see Eq. 8.15). In initial work, the average value for α determined from hydrophobicity profiles was found to be 0.517 ± 0.005 for 16 different proteins (Strait and Dewey, 1995). This is in excellent agreement with the value of 0.520 ± 0.005 determined from the bridge analysis of the hydrophilicity of an ensemble of proteins (Pande et al., 1994). The correspondence between the average of walks within a protein to ensemble average walks suggests that protein sequences are ergodic. The ergodicity of protein sequences has important implications for information theory approaches to molecular evolution and sequence statistics, because the central theorems of information theory assume ergodicity in the signal or message. Recently, more extensive data sets have been explored (Dewey and Strait, 1995; Strait and Dewey, 1995), and these gave essentially the same results as the earlier one. These results are discussed in the next section with reference to the ergodicity of protein sequences.

8.5 Information Theory and the Ergodicity of Protein Sequences

We return now to a consideration of the number of possible protein sequences. Much has been written about the size of this number (see Eigen, 1992; Kauffman, 1993); and its implication for molecular evolution. It has even been suggested that the vast protein sequence space should be systematically searched in the hope of finding new proteins with commercial or medicinal value. Yet it might be argued that the useful corners of sequence space may not be extensive. From an information theory perspective, the number of possible sequences is not as significant as the number of most probable sequences. An analogy can be made with the information content of languages. Certainly there are astronomical numbers of letter sequences that one can generate, but this has nothing to do with the structure or information content of the language. Should we be systematically generating random letter sequences in order to acquire more knowledge or greater literature?

A basic theorem in information theory relates the most probable number of messages, Ω, of length N to the information entropy, I, by (see Shannon and Weaver, 1962; Cover and Thomas, 1991):

$$\Omega = 2^{NI} \tag{8.16}$$

This theorem was derived for ergodic messages. An ergodic message is one in which the information statistics within the message is the same as the information statistics of an ensemble of messages. Equation 8.16 can be used to calculate the most probable number of protein sequences that populate sequence space. To calculate this number properly, one must assume (or attempt to establish) the ergodicity of protein sequences. Additionally, an accurate determination of I must be achieved.

TABLE 8.1 Walk exponents derived from multifractal spectra

	Size	α_w
Sets		
Concatenated set I	30,309 amino acids	0.516 ± 0.001
Concatenated set II	37,101 amino acids	0.512 ± 0.001
Average of individuals, set I	155 proteins	0.517 ± 0.007
Average of individuals, set II	190 proteins	0.517 ± 0.006
Representative proteins		
Cytochrome b_{562}	106	0.523
Hemagglutinin	328	0.521
Aconitase	754	0.516
Tryptophan synthase	248	0.512

To actually prove ergodicity is a difficult task and, at best, one can hope to establish consistency or inconsistency with an assumed ergodicity. To this end, a large, representative data set of protein sequences has been analyzed using multifractal analysis to determine the walk exponent, α (Dewey and Strait, 1995). Two such "representative" data sets were constructed using algorithms (Hobohm et al., 1995) the goals of which were to reduce redundancy in the set while maximizing sequence coverage. With these algorithms, representative sets of 155 (set I) and 190 (set II) different proteins sequences were examined; the average of the individual α values of the entire data sets (I and II) are given in Table 8.1. The statistics of the ensemble of proteins in the data set could be determined by concatenating the entire data set into a single sequence and determining the multifractal spectrum; these α values are also shown in Table 8.1. It can be seen that the mean of the individual proteins is well within a standard deviation of the value for the ensemble of proteins. This shows that the behavior of the hydrophilicity profile is consistent with ergodicity of protein sequences.

The estimation of the information entropy is also a difficult matter, and one is restricted to determining an upper limit for this parameter. The information entropy can be estimated from the distribution of k-tuplets in a sequence. This has been done for languages (Shannon and Weaver, 1962), for DNA sequences (Gatlin, 1972; Volkenstein, 1994; Mantegna et al., 1995), and for protein sequences (Dewey and Strait, 1995; Strait and Dewey, 1996). The protein sequence calculation has the largest uncertainty as a result of a much more limited "text." At the lowest order approximation, the entropy is defined as:

$$I = - \sum_{k=1}^{20} p_k \ln_2 p_k \tag{8.17}$$

where p_k is the probability of finding the kth amino acid and base 2 is used to represent the entropy in terms of bits. At the lowest order of approximation, I_0, each amino acid is equally probable and $I_0 = \ln_2 20 = 4.3219$. The first-order approximation accounts for the nonuniform amino acid composition, and gives $I_1 = 4.177$ for data set II.

Higher order approximations can be determined from the k-tuplet distribution using (Shannon and Weaver, 1962):

$$I_m = -\sum_{k=1}^{20} \sum_{s} p_s p(k \mid s) \ln p(k \mid s) \qquad (8.18)$$

where s represents a sequence that is $m-1$ units long and the inner sum is over all possible sequences of this length. The conditional probability, $p(k|s)$, is the probability of a kth amino acid following an s sequence. Doublet and triplet frequencies give $I_2 = 4.161$ and $I_3 = 3.988$, respectively. Results for higher order sequences drop precipitously. This is a result of the finite size of the text. The number of possible k-tuplets exceeds the number in the text and gives the appearance of an unusually low information content. From these data a value of 3.988 can be considered to be an upper limit. Recently, a Monte Carlo method was used to computer-generate very large sequences from existing sequence data bases. A k-tuplet analysis of these sequences suggests that the information content of an amino acid in a protein could be as low as 2.5 bits per amino acid. Interestingly, estimates of the information content of nucleic acid sequences (from an early and limited data base, (Gatlin, 1972) found an I value of 1.94. Thus, two bases have enough information in the genetic code.

With this lower estimate of the information entropy, Eq. 8.16 can be used to calculate the most probable number of protein sequences. First, it is recast into a more convenient form:

$$\Omega = 20^{0.2314\,NI} \qquad (8.19)$$

For a protein that contains 100 amino acids Ω is, at most, 20^{58}, which is over 57 decades less than the number of possible sequences. This indicates that there are significant regions of "sequence space" that have little or no probability of being populated. Nevertheless, the "most probable sequence space" is still vast and the number of proteins that have existed throughout evolution is still minute compared with the number of "most probable" proteins. One must bear in mind that the number of probable protein sequences is an upper estimate. The use of more sophisticated data compression methods may reduce this number further.

The number of proteins that could have existed during our history is difficult to estimate, and estimates ranging from 10^{35} to 10^{48} (see Eigen, 1992; Kauffman, 1993). Despite this wide range, these estimates are still small compared with the number of possible sequences. This suggests that the conditions for forming a protein from a random polymer cannot be all that stringent, and proteins can still be considered "slightly edited random polymers." Yet the concept of a "slightly edited random polymer" is extremely vague. The information estimates serve to define this condition better. A "most probable" protein sequence is approximately 10^{-57} as likely as a random protein sequence, thus suggesting that considerable editing has taken place. Yet, even with this, the sequence space for "most probable sequences" is still much larger than the space that has been explored to date. An existing sequence is roughly 10^{-25} as likely as a most probable sequence.

8.6 Summary

In this chapter we have looked at an unusual form of discrete dynamical processes. These are the trajectories encoded in the sequences of nucleic acids and proteins. Sequence data can readily be converted into a random walk trajectory, and there are a number of statistical techniques for analyzing such walks. Controversial work on DNA sequences reveals that coding DNA is random, while noncoding or junk DNA has long-range correlations. Two plausible models for this finding have been presented. Walks can also be created from protein sequences, and one such algorithm, the Brownian bridge, provides a facile means of distinguishing between random, persistent and antipersistent behavior. A quantitative connection can be made between encoded walks and the multifractal analysis of sequence data. With this connection, the ergodicity of protein sequences has been explored. The analysis of a large data set of protein sequences suggests that the statistics within a sequence is the same as the statistics within the library. This apparent ergodicity has important consequences for information theoretical considerations.

References

Balafas, J. S., and Dewey, T. G. 1995. Multifractal Analysis of Solvent Accessibilities in Proteins. *Phys. Rev. E.* **52**, 880–887.

Berthelsen, C. L., Glazier, J. A., and Skolnick, M. H. 1992. Global fractal dimension of human DNA sequences treated as pseudorandom walks. *Phys. Rev. A* **45**, 8902–8913.

Buldyrev, S. V., Goldberger, A. L., Havlin, S., Peng, C.-K., Stanley, H. E., Stanley, M. H. R., and Simons, M. 1993. Fractal Landscapes and Molecular Evolution: Modeling the Myosin Heavy Chain Gene Family. *Biophys. J.* **65**, 2673–2679.

Cover, T. M., and Thomas, J. A. 1991. *Elements of Information Theory.* Wiley, New York, pp. 50–59.

Dewey, T. G. 1993. Fractal Aspects of Protein Structure and Dynamics. *Fractals* **1**, 179–189.

Dewey, T. G., and Strait, B. J. 1995. Multifractals, Encoded Walks and the Ergodicity of Protein Sequences. In *Pacific Symposium on Biocomputing '96*, ed. L. Hunter and T. E. Klein. World Scientific, Singapore.

Doolittle, R. F. 1990. *Molecular Evolution: Computer Analysis of Protein and Nucleic Acid Sequence. Methods Enzymol* **183**, Academic Press, New York.

Eigen, M. 1992. *Steps Toward Life*. Oxford University Press, Oxford.

Feder, J. 1988. *Fractals*. Plenum Press, New York, pp. 66–103.

Gatlin, L. L. 1972. *Information Theory and the Living System*. Columbia University Press, New York, pp. 47–72.

Grosberg, A., Rabin, Y., Havlin, S., and Neer, A. 1993. Crumpled Globule Model of the Three-Dimensional Structure of DNA. *Europhys. Lett.* **23**, 373–378.

Hobohm, U., Scharf, M. Schneider, R., and Sander, C. 1995. Selection of Representative Protein Data Sets. *Protein Science* **1**, 409–417.

Karlin, S., Bucher, P., Brendel, V., and Altschul, S. F. 1991. *Annu. Rev. Biophys. Biophys. Chem.* **20**, 175–203.

Karplus, M., and Shakhnovich, E. I. 1992. Protein Folding: Theoretical Studies of Thermodynamics and Dynamics. *Protein Folding*, ed. T. E. Creighton. W. H. Freeman, New York.

Kauffman, S. A. 1993. *The Origins of Order.* Oxford University Press, New York, pp. 20–22, 121–172.

Klapper, M. H. 1977. The Independent Distribution of Amino Acid Near Neighbor Pairs into Polypeptides. *Biochem. Biophys. Res. Commun.* **78**, 1018–1024.

Korolev, S. V., Vin'a, A. R., Tumanyan, V. G., and Esipova, N. G. 1994. Fractal Properties of DNA Sequences. In *Fractals in the Natural and Applied Sciences (A-41)*, ed. M. M. Novak. Elsevier, Amsterdam, pp. 221–228.

Mantegna, R. N., Buldyrev, S. V., Goldberger, A. L., Havlin, S., Peng, C.-K., Simons, M., and Stanley, H. E. 1995. Linguistic Features of Noncoding DNA sequences. *Phys. Rev. Lett.* **73**, 3169–3172.

Pande, V. S., Grosberg, A. Y., and Taaka, T. 1994. Nonrandomness in Protein Sequences: Evidence for a Physically Driven Stage of Evolution? *Proc. Natl. Acad. Sci. USA* **91**, 12 972–12 975.

Peng, C.-K., Buldyrev, S. V., Goldberger, A. L., Havlin, S., Sciortino, F., Simons, M., and Stanley, H. E. 1992. Long-Range Correlations in Nucleotide Sequences. *Nature* **356**, 168–170.

Peng, C.-K., Buldyrev, S. V., Havlin, S., Simons, M., Stanley, H. E., and Goldberger, A. L. 1994. Mosaic Organization of DNA Nucleotides. *Phys. Rev. E* **49**, 1685–1689.

Shakhnovich, E. I., and Gutin, A. M. 1990. Implications of Thermodynamics of Protein Folding for Evolution of Primary Sequences. *Nature* **346**, 773–775.

Shannon, C. E., and Weaver, W. 1962. *The Mathematical Theory of Communication.* University of Illinois Press, Urbana, IL, pp. 7–33.

Stanley, H. E., Buldyrev, S. V., Goldberger, A. L., Hausdorff, J. M., Havlin, S., Mietus, J., Peng, C.-K., Sciortino, F., and Simons, M. 1992. Fractal Landscapes in Biological Systems: Long-range Correlations in DNA and Interbeat Heart Intervals. *Physica A* **191**, 1–12.

Stanley, H. E., Buldyrev, S. V., Goldberger, A. L., Goldberger, Z. D., Havlin, S., Mantegna, R. N., Ossadnik, S. M., Peng, C.-K., and Simons, M. 1994. Statistical Mechanics in Biology: How Ubiquitous are Long Range Correlations? *Physica A* **205**, 214–253.

Strait, B. J., and Dewey, T. G. 1995. Multifractals and Decoded Walks: Application to Protein Sequence Correlations. *Phys. Rev. E* **52**, 6588–6591.

Strait, B. J., and Dewey, T. G. 1996. The Shannon Information Entropy of Protein Sequences. *Biophys. J.* **71**, 148–155.

Tsonis, A. A., Elsner, J. B., and Tsonis, P. A. 1993. On the Existence of Scaling in DNA Sequences. *Biochem. Biophys. Res. Commun.* **197**, 1288–1295.

Volkenstein, M. V. 1994. *Physical Approaches to Molecular Evolution.* Springer-Verlag, Berlin.

Voss, R. F. 1992. Evolution of Long-Range Fractal Correlations and 1/*f* Noise in DNA Base Sequences. *Phys. Rev. Lett.* **68**, 3805–3808.

White, S. H. 1994. Global Statistics of Protein Sequences: Implications for the Origin, Evolution, and Prediction of Structure. *Annu. Rev. Biophys. Biomol. Struct.* **23**, 407–439.

9

Percolation

The liquid crystalline phase transition of lipid bilayers has been studied extensively and has important biological implications. There is an enormous literature in this area, particularly on one-component model systems. Recently, attention has shifted to two-and three-component systems and to protein–lipid systems. Simple two-component membranes are often nonideal and exhibit complicated phase behavior. Typically, each lipid component can exist in one of two states: solid (or gel) and liquid (or fluid). In the solid state the hydrocarbon chains of the lipid are immobilized and tightly packed in a two-dimensional lattice. The translational diffusion of the lipid is greatly restricted, as is rotational diffusion about the long axis. The tight packing also prohibits bond rotation in the hydrocarbon chains. At a specific "melting temperature," T_m, a highly cooperative, order–disorder transition results in a mobile, fluid phase. The cooperativity is a result of hydrocarbon bond rotation and subsequent disruption of the packed array. This fluid phase now has rapid translational motion as well as long-axis rotational motion.

The phase diagrams for a binary system (Figure 9.1) can become quite complicated because conditions arise where each component, regardless of phase, is immiscible in the other component (Almeida et al., 1993). For the cholesterol–phospholipid system shown in Figure 9.1, five different phase regions exist. At low cholesterol concentration, one has a homogeneous, liquid-disordered state above the T_m (region (a)), while a solid, uniform phase occurs below the T_m (region (b)). As the cholesterol concentration is increased, the high-temperature state undergoes a transition from the liquid-disordered state to an immiscible mixture of a liquid-disordered and a liquid-ordered state (region (d)). Likewise, in the low-temperature

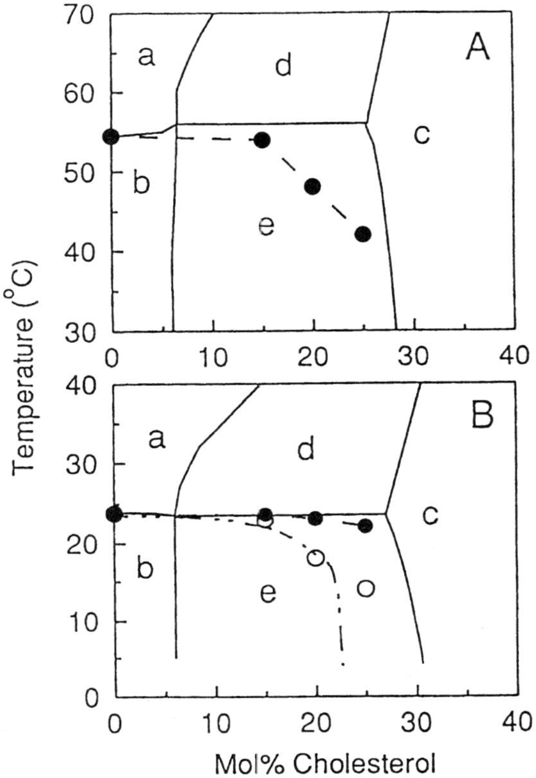

FIGURE 9.1 Phase diagrams for a cholesterol/distearoylphosphatidylcholine mixture (A) and a cholesterol/dimyristoylphosphotidylcholine mixture (B). (a) Disordered liquid; (b) solid; (c) ordered liquid; (d) a mixture of ordered and disordered liquid; (e) a mixture of solid and ordered liquid. (●) Percolation thresholds. (From Almeida et al., 1993. Reprinted by permission of Biophysical Society,)

regime, the solid state becomes a mixture of solid and liquid-ordered states (region(e)). As the cholesterol concentration is increased further, both the high- and low-temperature phase undergo a transition to a single, homogeneous, liquid-ordered state (region (c)).

Not only is the phase behavior of these systems intriguing, but the microscopic structure within a given phase has generated considerable interest. With the advent of tools such as fluorescence recovery after photobleaching (FRAP), it has become possible to study the diffusion of fluorescent molecules within a phase domain. From these experiments, domain structures have been observed in regions of coexisting phases. These structures may be important in a number of biological settings (for reviews, see Vaz and Almeida (1993) and Thompson (1992)). Figure 9.2 shows the phase diagram for a system in which domain structure has been investigated. It consists of a simpler, binary system containing two different types of phospholipid. In the region f_1 in Figure 9.2, both components exist in a miscible,

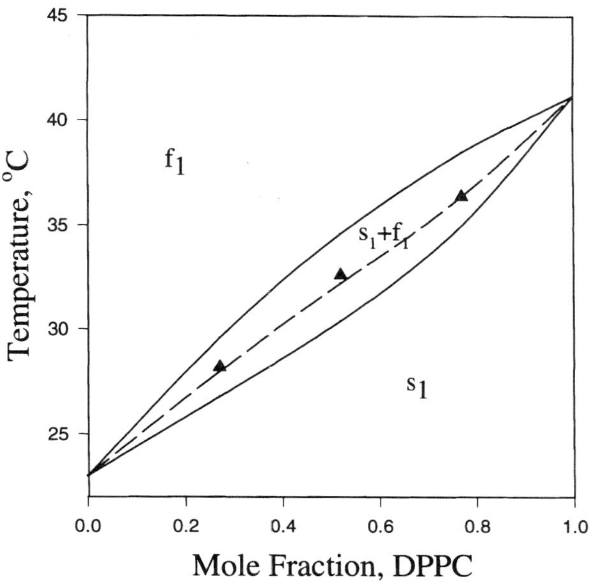

FIGURE 9.2 Phase diagram for a dipalmitoylphosphatidylcholine (DPPC)/dimyristoylphosphatidylcholine (DMPC) mixture. f_l, the miscible fluid phase; s_l, solid; $s_l + f_l$, the region of coexistence of fluid and liquid. (▲) Percolation thresholds. (Data from Almeida and Vaz, 1994.)

fluid phase. The region s_l represents a miscible solid phase. The region in between these phases (region $s_l + f_l$) consists of a mixture of immiscible, fluid and solid phases. At a given composition and at the low-temperature end of this mixed phase, the fluid regions form patches or islands within the connected solid regions. As the temperature increases, the percentage of area occupied by the fluid phase increases. This could be a result of formation of more islands and/or the growth of larger islands. Eventually, a temperature is reached at which the fluid domains merge, forming a connected region. The solid phase now becomes disconnected and forms islands. The transition from a disconnected domain to a continuous, connected one is called percolation. The condition for the onset of percolation, called the percolation point or threshold, is shown by the symbols in the phase diagrams in Figures 9.1 and 9.2. These points are determined by FRAP. As discussed in Chapter 5, such domain structures can strongly influence diffusion and reaction in membranes.

Percolation has been studied extensively by computer simulation as well as other methods. The phenomenon has been of great interest to physicists because it represents a "geometric" phase transition. In this chapter a modest introduction to this vast subject is presented. As will be seen, percolation clusters (the connected domain described above) are fractals. The statistical properties of percolation clusters are discussed, with emphasis on static behavior (Section 9.1). After presenting the basic scaling laws used to characterize such clusters, an exactly solvable model called the Cayley tree or Bethe lattice is considered in Section 9.2. Such lattice models have been used to describe antibody-receptor clustering in cell membranes. This clustering represents an initial response of the immune system. This discussion is followed by a description of the experimental and theoretical work on lipid domains and percolation in biological membranes (Section 9.3). Finite size effects must be considered in handling such systems, and this results in interesting scaling behavior. The hydration properties of lyophilized proteins have also be treated as percolation phenomena (Section 9.4). Water will adsorb on a protein surface and create hydrogen-bonded networks. At a critical hydration level, conducting networks will form as a result of proton transfer between water molecules. Experimental results on the dc conductivity of hydrated powders reflect the percolation of these conducting networks. Percolation in biological systems is generally not the random, geometric process studied in computer simulations; rather, thermodynamics enters into the phenomenon, causing percolation to be "correlated." In the Appendix, a simple model of two-dimensional, correlated percolation is presented that incorporates features of the order–disorder transition of the lipid membrane.

9.1 Introduction to Percolation Phenomena

9.1.1 Scaling Laws for Percolation

In this section, an introduction to percolation theory is presented, which is sufficient to discuss biophysical examples. The literature on percolation phenomena is vast, and there are a number of excellent treatises on the subject. Stauffer's book still serves as a good introduction to the subject (Stauffer, 1985). Physical applications can be found in the book by Sahimi (1994), and two excellent and wide-ranging reviews are recommended (Bunde and Havlin, 1991; Havlin and Bunde, 1991). The book chapters by Feder (1988) and Creswick et al. (1992) are also useful.

Percolation is a deceptively simple geometric problem that shows a geometric phase transition. This phase behavior accounts for much of the appeal of this subject. A good example of percolation is the growth of clusters on an infinite square lattice. Consider such a lattice where particles are allowed to "rain" down randomly onto the lattice sites. Initially, the particles will occupy sites that are dispersed throughout the lattice. A cluster is defined as a group of particles that forms contiguous nearest neighbors. (For a square lattice, each position has four nearest neighbor sites. Diagonal sites are considered to be next nearest neighbors.) As the "rain" continues, the density of filled sites increases and the size of the clusters grows. Clusters formed in this fashion are sometimes referred to as "lattice animals." As the process continues, regions between clusters will be filled in and clusters will merge into single, large clusters. At a critical density of sites, all the clusters will be connected into one infinite cluster. This is the percolation threshold. The resulting percolation cluster is a fractal. The probability of finding an occupied site anywhere on the lattice at the percolation threshold, p_c, will depend on both the dimensionality of the lattice and the lattice structure. Values of p_c for some simple

TABLE 9.1 Percolation parameters*

Parameter (Exponent)	Two-dimensions	Three-dimensions
Order parameter, P_∞ (β)	5/36	0.42
Correlation length ξ (ν)	4/3	0.88
Mean cluster size, s (γ)	43/18	1.80
Percolation threshold p_c		
Lattice:		
Triangular	0.5	2 sin (π/18)
Square	0.59	0.5
Honeycomb	0.70	0.5 – 2 sin (π/18)
Face centered cubic	0.20	0.12
Body centered cubic	0.25	0.18
Continuum (circles, spheres)	0.31	0.03

* Adapted from Bunde and Havlin (1991).

lattices are given in Table 9.1. At the percolation threshold, the vacant sites have now become discontinuous and the filled sites are fully connected. The problem described here is known as a site percolation. One can also define a bond percolation process. In this case, bridges are created between lattice sites, as opposed to the lattice sites themselves being occupied. This generates connected networks or clusters. The physical characteristics of percolation clusters can be described in a number of ways, with a seeming multitude of scaling laws. This simple geometric problem has a number of surprising nuances.

As a starting point for examining the percolation problem, consider the calculation of the distribution of clusters of different size. Let p be the probability of site occupation and $n_s(p)$ is the probability per lattice site of a cluster of size s occurring. With a total of N sites, the average number of clusters of s units is $Nn_s(p)$. For a "cluster" containing a single site on a square lattice, one has:

$$n_1(p) = p(1-p)^4 \tag{9.1}$$

which is the probability of a site being occupied times the probability of the surrounding four sites being vacant. For clusters of size 2, $n_2(p) = 4p^2(1-p)^6$, where the factor of 4 arises from the four possible orientations of the cluster on the lattice. With clusters of size 3, the configurational factor has already increased to 17 and by size 4 clusters of different "surface area" (number of boundary sites) appear. The enumeration of all possible clusters at higher sizes becomes a formidable task. In general, one has a sum:

$$n_s(p) = \sum_b g_{sb} p^s (1-p)^b \tag{9.2}$$

where the sum is over clusters of different number of boundary sites, b, and g_{sb} is the number of configurations for a cluster of s sites and b boundary sites. The quantity, g_{sb} is not easily calculated, and consequently $n_s(p)$ is not readily obtained.

Near the percolation threshold, p_c a scaling Ansatz has proven successful in describing $n_s(p)$. For large s, it is often assumed that $n_s(p)$ has the following functional form:

$$n_s\,(p) \sim s^{-\tau} f[(p-p_c)s^{\sigma}] \tag{9.3}$$

where τ and σ are critical exponents, and f is a function to be determined. Using this assumed form, the critical exponents describing a number of different physical processes can be derived. In our present application, all these exponents can be related to τ and σ in Eq. 9.3.

For instance, consider the calculation of the average number of sites in a cluster. The probability that a given site is in a cluster of size s is simply, $sn_s(p)$. The relative probability, $w_s\,(p)$, that among all the clusters available a given site belongs to an s cluster is:

$$w_s(p) = \frac{sn_s(p)}{\sum_{s'} s'n_{s'}(p)} \tag{9.4}$$

The average number of sites per cluster, S, is then given by

$$S = \sum_s sw_s(p)n_s(p) = \frac{\sum_s s^2 n_s(p)}{\sum_s sn_s(p)} = \frac{1}{p}\sum_s s^2 n_s(p) \tag{9.5}$$

where the rightmost equality uses the definition of p as the number of filled sites divided by the total number of sites. The scaling law for S can be derived by considering the integral form of Eq. 9.5 and by using Eq. 9.3. This gives:

$$S = \frac{1}{p}\int_0^{\infty} s^{2-\tau} f[(p-p_c)s^{\sigma}]\, ds \tag{9.6}$$

With a change of variables such that $x = |p-p_c|\, s^{\sigma}$, one obtains:

$$S = |p-p_c|^{-\left(\frac{3-\tau}{\sigma}\right)} \left\{\frac{1}{\sigma p}\int_0^{\infty} x^{-\left(\frac{3-\tau-\sigma}{\sigma}\right)} f[\mathrm{sgn}\,(p-p_c)x]\, dx\right\} \tag{9.7}$$

The integral in Eq. 9.7 is now merely a proportionality constant, and the scaling is contained in the lead term. Historically, the scaling of S has been defined by an exponent, γ, such that:

$$S = |p-p_c|^{-\gamma} \tag{9.8}$$

The following relationship between exponents is obtained: $\gamma = (3-\tau)/\sigma$. The average number of sites in a cluster, S, diverges as the percolation threshold is approached. This behavior is illustrated in Figure 9.3, where S is plotted as a function of p. At the percolation threshold, p_c, a single cluster is formed. Since these results hold for an infinite lattice, the "percolation cluster" has an infinite number of sites. Figure 9.3 shows the critical nature of the transition. Equation 9.8 appears to

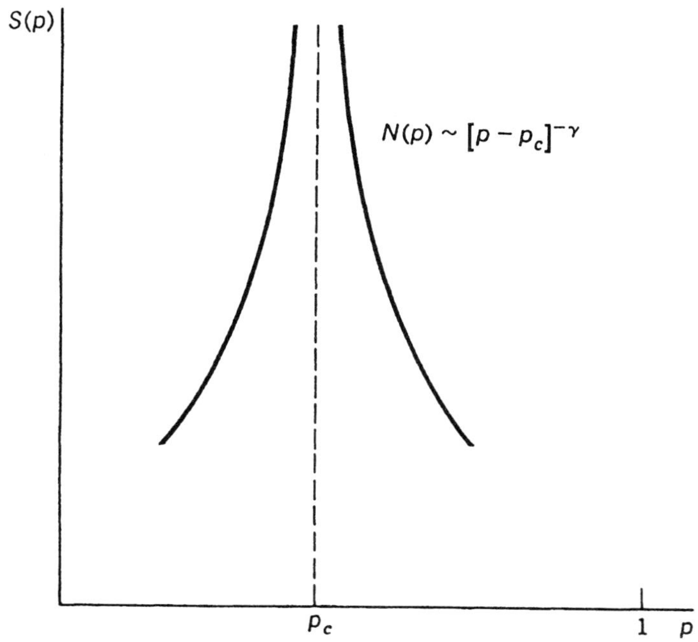

FIGURE 9.3 Plot of the average number of sites, S, in a finite cluster as a function of the site occupancy probability, p. At the percolation threshold, $p_{c'}$, the number of sites diverges according to Eq. 7.8.

be a general relationship, with specific values of γ applying for different geometries. Some of these values are tabulated in Table 9.1.

Other properties of the cluster distribution will give similar scaling laws near the percolation threshold. A parameter of particular interest is P_∞, the probability that a random site will be part of the percolation cluster. This is the order parameter for the system. It is defined by the relationship:

$$P_\infty + \sum_s sn_s\,(p) = p \tag{9.9}$$

which essentially says that a particle must either exist in a finite cluster or in the percolation cluster. Below the percolation threshold, there is no infinite cluster and $P_\infty = 0$. Above the threshold, a scaling relationship holds:

$$P_\infty \sim (p - p_c)^\beta \tag{9.10}$$

where β is the critical exponent. This exponent can be related to τ and σ by transforming the sum in Eq. 9.9 to an integral and using scaling arguments as before. Again, the specific value of β will depend on the geometry of the system. The behavior of P_∞ as a function of p is shown in Figure 9.4. The function becomes nonzero at the percolation threshold and increases to 1 when all the sites are filled ($p = 1$).

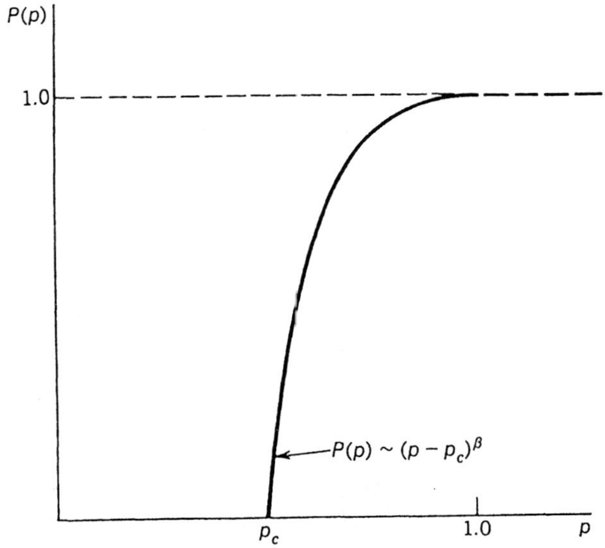

FIGURE 9.4 Plot of the probability of a site being in the percolation cluster, P_∞, versus the site occupancy probability, p. Below the percolation threshold, $p_{c'}$, the percolation cluster does not exist. At high occupancy all occupied sites are attached to the percolation cluster. Near the percolation threshold the order parameter obeys Eq. 7.10.

A parameter of particular interest, especially when considering diffusion problems, is the average radius of gyration of the clusters, ξ. This parameter is given by:

$$\xi^2 = \frac{\sum_s R_s^2 s^2 n_s(p)}{\sum_s s n_s(p)} \tag{9.11}$$

where R_s is the radius of gyration of an s cluster. Again, this parameter follows a scaling law in the vicinity of the percolation threshold. The scaling defines the exponent ν as

$$\xi^2 \sim |p - p_c|^{-2\nu} \tag{9.12}$$

As for the polymer case discussed in Chapter 2, the radius of gyration can be related to the number of units by the scaling law $R_s \sim S^{1/d_f}$, where d_f is the fractal dimension of the cluster. Inserting this scaling law into Eq. 9.11 and using the Ansatz of Eq. 9.3, one finds an expression of the form of Eq. 9.12. This provides the following relationship between exponents:

$$\nu = \frac{1}{\sigma d_f} \tag{9.13}$$

Note that this definition of ν differs from the result in Chapter 2 (Eq. 2.4) by the factor σ. Theoretical methods, identical to those used the polymer case, can be used to determine ν in percolation problems. The real-space renormalization techniques described in Chapter 3 can be employed to compute values of p_c and of ν.

9.1.2 Real-Space Renormalization

As a simple example of such a renormalization calculation, percolation on a square lattice is considered. The techniques used in Chapter 3 are more applicable to bond percolation rather than site percolation. However, site percolation is readily treated by an appropriate definition of the renormalized lattices. In the small-cell approach (Creswick et al., 1992), a cell of lattice points maps into a decimated cell. We consider a four-site, square lattice mapping to a single point. All configurations of sites that span the cell from left to right are taken to be percolating and will map onto a single point on the decimated lattice. Figure 9.5 shows the four-site lattices that span the cell and their statistical weights. The probability of finding a percolation cluster point on the decimated lattice, p', is now given by:

$$p' = p^4 + 4p^3(1-p) + 2p^2(1-p)^2 = 2p^2 - p^4 \tag{9.14}$$

This renormalization group equation is the analog of Eq. 3.11 discussed in Chapter 3. The percolation threshold is determined from the fixed point, i.e., where $p' = p$. Equation 9.14 has trivial fixed points at 0 and 1 and a nontrivial one that gives $p_c = 0.618$. Considering the small size of the renormalization, this compares favorably with the best computational estimate of 0.593 for percolation on a square lattice. The exponent, ν, is given by the equivalent of Eq. 3.15:

$$\nu = \frac{\ln b}{\ln (\partial p'/\partial p)|_{p^*}} \tag{9.15}$$

where b is the lattice decimation constant, which in this case is 2. Using Eq. 9.14, the value for the fixed point, and Eq. 9.15, a ν value of 1.635 is obtained. The accepted value of ν is 4/3, so the small cell does not do particularly well with this parameter. To improve this estimate, one must go to significantly larger cells (Creswick et al., 1992). Later in this chapter, this simple model will be extended to treat correlated percolation in biomembranes.

9.1.3 Percolation Clusters as Fractals

The exponent, ν, turns out to be a significant parameter because it will determine the characteristic length of the system, ξ (Eq. 9.12). This length is important in establishing the crossover point in the scaling of clusters from a fractal regime to a homogeneous one. It was Stanley (1977) who suggested that percolation clusters are inherently fractal. This can be seen visually in Figure 9.6, where a percolation cluster under different levels of magnification is shown. As can be seen, the distribution of holes in the cluster appears the same, regardless of scale. Not only are percolation clusters fractal, but their fractal dimension is a universal constant that is determined by the dimensionality of the embedding space, d. To see this, consider

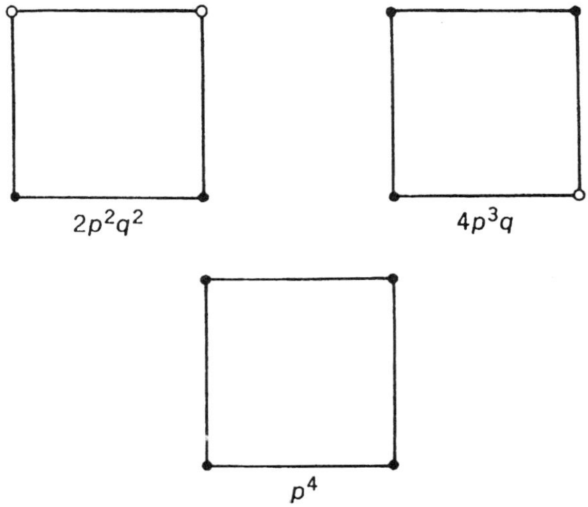

FIGURE 9.5 Percolating configurations of the four-site cluster on the square lattice. The statistical weights for each configuration are shown, with p being the probability of an occupied site and q the probability of an unoccupied site. Percolation is taken as connectivity from left to right. Numerical weights show the number of equivalent structures.

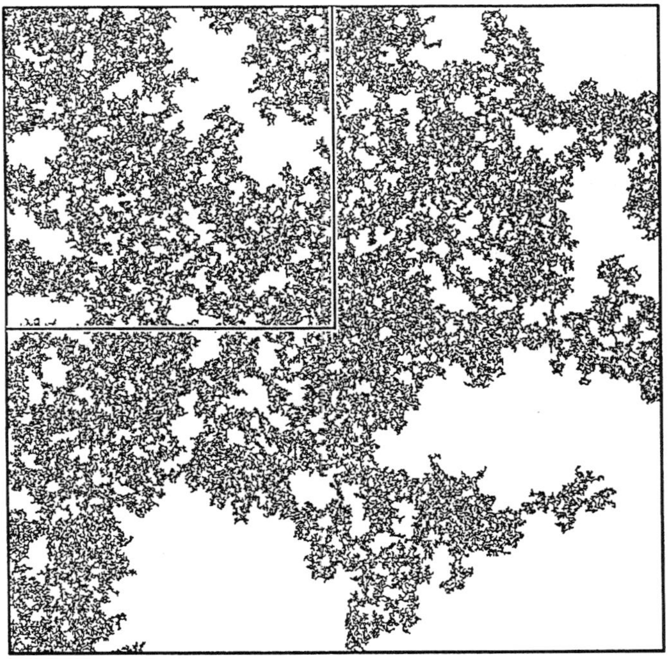

(a)

FIGURE 9.6 A percolation cluster at the critical concentration. Magnification of successive regions within the cluster shows its self-similar features. The region enclosed in the box in (a) is given in (b), and (c) is the boxed region in (b). (Percolation simulations courtesy of M. Saxton.)

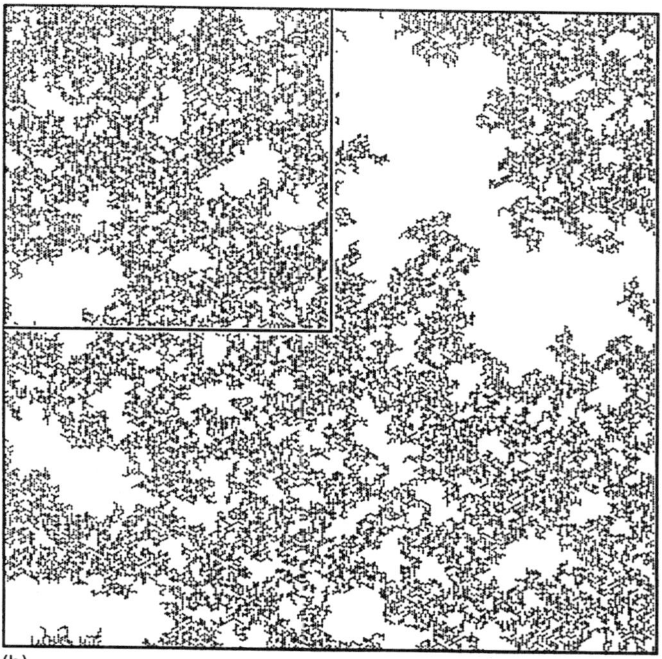

(b)

FIGURE 9.6b

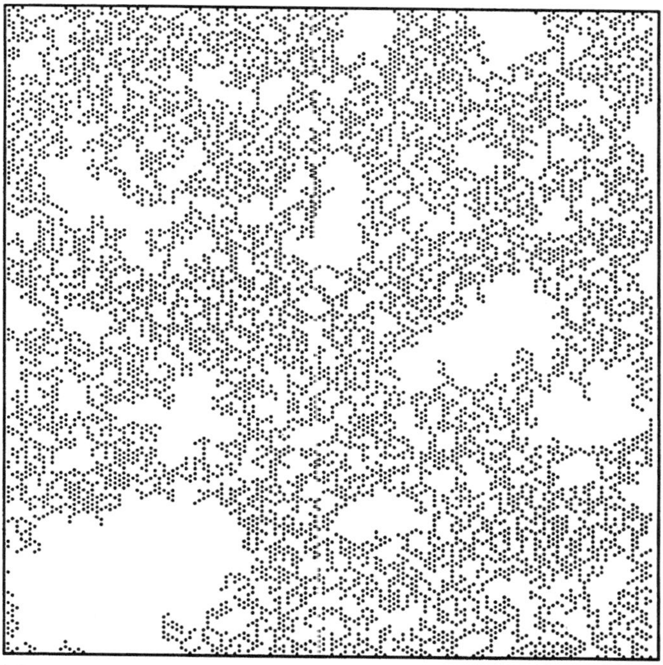

(c)

FIGURE 9.6c

the behavior of the infinite cluster probability function, P_∞. Instead of defining it by Eq. 9.9, an alternative definition is:

$$P_\infty = \frac{N_{cluster}}{N_{lattice}} = \frac{r^{d_f}}{r^d} \tag{9.16}$$

where $N_{cluster}$ and $N_{lattice}$ are the number of sites in the infinite cluster and in the lattice, respectively. The rightmost equality in Eq. 9.16 uses the standard scaling relationship for the radius, r, and can be used to define the fractal dimension of the cluster. Let the "ruler" used to measure the cluster be taken as ξ, so that one has $r = a\xi$, where a is an arbitrary constant. With Eq. 9.16, one finds: $P_\infty \sim \xi^{d_f - d}$. Using Eqs 9.12 and 9.10, one also has:

$$P_\infty \sim |p - p_c|^{-\nu(d_f - d)} \sim |p - p_c|^\beta \tag{9.17}$$

Equation 9.17 provides the relationship between the exponents, giving:

$$d_f = d - \frac{\beta}{\nu} \tag{9.18}$$

Since β and ν are universal constants determined by the dimensionality of the system, the fractal dimension of the percolation cluster is also a constant. All percolation clusters in two dimensions will have fractal dimensions of 91/48, an exact result. In three dimensions, the fractal dimension obtained from numerical simulations is 2.524 (Bunde and Havlin, 1991). Thus, percolation problems intrinsically create fractals.

The above arguments are used to describe the fractal properties of the infinite or percolation clusters. Yet the finite clusters that occur on either side of the percolation threshold have a fractal character as well. At distances much shorter than the correlation length, ξ, the clusters are fractal. At distances much longer than ξ, the fine structure of the clusters is obscured and the lattice appears to have a homogenous, "background" of clusters. The clusters are mass fractals, so that the mass of points, $M(r)$, within a radius, r scales as:

$$M(r) \sim r^{d_f} \qquad r \ll \xi \tag{9.19a}$$

$$M(r) \sim r^{d} \qquad r \gg \xi \tag{9.19b}$$

The system shows a crossover between two scaling regimes at $r \sim \xi$. These length scales have important implications for diffusional processes associated with such systems and will result in crossovers in the time dependence of diffusion.

As seen in this brief introduction to percolation, percolation implies fractality. When percolation occurs in a physical setting, such as biomembranes, one has to deal with the fractal nature of the problem. We now move from the tidy world of lattice geometry and computer simulations to the messy world of experiments on biological systems. The computer simulations will serve as a guidepost to compare experimental results too. However, before doing this, an exactly solvable model of percolation is discussed. This provides a concrete example of how the scaling laws in the previous section can be derived. Additionally, it has relevance to a biological problem, antibody–receptor clustering.

9.2 Antibody–Receptor Clustering and the Cayley Tree

9.2.1 Clustering and the Immune Response System

The immune system coordinates a complex interplay between molecular and cellular events. Percolation models have been employed to describe the immune response at several different levels. Because of the great complexity of these systems, it is difficult to provide a quantitative description of the variety of effects that can occur. In this section, we discuss one aspect, antibody–receptor clustering, and show how percolation on a Cayley tree fits this one aspect of the overall immune response. As a matter of convenience, we focus on the immune response of one type of cell, basophils (or mast cells). These cells contain granules of histamine that are released in response to an antigen. Receptors for the antibody, immunoglobulin E (IgE), are located on the cell surface. These receptors have two subunits and are membrane proteins. A typical cell has 5000–500,000 such receptors on its surface (Goldstein and Dembo, 1984). Each receptor can bind the antibody in such a fashion as to preserve the antibody's bivalent, antigen binding capability (Figure 9.7). The antigens can have multiple binding sites for the IgE, and the number of such sites is sometimes referred to as the antigen's functionality, f. With a bivalent "crosslinker" such as IgE and an antigen of functionality, f, it is possible to form a network of crosslinked receptors. (In Figure 9.7, different antigens with functionalities of 2 and 6 are shown.) It is this aspect of the response that can be treated as a Cayley tree and will be discussed presently. Percolation theory has also been applied to immune networks between cell clones, a higher cellular level process (see DeBoer and Hogeweg, 1989); this will not be discussed here.

The immune response, the dumping of histamine granules from the cell, occurs when a crosslinked structure of a critical size is reached. Figure 9.8 shows the amount of histamine released versus antigen concentration. As antigen is increased, the equilibrium favors antibody–antigen interactions, and larger crosslinked structures appear. Using monovalent antibodies, it has been established experimentally that cluster, and not just antibody–antigen, interaction is required for a response. Some authors have suggested that a critical-sized cluster is required for activation of the cell (see Perelson, 1984). This cluster, which consists of 10–20 receptors, is referred to as an "immunon." This effect accounts for the sharp initial rise in the curve in Figure 9.8. Interestingly, at high antigen concentrations the effects can be reversed, as seen by the decreasing portion of the curve.

This second phenomenon might occur in two ways. First, when the binding equilibrium is not too strong and when the antigens are multifunctional, excess antigen will compete with crosslinked structures, so that ultimately each antibody only has one antigen (with two binding sites). Alternatively, an extremely larger

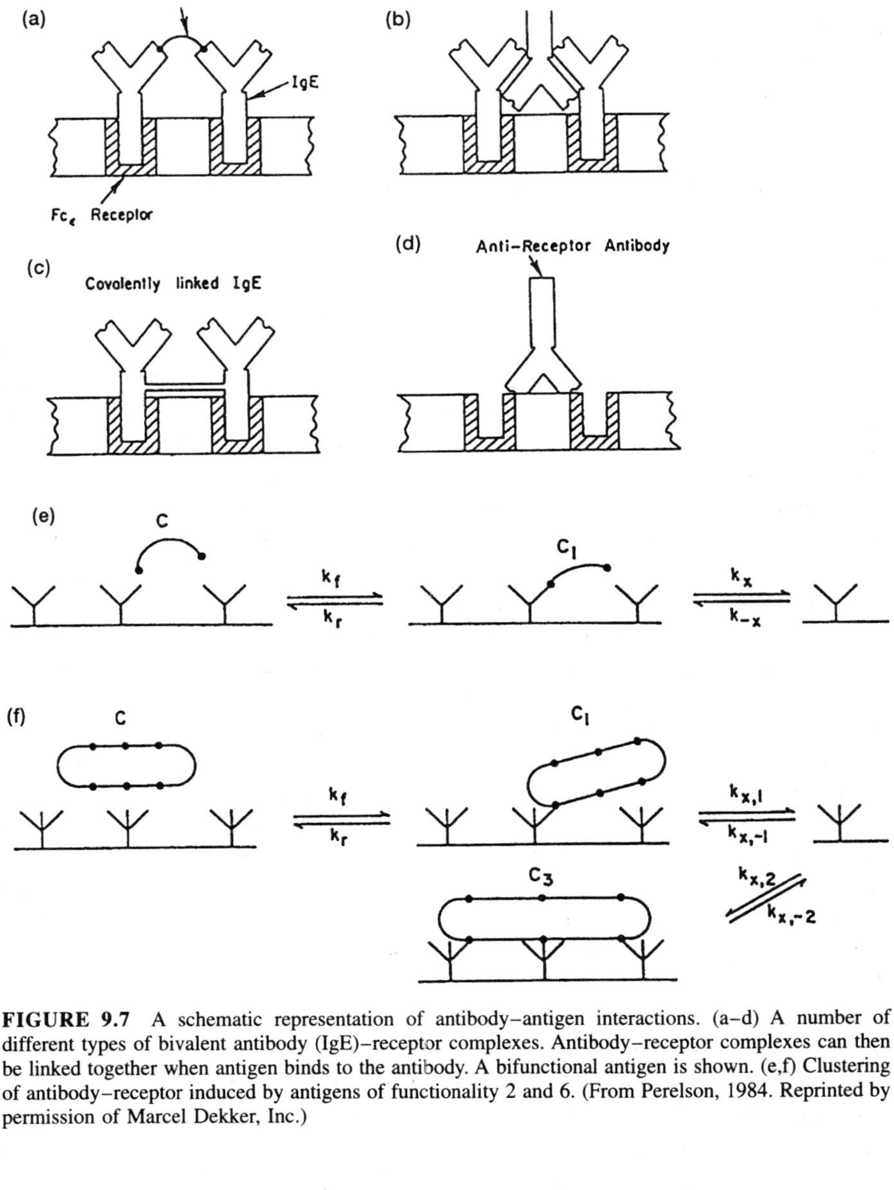

FIGURE 9.7 A schematic representation of antibody–antigen interactions. (a–d) A number of different types of bivalent antibody (IgE)–receptor complexes. Antibody–receptor complexes can then be linked together when antigen binds to the antibody. A bifunctional antigen is shown. (e,f) Clustering of antibody–receptor induced by antigens of functionality 2 and 6. (From Perelson, 1984. Reprinted by permission of Marcel Dekker, Inc.)

cluster can be formed at high antigen concentration and, for reasons that remain unclear, the response is inhibited. In this system, one does not want to reach the percolation threshold, rather it is better to have large, nonpercolating clusters.

Antibody–receptor clustering was treated in detail in the classic work of DeLisi and Perelson (1976). They essentially solved, in a very different setting, the problem of percolation on a Cayley tree. The Cayley tree is sometimes referred to as percolation on a Bethe lattice. An example of such a structure is shown in Figure 9.9. Continuing work on the antibody–receptor problem is aimed

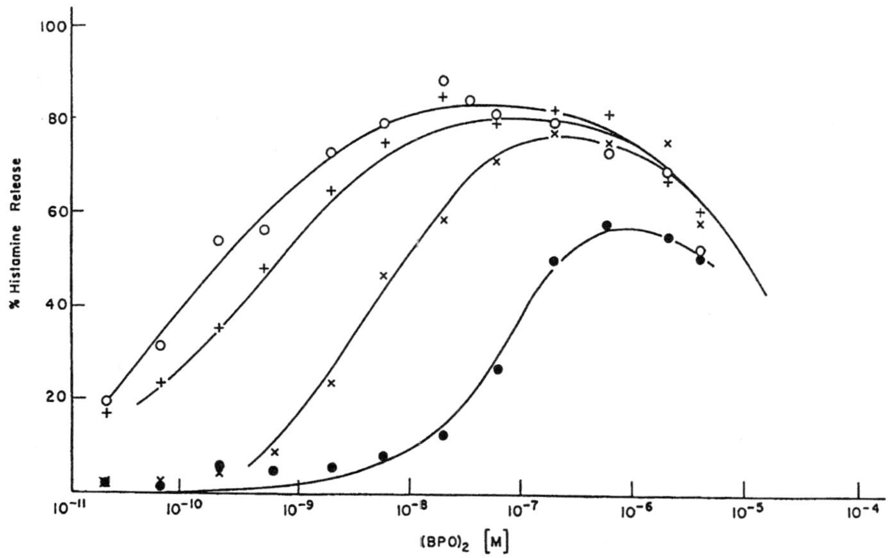

FIGURE 9.8 Plot of the percentage of histamine released from mast cells versus the molarity of a bifunctional antigen. Curves from top to bottom are for increasing amounts of a monovalent antigen. The monovalent antigen inhibits cluster formation and, consequently, a higher concentration of bifunctional antigen is required for cluster-induced release of histamine. At high bifunctional antigen concentration the effect is reversed. This may again be a result of inhibition of cluster formation. (From Perelson et al., 1984. Reprinted by permission of Marcel Dekker, Inc.)

at incorporating more physical and biological features in the model (see Perelson, 1984; Faro and Velasco, 1993). In this presentation, we restrict our attention to the simple problem and follow the approach of Bunde and Havlin (1991). Figure 9.9 illustrates a Cayley tree with a connectivity parameter, z, equal to 3. The tree is initiated with a central site from which z branches emanate. At the end of each branch, a new site is formed which than generates $z-1$ new, equal-length branches. This processes is iterated indefinitely. For the antibody–receptor problem a z value of 2 is a natural choice, because antibodies are bivalent. Notice that there is no steric hindrance between branches and that no loops can be formed. These unrealistic features are incorporated in the early antibody cluster models (DeLisi and Perelson, 1976).

9.2.2 *Percolation on the Cayley Tree*

The percolation properties of the Cayley lattice can be calculated directly using a combinatorial approach. In this problem, it is more meaningful to speak of a "chemical distance," l, rather than a Euclidean distance. The chemical distance is the distance traveled along the lattice. The number of sites at a distance, l, from the central site is given simply by $z(z-1)^{l-1}$. The correlation function, $g(l)$, is defined

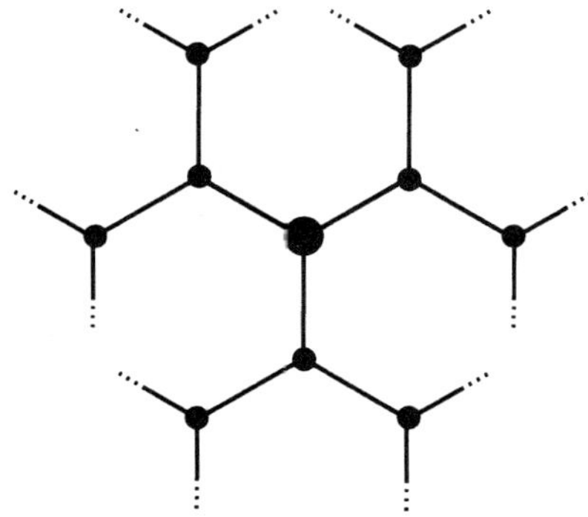

FIGURE 9.9 A Cayley tree with a connectivity of 3.

as the mean number of sites on a cluster at a distance l from a second point on the cluster. If p is the probability of finding a lattice point then:

$$g(l) = z(z-1)^{l-1} p^l \tag{9.20}$$

When $p(z-1) < 1$, the correlation decays rapidly to zero. At the point $p(z-1) = 1$, it stays finite for infinite, l, thus providing the relationship for the percolation threshold:

$$p_c = \frac{1}{z-1} \tag{9.21}$$

For a linear lattice, $z = 2$, percolation only occurs when the lattice is completely covered, $p_c = 1$.

Using the correlation function, exact expressions for a number of parameters can be derived. The correlation length, ξ_l, in chemical distance space is obtained from:

$$\xi_l^2 = \frac{\sum_{l=1}^{\infty} l^2 g(l)}{\sum_{l=1}^{\infty} g(l)} = \frac{\sum_{l=1}^{\infty} l^2 (z-1)^l p^l}{\sum_{l=1}^{\infty} (z-1)^l p^l} \tag{9.22}$$

For $p < p_c$, the summations on the right-hand side of Eq. 9.22 can be evaluated exactly from the binomial theorem or derivatives of it. Along with Eq. 9.21, this gives:

$$\xi_l^2 = p_c \frac{p_c + p}{(p_c - p)^2} \tag{9.23}$$

Referring to Eq. 9.12, the critical exponent ν for the Cayley tree is equal to 1, regardless of the value of z. The mean mass of the finite clusters, S, can be derived in a similar fashion. It is given by:

$$S = 1 + \sum_{l=1}^{\infty} g(l) = p_c \frac{1+p}{p_c - p} \qquad (9.24)$$

Referring to Eq. 9.8, it is seen that the exponent γ is equal to 1. The other exponents can be determined in a similar fashion. The exponent for the order parameter, P_∞, in Eq. 9.10 is found to be $\beta = 1$. Thus, the scaling relationships in Section 9.1 are obtained exactly for the Cayley tree. More complicated forms of this model have been investigated (Coniglio, 1976; Coniglio et al., 1979, 1982), and have been particularly useful in describing polymer gelation.

9.3 Percolation in Biomembranes

The "fluid mosaic" model of biological membranes was proposed over 20 years ago and has been one of the most successful paradigms of biochemistry (Singer and Nicholson, 1972). The original tenets of the model are that cell membranes consist of a lipid bilayer, that the membrane is fluid in nature, and that membrane-spanning proteins are adrift in this sea of lipids. Peripheral membrane proteins can also be adsorbed on the surface of the membrane. In the intervening years, there has been a wealth of work, on both native and model systems, that confirms the main features of the model. Current work suggests a slight modification in terms of what constitutes the "mosaic." Originally, the mosaic was conceived to be membrane proteins. Now, it is apparent that there can be a variety of large, relatively immobile "islands" in the lipid fluid. These can be protein aggregates, lipid-phase domains, or combinations of the two. As discussed in the introduction to this chapter, even simple lipid mixtures can show complex phase behavior, with coexisting immiscible states. A native membrane could doubtless be in an extremely, heterogeneous physical state with a variety of microdomains.

9.3.1 Fluorescence Recovery after Photobleaching

Initially, a large effort was directed at understanding the phase behavior of biological membranes. These experimental efforts generally employed calorimetry and a variety of spectroscopic techniques. This work established the bulk thermodynamic properties of the system. It allowed for the construction of phase diagrams such as those shown in Figures 9.1 and 9.2. More recent efforts have been aimed at probing the microscopic domain structure of the lipid bilayer. This became possible principally with the advent of the FRAP technique. Figure 9.10 shows a simple schematic that illustrates the technique known as spot FRAP (see Cardullo et al., 1991). In this technique a fluorescent molecule is embedded in a membrane and a micron-size region is illuminated. The fluorescence from the molecules in the region is monitored through a light microscope. The illumination source is usually a laser. At a given instant, the illuminating intensity is increased by 10^3–10^4 orders of magnitude so as to photobleach the fluorescent probes. Typically, 70% of the molecules are bleached. The photobleaching is required to be irreversible and not result in any adverse chemical reactions (such as crosslinking) by the photolysis

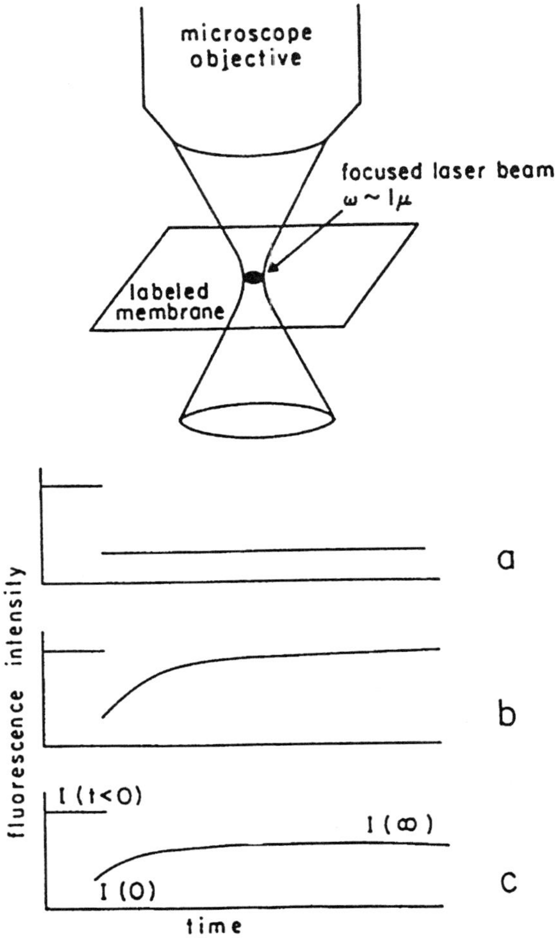

FIGURE 9.10 Schematic diagram of spot FRAP and three types of system response. Top schematic shows optical arrangement for bleaching a spot on a lipid film. Bottom schematic shows observed time courses: (a) fluorescence response for a uniform, solid system–no recovery; (b) response for a fluid system–complete recovery; (c) response in a system that contains fluid microdomains–partial recovery. (From Cardullo et al., 1991. Reprinted with permission of Plenum Publishing Corp.)

product. After the bleaching pulse, the illumination is returned to its initial level and the fluorescence of the region is monitored. To study diffusion in the membrane, there must be no competing external diffusional processes or bulk flow. For cellular systems, this is achieved by immobilizing cells by adhering them to a microscope slide. To study model lipid systems, a hydrated film of multibilayers is formed on a slide.

Figure 9.10 shows three different scenarios that have been observed experimentally. In (a), the probes are immobile in the membrane and a hole is "per-

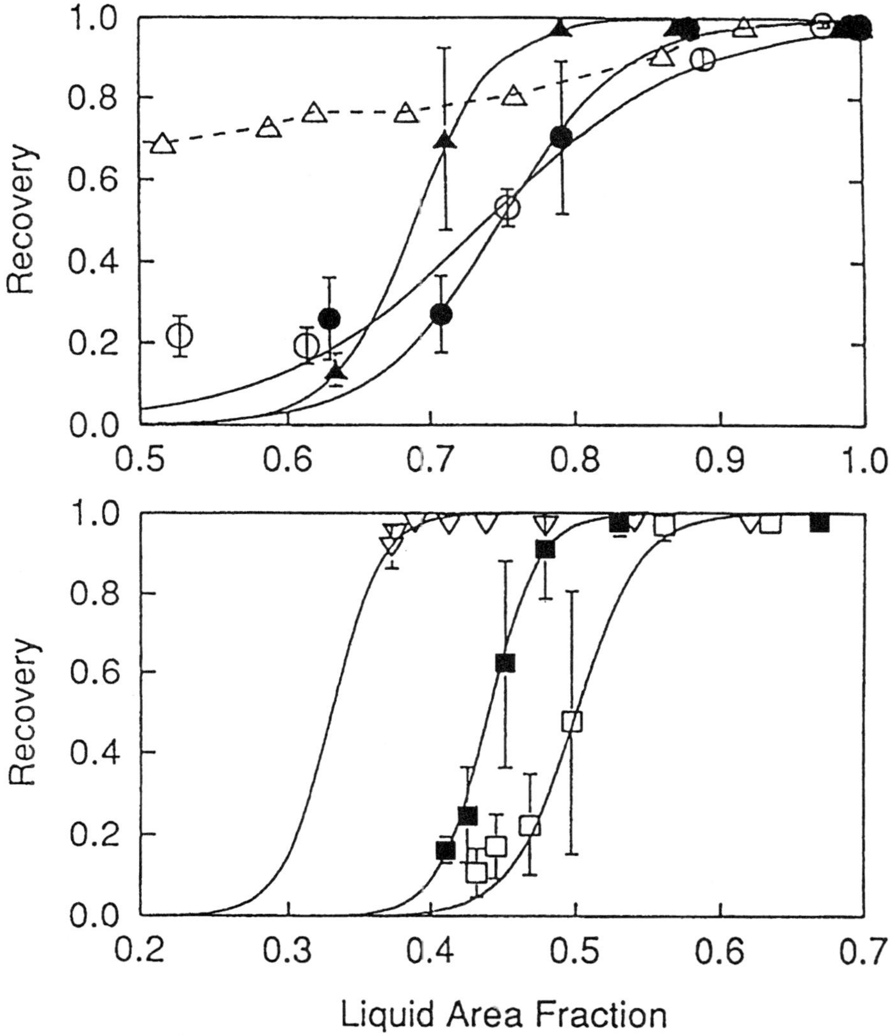

FIGURE 9.11 Plot of the fraction of fluorescence recovered in a FRAP experiment versus the area fraction of the liquid. The system is a 1 : 1 mixture of DMPC/DSPC (see Table 9.2) with varying amounts of cholesterol. The percolation point is taken as the inflection point of the curve. Mole fraction of cholesterol: (top) – (Δ) 0, (○) 4, (●), (Δ) 10; (bottom) – (□) 15, (■) 18, (∇) 20. (Data from Almeida et al., 1993.)

manently" burned into the region. This results in an instantaneous decrease in fluorescence with no subsequent recovery. In (b), the membrane is fluid and the photolyzed molecules diffuse out of the illuminated region, while unbleached, fluorescent molecules diffuse into it. The sample is initially bleached, but the fluorescence will recover to its original level. The third example (c) shows the response when nonpercolating, microscopic domains exist and the fluorescent molecule is confined to a specific phase domain. The illuminated area will contain

TABLE 9.2 Percolation thresholds of lipid membranes* and computer simulations†

System[1,2]	$T(p_C)$	p_C
DMPC/DPPC		
75 : 25	27.5	0.52
50 : 50	32.0	0.52
25 : 75	36.5	0.49
DMPC/cholesterol		
85 : 15	23.6	0.42
80 : 20	23.5	0.65
75 : 25	22.5	0.85
LigGalCer/DPPC		
20 : 80	42.0	0.30
33 : 67	42.0	0.26
50 : 50	46.0	0.25
DMPC/DSPC/cholesterol		
48 : 48 : 4	38.0	0.74
45 : 45 : 10	37.0	0.63
41 : 41 : 18	26.0	0.38
40 : 40 : 20	20.0	0.29
Fluid tracer/gel obstacle in computer simulations		
Point/point		0.4990
Hexagon (R = 1)/point		0.1156
Hexagon (R = 2)/point		0.04823
Point/hexagon (R = 1)		0.5852
Point/equilateral triangle		0.5256
Point/dimer		0.4863
Point/linear trimer		0.4609

* Data from Almeida and Vaz (1994).
† Data from Saxton (1993).
[1] Ratios are of mole fractions. DMPC, dimyristoylphosphatidylcholine; DPPC, dipalmitoylphospatidylcholine; DSPC, distearoylphosphatidylcholine; LigGalCer, N-lignoceroyldihydrogalactosylceramide.
[2] R = 1 corresponds to an object of unit radius.

many microdomains. Those domains that are entirely within the illuminated area will not recover, and exhibit behavior as in (a). Those domains that extend beyond the edge of the illuminated region will contain unbleached molecules that will eventually diffuse into the photolyzed region. These regions will show partial recovery.

Traditionally, two experimental parameters have been determined in a FRAP experiment: the time constant associated with the recovery process, and the fraction of fluorescence recovered. From the time constant, one can, in principle, determine the diffusion constant for the fluorescence probe. However, the analysis of the time course is not always simple, and there is now an increased awareness of the nonexponential behavior of recovery curves (Nagle, 1992) (see Chapter 5). The percentage recovery can be used to determine the percolation threshold for a system. Figure 9.11 shows the fraction of fluorescence recovered versus the area fraction of the solid phase. These experimental results are for one of the binary lipid mixtures described in Figure 9.2. Regions that show 100% recovery are connected. As the solid phase increases, it reaches a percolation threshold, which in this case is defined as the inflection point of the curve. When the system becomes completely solid, there is no recovery of fluorescence. Table 9.2 summarizes the results on percolation thresholds in some model lipid systems.

9.3.2 Finite Size Scaling

If the system were infinite, the transition curve in Figure 9.11 would be a step function. The experimental constraint on the system is the size of the illuminated area, which is approximately 2.5 μm. The effect of this finite size is to broaden the transition. The analysis of the transition curve can yield the percolation threshold and a "characteristic length" for the system. Stauffer's book provides a good introduction to finite size scaling effects (Stauffer, 1985). To account for such effects, typically a scaling law as in Eq. 9.3 is used. Such laws can be assumed for the parameter of interest and, often, the exponents can be fixed from the limiting behavior of the parameters. As an example, consider the probability, P, that a random site belongs to a cluster that percolates across the finite length, L. This parameter is the finite analog of P_∞ (see Eq. 9.10). Following Stauffer (1985), the scaling Ansatz gives:

$$P = L^{-A}F[(p - p_c)L^{B}] \tag{9.25}$$

where A and B are exponents to be determined, and F is an unspecified function. In the limit of large L and $p > p^c$, Eq. 9.10 should be recovered $P \sim (p - p_c)^{\beta}$, and there should be no dependence on length. In this limit $F[x] \sim x^{\beta}$, and to eliminate the length dependence $\beta = A/B$. At the percolation threshold $P = L^{-A}F(0)$, where $F(0)$ is now just a proportionality constant. Under these conditions, where a single cluster spans the length L, one has:

$$P \sim \frac{N_{sites}}{N} \sim \frac{L^{d_f}}{L^{d}} = L^{d_f - d} \tag{9.26}$$

where N_{sites} is the number of lattice sites in the spanning cluster and N is the total number of sites in the d dimension lattice of length L. Recalling Eq. 9.18, Eq. 9.25 now becomes:

$$P = L^{-\beta/\nu}F[(p - p_c)L^{1/\nu}] \tag{9.27}$$

The parameter P is closely related to the fluorescence recovery parameter, R, shown in Figure 9.11. R is the probability that a lattice of length L has a cluster that percolates. For an infinite system, R is a step function, being 0 for $p < p_c$ and 1 for $p > p_c$. The value of R can be calculated directly from P because $R \sim N_{cluster}\, P$, where $N_{cluster}$ is the number of clusters. The number of clusters is proportional to N/N_{sites} Using Eqs 9.26 and 9.27, one obtains:

$$R \sim F[(p - p_c)L^{1/\nu}] \tag{9.28}$$

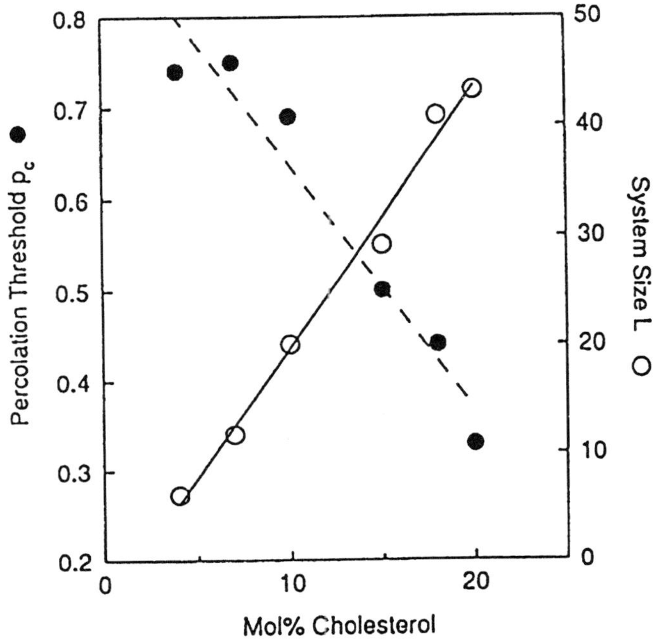

FIGURE 9.12 Plot of the percolation threshold and system length versus the mole fraction of cholesterol. Data taken from computer fits to the curves in Figure 9.11. (Data from Almeida et al., 1993.)

If one knew the function, $F(x)$, the parameters L and p_c could be determined from a statistical fit of the data in Figure 9.11. The slope of the curve is given by:

$$\frac{dR}{dp} = L^{1/\nu} \; F[(p - p_c)L^{1/\nu}] \tag{9.29}$$

At the percolation threshold this slope is given by $dR/dp = L^{1/\nu} \; F(0)$, and is, consequently, a direct measure of the length of the system. As the system becomes larger, this slope gets steeper, and for an infinite system the slope becomes infinite, reflecting the step function.

In previous work, a specific function form was assumed for $F(x)$, and FRAP recovery data were fitted to determine L and p_c (Almeida et al., 1993). The values for these parameters in multibilayers consisting of a ternary mixture are shown in Figure 9.12 as a function of one of the components. As the cholesterol is varied in the ternary mixture of dimyristoylphosphatidylcholine (DMPC)/distearoyl phosphatidylcholine (DSPC) cholesterol, the fraction of fluid-phase molecules is also varied. Although it is difficult on theoretical grounds to justify a specific form for the function, $F(x)$, this will not represent a serious problem. In a sense, all one needs is a curve-fitting device that consistently provides a measure of the transition point and its steepness. In this previous work, L was treated as a dimensionless parameter and is defined as $L = \omega/\phi$, where ω is the radius of the bleached spot and ϕ is a characteristic length of the system. Typically, the characteristic lengths determined in this method will vary from 50 to 500 nm. Assuming these lengths are related to the size of the domains, then each domain will contain approximately 10^4–10^5 lipid molecules.

From the data on the ternary system, it is interesting to note the behavior of L as it approaches the percolation threshold. In previous work, a number of scaling assumptions were made to determine the dependence of ϕ on p_c. In this presentation, we follow a more direct approach, as described previously (Stauffer, 1985). Since the experimental observation is made on a multibilayer, one is measuring an average response of many systems. To calculate this average, Eq. 9.29 is employed, since dR/dp gives the probability that the lattice percolates upon increase of the fluid species concentration from p to $p + dp$. The average probability, $\langle p \rangle$, for the onset of percolation is then given by (Stauffer, 1985):

$$\langle p \rangle = \int_0^1 p\left(\frac{dR}{dp}\right) dp = \int_0^1 pL^{1/\nu}F[(p - p_c)L^{1/\nu}]\, dp \qquad (9.30)$$

where Eq. 9.29 was used in the rightmost equality. Making the substitution $x = (p - p_c)L^{1/\nu}$ and employing the normalization condition, $\int_0^1 (dR/dp)\, dp = 1$, one obtains:

$$\langle p \rangle - p_c = L^{-1/\nu} \int_0^1 xF(x)\, dx \qquad (9.31)$$

where the integral is now just a proportionality constant. Equation 9.31 could be used to determine both ν and the percolation threshold for the infinite system. In practice, such a determination is problematic because to change $\langle p \rangle$ the composition is changed. Since percolation is not strictly geometric in these systems, p_c will also change with composition. Thus, both parameters on the left-hand side of Eq. 9.32 will be changing. It is also difficult to obtain an independent measure of p_c. Nevertheless, the length dependence can be fitted using empirical models (Almeida et al., 1993).

9.3.3 Continuum and Correlated Percolation Thresholds

Using the FRAP technique, percolation thresholds in a variety of lipid systems can be measured, and some of these data are listed in Table 9.2. It is interesting to compare these results with those of computer simulations (also given in Table 9.2). For the system DMPC/dipalmitoyl phosphatidyl choline (DPPC), the phase diagram indicates ideal behavior, and the percolation threshold is 0.5, independent of composition. At different compositions a temperature must be reached where there is a 50:50 mix of solid and liquid before percolation occurs. This gives a constant "lever" ratio for the percolation point on the phase diagram (see Figure 9.2). This value is consistent with percolation on two-dimensional lattices. There are a number of systems that show percolation thresholds as low as 0.3, and some that are as high as 0.8. In these systems the percolation thresholds change with composition. This change with composition means that either the geometry of the percolation cluster is changing or that the percolation process is not random. Nonrandom or correlated percolation has not received as much attention as random, geometric percolation. Generally, it is much easier to explain a low percolation threshold than a high one. Saxton (1993) investigated site percolation in a triangular lattice in which the liquid phase tracer (a "computer analog" to the fluorescent probe) and the solid phase obstacles had specific structures. His results are shown in Table 9.2. They show that random percolation involving objects of different geometry can significantly decrease the percolation threshold below 0.5. However, such effects never result in an increase in thresholds.

Continuum percolation is a process that can also yield variable threshold values. In a continuum percolation problem, objects of a given shape are "punched" out of a region at random. The punched-out regions are allowed to overlap, giving the resulting region a Swiss cheese appearance. Percolation occurs when the vacancies connect the edges of the region. For circular punches in a two-dimensional region, the percolation threshold is 0.312 (Bunde and Havlin, 1991). Using ellipses as the hole punch, the following empirical relationship is obtained (Xia and Thorpe, 1988):

$$p_c = \frac{1 + 4y}{19 + 4y} \tag{9.32}$$

where $y = (a/b + b/a)$ and the aspect ratio of the ellipse is b/a, with $2a$ and $2b$ being the length of the major and minor axes, respectively. As pointed out by Saxton (1993), when the ellipses assume an extreme, needle-like structure, the percolation threshold can become very high, approaching 1. If the solid phase showed such a needle-like structure, randomly criss-crossing the membrane, the percolation threshold could be extremely high. However, this model alone cannot account for observations of high thresholds, because it does not predict a change in p_c with temperature or composition. In such instances, a correlated percolation model must be used that incorporates the thermodynamic features of lipid–lipid interactions. Such a model would account for interactions between solid and liquid phase molecules at the boundary of the cluster. A simple, heuristic model based on small-cell renormalization is considered in the Appendix to this chapter.

Correlated percolation models with thermodynamic features can show high percolation thresholds (Napiorkowski and Hemmer, 1980). This can be nicely seen by considering the case where neighboring occupied sites have extremely strong repulsive energies. As the occupancy of the lattice approaches 0.5, the occupied sites will arrange themselves at every other site, so as to avoid the strong repulsive energy of another site (see Figure 9.13). This creates a sublattice that contains no filled sites and is a decimated form of the original lattice. As the site coverage increases, the addition of each site is random because each site has an identical energy for occupancy. Thus, the filling of the sublattice will be a random percolation problem. After filling half the sites, the system will then

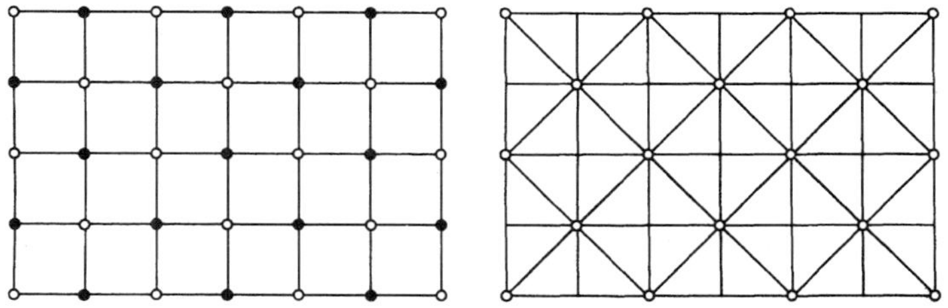

FIGURE 9.13 Percolation problem on a square lattice with strongly repulsive interactions between neighbors. (a) Lattice at occupancy (black sites) $p = 0.5$, which creates a sublattice (b) in which addition of sites will be uncorrelated.

percolate at p_c of the sublattice. The percolation problem for strong repulsion between sites will have a percolation threshold give by:

$$p_c(-\infty) = \frac{1}{2}\ [1 + p_c\ (sublattice)] \tag{9.33}$$

where $p_c(-\infty)$ is the percolation threshold for the lattice with infinitely repelling sites. The first term on the right-hand side of Eq. 9.33 represents the filling of every other site, and the second term represents percolation on the sublattice. Taking $p_c(sublattice) = 0.50$ for a random process, the result for the system is 0.75. As seen in Table 9.2, numbers this high have been observed experimentally.

The results from FRAP in conjunction with calorimetric information convincingly show that percolation clusters occur in model, multibilayer lipid systems. But how general are these results? Do similar effects occur in native membranes where the composition is even more heterogeneous? Not surprisingly, the situation is not as clear. FRAP experiments on a wide variety of different cell types show an immobilized phase (for a review, see Tocanne et al., 1989). Clearly, heterogeneous microdomains exist in cell membranes. However, in these cases the heterogeneity is probably more a consequence of the large protein concentrations in the membrane rather than lipid phase domains. Estimates have been made of the fraction of area covered by proteins in a variety of native membranes (Saxton, 1989). Some membranes, such as the mitochondrial inner membranes, can show exceedingly high fractional area coverage by protein (0.4–0.5). Such membranes presumably have percolating protein networks. In general, the protein coverage of native membranes is considerably lower than this, with typical values of approximately 0.25. It is not obvious whether these systems will be percolating and, unfortunately, membrane composition in a native system is not easily manipulated. Thus it would be difficult to perform a FRAP experiment at different coverage levels. An additional complication is that native membranes often have a nonrandom distribution of proteins. This is especially true for membrane components of specialized cells such as spermatozoa, eggs, epithelia, retinae, and chloroplasts, in which proteins are segregated in well-defined regions. The membrane organization in such cells clearly has a biological function. The elucidation of the topography of the microdomains in such settings will certainly offer a continuing challenge to the field.

9.4 Percolation of Proton Transport in Hydrated Proteins

Proteins in their native aqueous environment have polar residues at the surface and hydrophobic residues tucked inside. Because of this, the protein surface will naturally have an affinity for water. If a dehydrated protein powder is exposed to an atmosphere containing water, the water will begin to adsorb to the surface of the protein. From capacitance measurements on protein powders (Careri et al., 1985, 1986, 1987; Rupley et al. 1988), it appears that the adsorption process is a percolation phenomenon. The dc conductivity of the protein powders can be inferred from frequency-dependent capacitance measurements. At low levels of hydration, this conductivity is low. It increases dramatically at a threshold, hydration level, and this increase shows power law scaling. The observation of a hydrogen isotope effect was crucial in unraveling the nature of this phenomenon, because it implicates proton transfer reactions. The dielectric dispersion shows three relaxation regimes. All three show a deuterium isotope effect that would be anticipated for proton transfer reactions. This suggests that water bound at the protein surface is responsible for the major component of the relaxation processes. Interestingly, the threshold hydration levels are independent of isotope. Relaxation in the high-frequency regime has been attributed to proton hopping through water networks on the surface. The threshold effect is thought to occur when the bound water on the surface of a single protein percolates and provides a conducting dc network.

The frequency-dependent capacitance, $C(\omega)$, of a protein powder in a hydrated atmosphere is given by:

$$C(\omega) = C^* \left[1 + \left(\frac{\varepsilon^*}{\varepsilon} \right) \frac{\omega + \tau}{1 + \omega + \tau^2} \right]^{-1} \tag{9.34}$$

where C^* and ε^* are instrumental parameters, and $\tau = \varepsilon_0(\varepsilon/\sigma)$, with (ε/σ) being the vacuum permittivity, ε the dc dielectric constant, and σ the dc conductivity. Figure 9.14 shows changes in the capacitance as the hydration level of a lysozyme powder is changed. These curves are similar in shape to those shown in Figure 9.11 for percolation of membrane domains. In principle, finite size scaling theory could be used to determine the characteristic length of the system. As seen by the change in the slope at the inflection point, different frequencies will probe different length scales.

While the threshold behavior of the capacitance is apparent at each of the three frequencies shown, the threshold is more dramatically shown when the conductivity is considered. The conductivity calculated from the experimental data using Eq. 9.34 is displayed in Figure 9.15. This quantity shows the following scaling behavior:

$$\sigma(p) \sim (p - p_c)^{\mu} \sim (h - h_c)^{\mu} \tag{9.35}$$

where μ is the critical exponent for dc conductivity, h is the hydration level (weight adsorbed water per weight protein), and h_c is the hydration level at the percolation threshold. At low levels of hydration, all the water will be surface bound and the probability of a site being occupied will be proportional to the hydration level. The

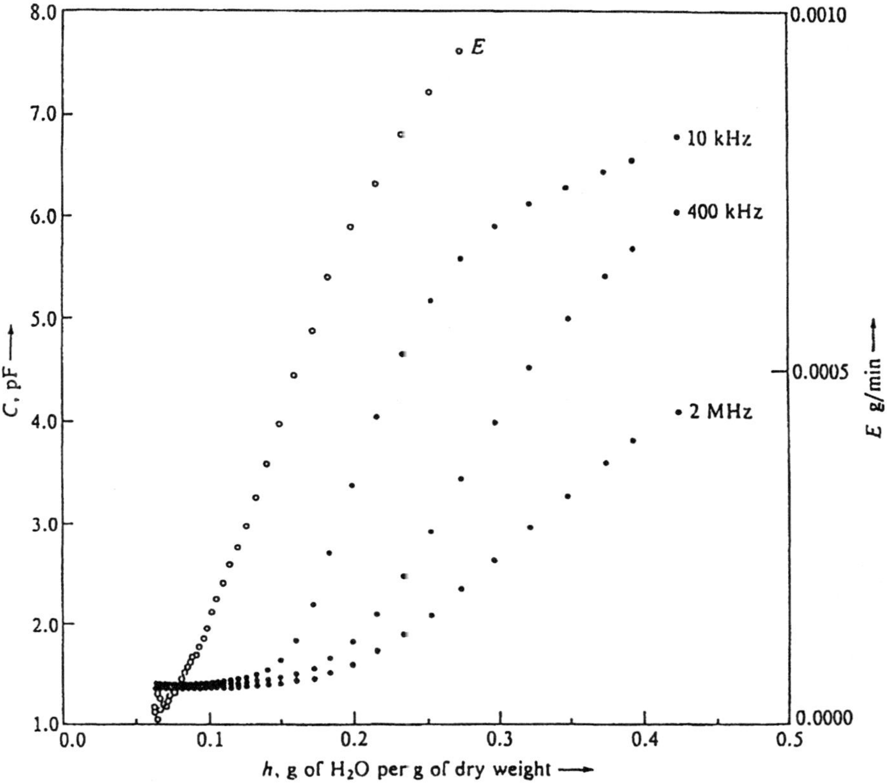

FIGURE 9.14 Plot of capacitance versus the hydration level of a lysozyme powder. Capacitance was measured at the indicated frequency. Curve E shows the rate of hydration. (From Careri et al., 1986. Reprinted by permission of John A. Rupley.)

exponent, μ, differs from the exponents discussed previously in that it is a dynamical quantity rather than a static one. In general, dynamical scaling is not as well understood as static scaling. Dynamical effects in percolation are discussed in the excellent review by Havlin and Bunde (1991) and will not be treated here. From a variety of theoretical and computational determinations, μ is found to be approximately 1.3 in two dimensions and 1.9 in three dimensions.

The data for lysozyme show two scaling regimes. At hydration levels in a limited regime close to the threshold, a μ value of 1.30 ± 0.09 was determined (Careri et al., 1988), consistent with two-dimensional percolation. Within experimental error, there was no effect when deuterium oxide was used instead of water. This is expected if the phenomenon is due strictly to site occupancy. Experiments were also performed when the lysozyme had a substrate bound at the active site. Having bound substrate will affect the dynamics of the protein, but probably does not alter its gross morphology (see Chapter 5). Again, no change in scaling exponent was observed. At higher levels of hydration the scaling exponent shifts and becomes 2.20 ± 0.19. This higher value is anticipated for three-dimensional percolation. At

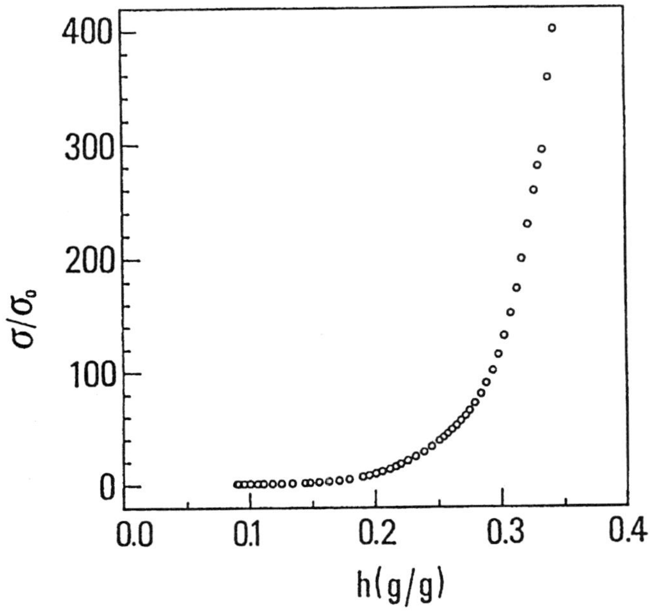

FIGURE 9.15 Plot of normalized dc conductivity versus the hydration level of lysozyme powder. Conductivity was determined from an analysis of data similar to those in Figure 7.15 (see Eq. 9.34). Scaling behavior as in Eq. 9.35 is seen. (From Careri et al., 1988. Reprinted by permission of John A. Rupley.)

these higher hydration levels, there may be more than a monolayer of water. If connections are made between proteins by bridging water molecules, the system could become three dimensional. Alternatively, these changes may be simply due to the breakdown of Eq. 9.35 in a regime far from the threshold.

The percolation threshold can be determined from the values of h_c. As in all finite systems, there is a certain arbitrariness in defining h_c. In the lysozyme work, h_c is defined by extrapolating a curve of dC/dh versus h to zero (Careri et al., 1986). Typical values for lysozyme are of the order of 0.15 g water/g protein. Fully hydrated protein was measured to have a value of 0.38 g water/g protein. Assuming that this represents a monolayer, the percolation threshold is given by the ratio of these two numbers. A value for p_c of 0.39 is obtained. Referring to Table 9.2, it is seen that this value is lower than expected for two-dimensional lattices and higher than for continuum percolation. This threshold does not show a hydrogen isotope effect. However, the presence of a bound substrate does change the threshold, giving a p_c value of 0.61. Also, at high pH a similar high value is obtained. Since these two conditions probably do not grossly alter the surface morphology of the protein, such effects argue against a random percolation process. The pH effect, in particular, indicates a specific role of protein surface residues, either by direct hydrogen bonding or by providing sites for bound water. Percolation of bound water on protein surfaces may well represent another example of correlated percolation.

Protonic percolation has been measured in arrays of the membrane protein bacteriorhodopsin (Rupley et al., 1988) and in nucleic acids (see Pethig, 1979; Bonincontro et al., 1988). Bacteriorhodopsin is a membrane protein that forms dense, closely packed arrays. These aggregates form membrane sheets known as "purple membrane." The experimental results in this system mirror those for lysozyme and support a two-dimensional percolation model. However, unusually low thresholds for percolation are obtained ($p_c \sim 0.18$). This is a more complicated system because of the aggregated state of the protein. The percolative process could be along the membrane surface or through the membrane. Regardless, the unusual low value is suggestive of a specific route or correlated percolation. The work on DNA is quite controversial (see Pethig, 1979), for both experimental and theoretical reasons. It will not be discussed here.

9.5 Summary

Percolation is a site-occupancy phenomenon that shows geometric phase transitions. Percolation occurs when occupied lattice sites form a region that spans the lattice. The resulting percolation cluster is intrinsically a fractal structure. There are a number of interesting scaling relationships that characterize the structure of these clusters. Percolation occurs in a number of biological settings, including antibody–receptor clustering, microdomains in biomembranes, and hydration of protein powders. Work on biomembranes provides the most complete characterization of percolation phenomena in biological systems. Because of the domain structures formed and the experimental constraints of the FRAP technique, finite size scaling models must be used to analyze the biomembrane data. These data show percolation thresholds that depend strongly on composition and temperature. Some of these thresholds are extremely high, and correlated percolation models are required to explain these results. Ultimately, a correlated model that incorporates the thermodynamics of the system is required to understand phase-separated microdomains in biomembranes.

Appendix: A Correlated Percolation Model for Biomembranes

As seen in the previous sections, most examples of percolation in biological systems are not "purely geometrical." Rather, the percolation threshold will depend on thermodynamic variables of the system. This is particularly evident in the work on biomembranes. As can be seen in Table 9.2, the percolation threshold will depend on such quantities as temperature and composition. Clearly, one has to incorporate thermodynamics into the percolation formalism to account for such phenomena. Such a melding of theories will result in correlated percolative processes. In these cases, occupancy of sites will not be random, but rather will depend on interactions between neighbors. Theories of correlated percolation have arisen in considering the site–bond percolation problem that have applications to gelation phenomena in polymer systems (see Coniglio, 1976; Nakanishi and Reynolds, 1979; Tanaka et al., 1979; Coniglio et al., 1979, 1982).

In the site–bond problem, one considers a lattice that is partially covered with monomers at specific sites. When monomers are neighbors, there is a fixed probability that they will form a bond between each other, thus creating dimers, trimers, and so forth. Thus, in addition to the probability of a site being occupied, one must also consider the probability of bond formation. Percolation occurs when a "polymer" formed in this manner traverses the lattice. The percolation point represents the sol–gel transition of the polymer system. Here, a simple model of biomembranes is considered. In this model the lipids are considered to exist in one of two phases: fluid or solid. The fluid phase will have a different energy per site than the solid. This determines the probability of site occupancy. Additionally, there is an unfavorable energy associated with a fluid and solid site being next to each other. Thus, fluid–fluid neighbors will have a stronger "bond" than will fluid–solid ones. This problem is then closely related to site–bond percolation. One big difference is that in the site–bond model the total number of sites is fixed by experimental conditions, while in the fluid–solid membrane the number of fluid molecules will change with temperature.

To illustrate how the thermodynamics of the lipid phase transition might be incorporated into a percolation model, a small-cell renormalization approach is considered. This approach is similar in spirit to a previous treatment of percolation in an Ising model (Napiorkowski and Hemmer, 1980) and will give essentially the same results. The square lattice used in the renormalization calculation in Section 9.1 is used with added features. For consistency, the notation used for helix–coil transitions of biopolymers (Chapter 3) is followed. The Boltzmann weight for a fluid molecule is s, while solid or gel molecules are assigned a weight of I. Boundary effects between a fluid and a solid are given a statistical weight of $\sigma^{1/2}$ (in correspondence to the helix – coil boundary in biopolymers). In Figure A9.1 all possible sites for the 2×2 lattice are shown along with their probabilities. The renormalization of this to a single-site lattice is considered. To achieve this, the probability of percolation is first considered. We seek a relationship of the form:

$$p' = f(p) \tag{A9.1}$$

where p' and p are the probabilities of percolation of the decimated (single site) and original (2×2) lattice, respectively. This requires that probabilities be considered and the Boltzmann weights must be divided by the respective partition functions. The partition function for the single site, Z', is given by: $Z' = 1 + s'$. The partition function for the original lattice is:

$$Z = s^4 + 4s^3\sigma + 2s^2\sigma(2 + \sigma) + 4s\sigma + 1 \tag{A9.2}$$

The lattice configurations that percolate are indicated in Figure A9.1 and are used to calculate the probability of occupancy for the 2×2 lattice. This allows the

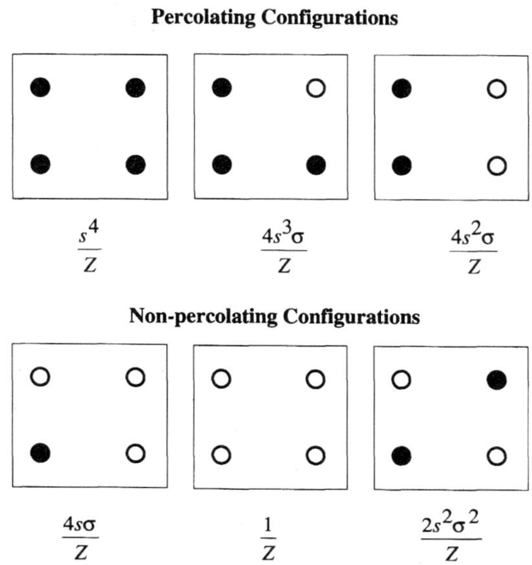

FIGURE A9.1 Renormalization model for correlated percolation. Configurations and statistical weights for four-site clusters are shown. Percolating configurations are shown in the top. The statistical weights for each configuration are determined with a weight of s for a filled site and a weight of 1 for an empty site. The boundary between filled and empty sites has a weight of $\sigma^{1/2}$.

calculation of the explicit form of Eq. A9.1 and gives the first of our renormalization equations:

$$p' = \frac{s'}{Z'} = \frac{s^4 + 4s^3\sigma + 4s^2\sigma}{Z} \tag{A9.3}$$

For a more rigorous approach, one would consider the probability of occupancy and the probability of percolation separately. This would provide a critical surface in which s and σ are variables. For simplicity, we restrict ourselves to the present approach.

At this point, one could solve Eq. A9.3 for the fixed point and calculate the scaling exponent as in Eq. 9.15. However, this would require that σ be considered a fixed constant, which it is not. To obtain a second relationship, the energy of the system must be renormalized, as was done in the example in Chapter 1. The energy of the system with N lattice sites is given by:

$$N\varsigma = \ln Z \tag{A9.4}$$

where ς is the energy per lattice site divided by kT. Since ς should not depend on the size of the systems, its value for the original and decimated lattice should be the same. This condition, along with the respective partition functions, provides a second renormalization equation:

$$4 \ln Z' = \ln Z \tag{A9.5}$$

where the factor of 4 is a result of a fourfold reduction in number of sites upon decimation. Substituting the respective partition functions into Eq. A9.5 gives:

$$(1 + s')^4 = 1 + 4s\sigma + 2s^2(2\sigma + \sigma^2) + 4s^3\sigma + s^4 \tag{A9.6}$$

Equations A9.3 and A9.6 constitute the two renormalization equations. The fixed point is determined by setting $s' = s$. Normally, one would have a comparable relationship for σ, but because of the small cells considered σ' never appears. Equation A9.6 can be solved directly for σ, and it is seen that the only physical solution occurs when $\sigma = 1$. This corresponds to zero energy of interaction between occupied and unoccupied neighbors, and the fixed point will represent random percolation. The fixed point for s is 0.618, which happens to give a percolation threshold for the solid phase of 0.618. This result is identical to that obtained for the Ising model (Napiorkowski and Hemmer, 1980).

Because the fixed point for this correlated percolation problem corresponds to random percolation, the critical exponent, υ, remains unchanged, even for correlated situations. However, the percolation point will change and the specific value it assumes will depend on experimental conditions. The critical surface will be a trajectory of s, σ pairs determined from the renormalization equations that lead to (or from) the fixed point. This critical surface separates regions of nonpercolating structures from percolating structures. Different points along the critical surface will have different thresholds. The change in percolation threshold with a given thermodynamic variable can, in principle, be related to changes in s and σ. This approach could lead to an analysis of the effects of temperature in real systems, as seen in Table 9.2. However, composition must be included for a complete model of percolation appropriate to biomembranes. Such models have been considered for gel–sol problems (Coniglio et al., 1979, 1982), and could be adapted to the present case.

References

Almeida, P. F. F., and Vaz, W. L. C. 1994. Lateral Diffusion and Membranes. In *Membranes: Their Structure and Conformation*, ed. R. Lipowsky and E. Sackmann. Elsevier, Amsterdam.

Almeida, P. F. F., Vaz, W. L. C., and Thompson, T. E. 1993. Percolation and Diffusion in Three-Component Lipid Bilayers: Effect of Cholesterol on an Equimolar Mixture of Two Phosphatidylcholines. *Biophys. J.*, **64**, 399–412.

Bonincontro, A., Careri, G., Giansanti, A., and Pedone, F. 1988. Water-Induced dc Conductivity of DNA: A Dielectric–Gravimetric Study. *Phys. Rev. A* **38**, 6446–6447.

Bunde, A., and Havlin, S. 1991. Percolation I. In *Fractals and Disordered Systems*, ed. A. Bunde and S. Havlin. Springer-Verlag, Berlin, pp. 55–95.

Cardullo, R. A., Mungovan, R. M., and Wolf, D. E. 1991. Imaging Membrane Organization and Dynamics. In *Biophysical and Biochemical Aspects of Fluorescence Spectroscopy*, ed. T. G. Dewey. Plenum Press, New York, pp. 231–260.

Careri, G., Geraci, M., Giansanti, A., and Rupley, J. A. 1985. Protonic Conductivity of Hydrated Lysozyme Powders at Megahertz Frequencies. *Proc. Natl. Acad. Sci. USA* **82**, 5342–5346.

Careri, G., Giansanti, A., and Rupley, J. A. 1986. Proton Percolation on Hydrated Lysozyme Powders. *Proc. Natl. Acad. Sci. USA* **83**, 6810–6814.

Careri, G., Giansanti, A., and Rupley, J. A. 1988. Critical Exponents of Protonic Percolation in Hydrated Lysozyme Powders. *Phys. Rev. A* **37**, 2703–2705.

Coniglio, A. 1976. Some Cluster-Size and Percolation Problems for Interacting Spins. *Phys. Rev. B* **13**, 2194–2207.

Coniglio, A., Stanley, H. E., and Klein, W. 1979. Site-Bond Correlated-Percolation Problem: A Statistical Mechanical Model of Polymer Gelation. *Phys. Rev. Lett.* **42**, 518–522.

Coniglio, A., Stanley, H. E., and Klein, W. 1982. Solvent Effects on Polymer Gels: A Statistical Mechanical Model. *Phys. Rev. B.* **25**, 6805–6821.

Creswick, R. J., Farach, H. A., and Poole, Jr., C. P. 1992. *Introduction to Renormalization Group Methods in Physics*. Wiley, New York, pp. 69–107.

DeBoer, R. J., and Hogeweg, P. 1989. Idiotypic Networks Incorporating T–B Cell Co-operation. The Conditions for Percolation. *J. Theor. Biol.* **136**, 17–38.

DeLisi, C., and Perelson, A. 1976. The Kinetics of Aggregation Phenomena. I. Minimal Models for Patch Formation on Lymphocyte Membranes. *J. Theor. Biol.* **62**, 159–210.

Faro, J., and Velasco, S. 1993. Crosslinking of Membrane Immunoglobulins and B-Cell Activation: A Simple Model Based on Percolation Theory. *Proc. R. Soc., London, ser. B* **254** 139–145.

Feder, J. 1988. *Fractals.* Plenum Press, New York, pp. 104–148.

Goldstein, B., and Dembo, M. 1984. The IgE Mediated Activation and Desensitization of Human Basophils. In *Cell Surface Dynamics – Concepts and Models*, ed. A. S. Perelson, C. DeLisi, and F. W. Wiegel. Marcel Dekker, New York, pp. 331–382.

Havlin, S., and Bunde, A. 1991. Percolation II. In *Fractals and Disordered Systems*, ed. A. Bunde and S. Havlin. Springer-Verlag, Berlin, pp. 96–149.

Nagle, J. 1992. Long Tail Kinetics in Biophysics? *Biophys. J.* **63**, 366–370.

Nakanishi, H., and Reynolds, P. J. 1979. Site–Bond Percolation by Position–Space Renormalization Group. *Phys. Lett.* **71A**, 252–254.

Napiorkowski, M., and Hemmer, P. C. 1980. Correlated Percolation on a Quadratic Lattice. *Phys. Lett.* **76A**, 359–361.

Perelson, A. S. 1984. Some Mathematical Models of Receptor Clustering by Multivalent Ligands. In *Cell Surface Dynamics – Concepts and Models*, ed. A. S. Perelson, C. DeLisi, and F. W. Wiegel. Marcel Dekker, New York, pp. 223–276.

Pethig, R. 1979. *Dielectric and Electronic Properties of Biological Materials.* Wiley, New York.

Rupley, J. A., Siemankowski, L., Careri, G., and Bruni, F. 1988. Two-Dimensional Protonic Percolation on Lightly Hydrated Purple Membrane. *Proc. Natl. Acad. Sci. USA* **85**, 9022–9025.

Sahimi, M. 1994. *Applications of Percolation Theory.* Taylor & Francis, London.

Saxton, M. J. 1989. Lateral Diffusion in an Archipelago: Distance Dependence of the Diffusion Coefficient. *Biophys. J.* **56**, 615–621.

Saxton, M. J. 1993. Lateral Diffusion in an Archipelago: Dependence on Tracer Size. *Biophys. J.*, **64**, 1053–1062.

Singer, S. J., and Nicholson, G. L. 1972. The Fluid Mosaic Model of the Structure of Cell Membranes. *Science* **175**, 720–731.

Stanley, H. E. 1977. Cluster Shapes at the Percolation Threshold: An Effective Cluster Dimensionality and its Connection with Critical-Point Exponents. *J. Phys. A* **10**, L211–L220.

Stauffer, D. 1985. *Introduction to Percolation Theory.* Taylor & Francis, London.

Tanaka, T., Swislow, G., and Ohmine, I. 1979. Phase Separation and Gelation in Gelatin Gels. *Phys. Rev. Lett.* **42**, 1556–1559.

Thompson, T. E. 1992. Biological Membrane Domains: Functional Significance. *Comments Mol. Cell. Biophys.* **8**, 1–15.

Tocanne, J.-F., Dupou-Cézanne, L., Lopez, A., and Tournier, J.-F. 1989. Lipid Lateral Diffusion and Membrane Organization. *FEBS Lett.* **257**, 10–16.

Vaz, W. L. C., and Almeida, P. F. F. 1993. Phase Topology and Percolation in Multi-Phase Lipid Bilayers: Is the Biological Membrane a Domain Mosaic? *Curr. Opinion Struct. Biol.* **3**, 482–488.

Xia, W., and Thorpe, M. F. 1988. Percolation Properties of Random Ellipses. *Phys. Rev. A* **38**, 2650–2656.

10

Chaos in Biochemical Systems

Chaos is a companion discipline to fractals. In the popular science literature, chaos and fractals are used almost interchangeably. Yet they are distinct topics that borrow heavily from each other. Fractals are used to describe complex topographies, whether they be temporal or spatial. Chaos, on the other hand, is concerned with the dynamical processes that show an exquisite sensitivity to initial conditions and which can generate complex fractal topographies in phase space. The language of fractals provides a natural descriptor for chaotic systems. Chaotic behavior has appeared in a number of biological settings (see May, 1976; Bassingthwaighte et al., 1994) and may confer specific physiological advantages. The goal of the present chapter is to provide the reader with a nonmathematical entrée into the subject and to discuss how fractal patterns arise naturally in chaotic systems. The emphasis will be on biochemical systems the chemical kinetics of which have been reasonably well defined. There are now a large number of introductory treatments of chaos, but Baker and Gollub's book (1990) remains one of the more attractive and approachable. For more mathematical treatments, see Rasband (1990) and the rigorous treatment by Lichtenberg and Lieberman (1983). For discussions of chaos in chemical and biochemical systems see the edited volume by Field and Gyögyi (1993), the monograph by Bassingthwaighte et al. (1994), and the introductory texts by Scott (1991) and Gray and Scott (1990).

Chaos arises from a deterministic system and gives seemingly random behavior. Consequently, one must focus on the difference between chaotic and random systems. There are a number of deep issues here. Operationally, chaotic systems are marked by an extreme sensitivity to initial conditions and a sharp divergence of trajectories in phase space. Often, the trajectories of chaotic systems are drawn to an "attractor" in phase space that is a fractal. Such an object is known as a "strange attractor." Because chaotic systems are deterministic, they are often thought to be amenable to deeper analysis and interpretation. In a sense, they are more "mechanistic." Despite these mechanistic laws that underlie chaos, there must be new descriptive formulations. This is where fractals have been so useful. Is our understanding of chaotic systems really more mechanistic than for random systems? As deeper understanding of the origins of different types of noise grows, random processes become open to mechanistic interpretation. The scaling of stochastic differential equations can be very closely linked to the types of fluctuation that drive the system. The physical origins of these fluctuations provide mechanistic grounds for interpretation. The chemical kinetics of biochemical system provide particularly fertile ground for exploring issues of fluctuations, randomness, and chaos.

In the Section 10.1 we provide a description of chaos, with special reference to chemical reaction systems. This has the advantage that the terminology of chaos is introduced in a setting that has direct physical significance. In the same section, methods for distinguishing chaotic behavior from random behavior are discussed. It is seen that fractal geometry becomes a natural tool for developing such analyses. In Section 10.1, the conditions required for chaos to exist in a chemical system are enumerated. Chaos and oscillatory behavior in biochemical systems have already developed a surprisingly long history. Chaos was observed in 1977 in the reaction catalyzed by the enzyme horseradish peroxidase (Olsen and Degn, 1977), shortly before it was discovered in the now famous, Belousov–Zhabotinsky reaction. Interestingly, chaotic enzymatic systems represent an area where theory and experiment have been closely tied and have developed in parallel. In Section 10.2 we discuss theoretical models of enzyme systems and explore what features are required for oscillations and chaos. While there is a maturing of our understanding of these systems, only a limited number of reactions have actually been investigated to date. Section 10.3 contains a discussion of two experimental systems in which chaos has been observed. These systems involve: phosphofructokinase from the glycolytic pathway and the enzyme horseradish peroxidase. In a final, highly speculative section, possible physiological advantages of chaos are discussed.

10.1 Chaos in Chemical Systems

10.1.1 Chemical Kinetics in Simple Systems

Traditionally, the goal of chemical kineticists has been to determine the mechanism of a reaction, i.e., to identify the elementary steps, and to measure quantities such as rate constants and activation parameters associated with these elementary steps. Even simple classical reaction schemes can provide extremely complicated rate laws that quite often are "underdetermined." In such cases, the model has more parameters than the data permit one realistically to determine. Consequently, the primary aim of kineticists has been to simplify. Often ingenious experiments are devised to study elementary steps in isolation and to measure

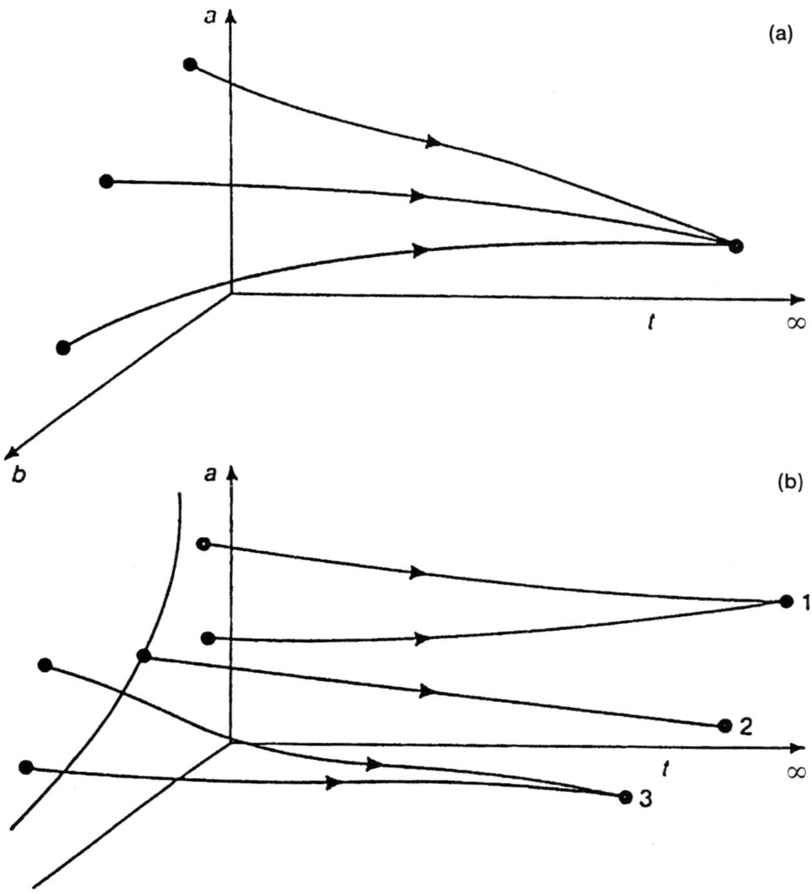

FIGURE 10.1 A phase space representation of the evolution of a chemical reaction. Coordinates a and b are reactant concentrations and t is time. For a reaction in a closed system, the trajectories will approach equilibrium in an exponential fashion. There is a net compression of points in phase space, as illustrated in (a) where all points converge to a single attracting point. (b) A system with two stable points (1 and 3) and an unstable regime (2). (From Scott, 1991. Reprinted by permission.)

their rate constants. Reactions in closed systems near equilibrium are ideally suited for this purpose. Such systems are known as dissipative systems. In these closed systems, a wide variety of initial conditions lead to a single set of final, equilibrium conditions.

One can create an n-dimensional space (called phase space) that contains the values of the n-independent concentrations of the reaction as parametric functions of time. Many different starting points (at time $t = 0$) in this phase space lead, as the reaction proceeds, to a single equilibrium point (at $t = \infty$). In mathematical parlance, this equilibrium point is known as an attractor. The attractor results in a compression of phase space, and this is why the system is said to be dissipative. This effect is illustrated in Figure 10.1. The key point for the chemical kineticist is that the approach to equilibrium of such systems always becomes exponential in time. This allows for a facile determination of rate parameters. To sort out different parameters, a kineticist will explore different regions of phase space by varying initial concentrations over an extremely wide range. This allows more than a single equilibrium point to be studied and assists in establishing the reaction order. Regardless of the "volume" of initial starting conditions, there is always a net compression in phase space as the equilibrium region is approached.

For most chemical reactions the final equilibrium state is achieved via a single reaction path and the reactions are relatively insensitive to conditions such as initial concentration, temperature, and pressure. However, there are reactions in closed systems that show a great sensitivity to these conditions. A common example would be the oxidation of hydrocarbons. For oxygen–hydrocarbon mixtures below the flashpoint, reaching equilibrium is an immeasurably slow process. Yet a slight increase in temperature to the flashpoint results in the rapid combustion of the mixture. This reaction is far from equilibrium and may initially not display the nice exponential approach to the final state. For reactions that are far from equilibrium in a closed system, oscillatory behavior can occur, but it is always against a background of changing reactant concentrations. Thus, sustained periodic behavior will not occur in such environments, but, ultimately, as the system nears equilibrium, exponential behavior will again appear.

10.1.2 Open Systems, Oscillations, and Chaos

Recently attention has turned from the examination of elementary reaction steps near equilibrium to collections of reactions in open systems. These systems do not exhibit the constraints of closed systems and, while perhaps more interesting, are also more difficult to study. Often such open systems consist of a continuously stirred tank reaction (CSTR), a device that allows a constant rate of flow of a reactant through a reaction vessel. Such reactions have a rich phenomenology. Figure 10.2 shows experimental results obtained with phosphofructokinase (PFK), an enzyme in the glycolytic pathway. As can be seen, a variety of complicated behavior is exhibited by one of the reactants as a function of time. This behavior is extremely sensitive to the flow rate for the CSTR, and can show periodic, aperiodic, and chaotic behavior, all as a result of slight changes in the flow rate.

For such flow reactions, behavior about a steady state is considered rather than that about the equilibrium state. Removing the equilibrium conditions permits a range of new phenomena. As can be seen from Figure 10.2, very regular oscillatory behavior is possible. The two-dimensional phase space representation of such oscillatory behavior is shown in Figure 10.3. The attractor, in this case, is a simple cycle. Conditions can exist where the same oscillatory behavior is seen regardless of the initial conditions. Such oscillations are said to occur in a limit cycle (see Figure 10.3). With a slight change in a parameter, seemingly random patterns can be generated. However, these patterns are not random at all. Rather, they are chaotic.

In contrast to the behavior of random systems and of systems with limit cycles, chaotic processes are characterized by an extreme sensitivity to initial conditions.

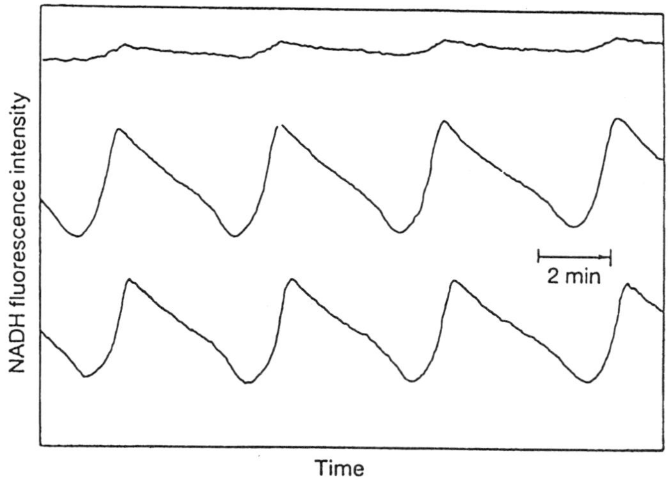

FIGURE 10.2 Example of sustained oscillations in a biochemical reaction. Oscillations in the nicotinamide adenine dinucleotide (NADH) NADH concentration in glycolysis are monitored by fluorescence changes. (From Hess and Boiteux, 1971. Reprinted by permission of the Biophysical Society.)

Trajectories that originate at points separated by minute concentration differences will ultimately diverge into different regions of phase space. This divergence is exponential in time and, consequently, is very fast. This divergence is illustrated in the trajectory in Figure 10.4. Such a condition creates a paradoxical situation. Chaotic systems are fully deterministic, yet one cannot experimentally reproduce a given time course. It is simply impossible to recreate the precise initial condition, regardless of the accuracy of the instrumentation. A second unusual condition results from this extreme sensitivity. One cannot predict the long-time behavior of this fully deterministic system. This is merely because the initial conditions cannot be represented accurately by finite numbers. We can continue to add more and more significant figures to specify our initial conditions, yet the exponential divergence will always catch up and prevent accurate long-time prediction.

Another important characteristic of a chaotic system is the topography of the attractor. Chaotic attractors are often fractals, and are called "strange attractors." A strange attractor is shown in Figure 10.4. When the trajectory of a chaotic system fills phase space it occupies the strange attractor. This is in sharp contrast to a random system which will uniformly occupy phase space. Thus, random and chaotic systems often have different topographies in phase space and this difference can be determined by measuring the fractal dimension. Note that the topography of the attractor is not a foolproof means of establishing chaos. There are situations where a chaotic system can have a uniform phase space and a random system can give a strange attractor. Barring these unusual cases, a random system typically will have an integer dimension equal to the dimensionality of phase space. A chaotic system will have a fractal dimension that is less than the dimensionality of phase space.

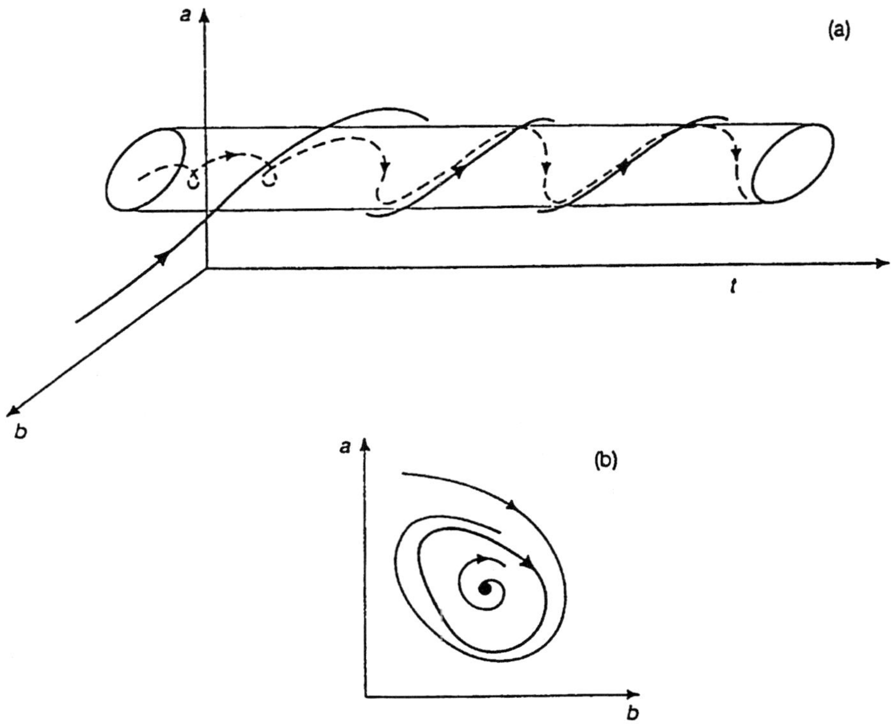

FIGURE 10.3 A phase space representation of the evolution of a chemical reaction in an open system that shows limit cycle behavior. Coordinates *a* and *b* are as in Figure 10.1. Trajectories are "attracted" to the limit cycle regardless of initial conditions. (a) The trajectory of the reaction coordinate over time; (b) a cross-section showing how all trajectories move into the limit cycle. (From Scott, 1991. Reprinted by permission.)

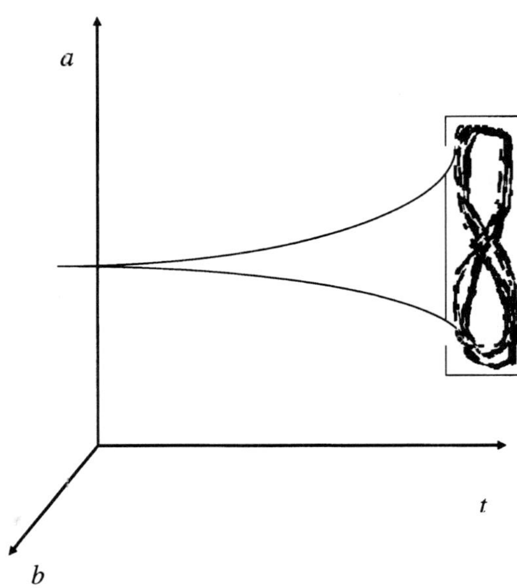

FIGURE 10.4 A phase space representation of the evolution of a chemical reaction that shows chaos. Trajectories diverge rapidly from each other and are extremely sensitive to initial conditions. Nevertheless, they will all migrate toward the "strange attractor," a fractal structure in phase space.

10.1.3 Origins of Chaos

Given an observation such as in Figure 10.2, how does one know if it is chaotic or random? The answer to such a question has a direct bearing on whether the underlying nature of the observed phenomena is stochastic or deterministic. Most of the mathematical work on chaos has focused on two different types of deterministic system: those generated by low-dimensional difference equations, and those that involve coupled nonlinear differential equations. Difference equations have numerous applications in biological and ecological problems, as they are ideal for treating seasonal phenomena. A common example of such an equation is the logistic equation (May, 1976):

$$x_{n+1} = ax_n\,(1 - x_n) \tag{10.1}$$

where x is a variable (such as the number of organisms in a population) in the nth generation, and a is a proportionality constant known as a control parameter. The logistic equation has been studied extensively and its chaotic properties are well understood. It is an example of a process that gives a "period-doubling route to chaos." Equation 10.1 is amazing in that such a simple equation can lead to such complex behavior. Yet, in its differential form, the logistic equation is exactly integrable and does not exhibit pathological behavior. This is an interesting observation that differential equations that are integrable and well behaved will have difference-equation counterparts that are chaotic.

Difference equations such as Eq. 10.1 are generally not found in chemical kinetic applications. The exception is instances of reactions in very small systems where differential equations do not apply. Reactions in small volumes often occur in biological systems as a result of the compartmentalization of the cell. Also, processes involved with gene regulation will also involve an extremely small number of species. There has been very little work on the difference equations of biochemical processes. Most chemical and biochemical investigations have focused on macroscopic systems that follow conventional mass action rate laws. Such systems that achieve chaos are governed by coupled, nonlinear differential equations. The mathematical conditions required for oscillatory behavior and chaos have been well established. For instance, a coupled linear two variable system can show oscillations, but not chaos. To achieve such oscillations a nonlinear term such as a cubic or exponential term must be present. A quadratic term in a two-variable system is not sufficient to produce oscillations.

The conditions for chaotic behavior are more stringent than for oscillatory behavior. The following rate equations show the simplest chaotic system investigated to date (Rössler, 1976):

$$\frac{dx}{dt} = -(y + z) \tag{10.2a}$$

$$\frac{dy}{dt} = x + ay \tag{10.2b}$$

$$\frac{dz}{dt} = b + xz - cz \tag{10.2c}$$

where x, y and z are variables, and a, b and c are constants. A minimum of three variables is required to achieve chaos. As seen in the above example, quadratic terms in a three-variable system are now sufficient to produce oscillations and chaos. Unfortunately, the coupled equations in Eq. 10.2 are not appropriate for a chemical reaction mechanism and much of the behavior of the trajectories involve negative values of the variables, i.e., negative rate constants. More complicated equations are required for a chemical mechanism. One such scheme that is compatible with chemical kinetics (Willamowski and Rössler, 1980) is:

$$\begin{aligned} &A_1 + X \leftrightarrow 2X \\ &X + Y \leftrightarrow 2Y \\ &A_5 + Y \leftrightarrow A_2 \\ &X + Z \leftrightarrow A_3 \\ &A_4 + Z \leftrightarrow 2Z \end{aligned} \tag{10.3}$$

where X, Y, and Z are chemical species that vary, and A_i are species in such high concentrations that they are essentially constants. Note that chaos is created with three variables, with the only nonlinearity arising from quadratic steps. The initial two steps and the final step in the mechanism are autocatalytic. This is a common feature in reactions that exhibit chaos. A feedback mechanism with an associated nonlinearity will allow for chaotic behavior.

Numerical computations on differential equations such as in Eq. 10.2 or Eq. 10.3 show how chaotic behavior arises from a specific mechanism. Such calculations have been very powerful in ascertaining the conditions required to produce chaos. An experimentalist is often concerned with the inverse problem: For a given time course of an experimental variable, how does one distinguish chaos from random noise? In observations where parameters are readily varied, perhaps the easiest way to establish chaos is to follow the example given in Figure 10.5. The "period doubling route to chaos" is a much studied process by which chaos arises. If an experimental parameter can be slightly altered and send the system first into periodic oscillations and then into seemingly random oscillations, then one has strong evidence of a chaotic system. This extreme sensitivity to experimental conditions is illustrated in Figure 10.5 for a reaction catalyzed by the enzyme peroxidase. In many physiological settings, the crucial system parameters are either unknown or are not readily manipulated. In such instances, observation of a time series is made and one must resort to statistical methods to establish chaos.

Time series in chaotic systems may appear random to the eye, but are actually quite different from and more complicated than a random system. This is because the time series reflects the underlying strange attractor. Conventional statistical methods have been used to distinguish random and chaotic time series. These include power spectrum analysis and autocorrelation techniques. For low-dimensional systems, one can attempt to reconstruct the phase space and to measure the fractal dimension of the

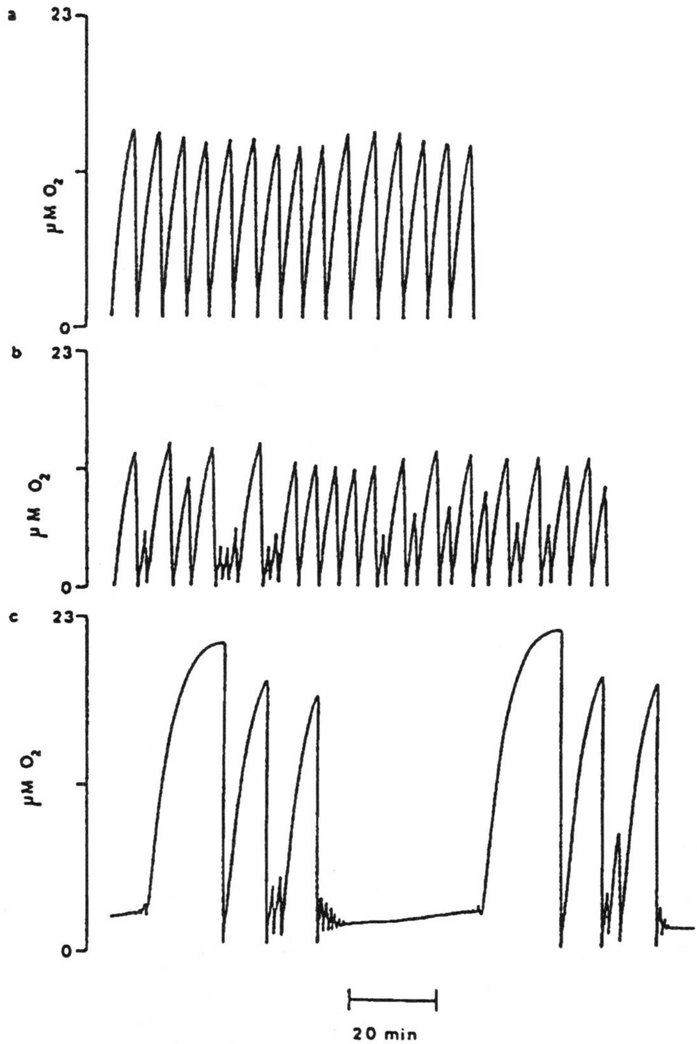

FIGURE 10.5 Experimental results showing chaos in the peroxidase reaction. Slight changes in reaction conditions create very different responses. The change in oxygen concentration versus time is shown. Enzyme concentration: (a) 0.90 μM, (b) 0.55 μM, (c) 0.45 μM. (From Olsen and Degn, 1977. Reprinted by permission of Nature Macmillan Magazines Ltd.)

strange attractor. This approach emphasizes the fractal nature of chaos and is considered here. Reconstructing phase space is not trivial, and there can be a number approaches as well as pitfalls (see Bassingthwaighte et al., 1994). For a time series or profile, as shown in Figure 10.2 or 10.5, one has only a single parameter. Computationally, one could calculate the parameter and its time derivative and create a two-dimensional phase space from these two parameters. This would provide a space that is analogous to the momentum-position phase space in classical statistical

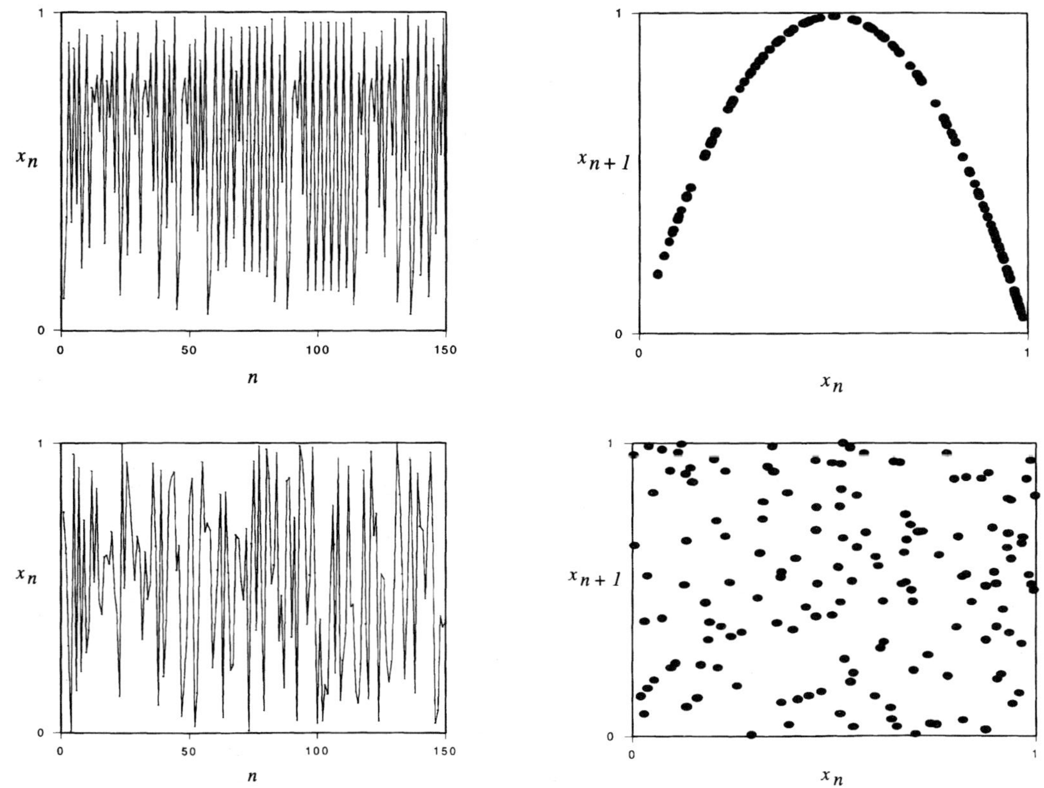

FIGURE 10.6 Analysis of a time series using delay plots. (Bottom left) A sequence of random numbers. (Bottom right) The resulting delay plot shows a uniform distribution. (Top left) The sequence of numbers generated by the logistic map and the resulting delay plot (top right). Chaotic behavior gives a delay plot with nonuniform distribution of points. This distribution can be characterized by a box-counting dimension. (Figure adapted from Figure 7.9 and Figure 7.10 of Bassingthwaighte et al., 1994.)

mechanics. Often it is easier to create a phase space from a delay plot, that is a plot of $x(t)$ versus $x(t + \Delta t)$ or of x_n versus x_{n+j}. Such a plot is shown in Figure 10.6 for a random set of points and for points generated by the logistic map (Eq. 10.1). The main problem in such analyses is how to choose the lag time, Δt or j. In practice, it is easy to try a number of different times and see which one generates an image that looks most like a chaotic attractor. For the logistic map in Eq. 10.1, $j = 1$ is the natural choice, and this attractor is shown in Figure 10.6. The random data set will show a uniformly distributed delay plot, regardless of the delay time.

Once a collection of points has been generated from the delay plot, the fractal dimension of the image can be determined. There are a number of different dimensions and algorithms that can be used for this purpose. It has been argued that the simplest and most useful one is the box-counting dimension described in Chapter 1 (Bassingthwaighte et al., 1994). With knowledge of the fractal dimension of strange attractor, one can return to the differential equations and try to generate a mechanism that will have an attractor of that dimension. The kineticist then returns to the old problem of sorting through a number of different mechanisms to find the one most consistent with the data.

10.2 Theoretical Models of Enzymatic Chaos

In the previous section, a number of necessary conditions required to achieve chemical chaos were described. We now focus this discussion on enzyme systems. As a minimal requirement for chaos, the system must be open and have nonlinear feedback. Indeed, feedback mechanisms are central to the control of metabolic processes at both the global and local level. Most metabolic reaction sequences have a number of control points. These commonly occur at rate-limiting steps (often referred to as the committed step) and before and after branching steps. There are a number of ways of regulating enzymes at this point. These include allosterism, covalent modification, and genetic control of enzyme expression. To date, allosteric mechanisms have received the most attention in model explorations of chaotic dynamics. This is perhaps due to the extensive work on glycolytic oscillations, where the primary source of nonlinearity is the allosteric enzyme phosphofructokinase. The focus of the discussion in this section is simple allosteric control models that result in chaotic behavior.

10.2.1 Mechanisms involving Michaelis–Menten Enzymes

To begin, we consider the difference between a simple, Michaelis–Menten enzyme and an allosteric enzyme. Many of the enzymes that are not at metabolic control points follow the Michaelis–Menten scheme. For a single substrate reaction, this scheme is:

$$E + S \leftrightarrow ES \rightarrow E + P \tag{10.4}$$

where E is the enzyme, S is the substrate (or reactant) and P is the product. The species ES is the enzyme with the substrate bound at the catalytic site. The

Michaelis–Menten equation relates the steady state velocity of the reaction, v, to the substrate concentration. It makes a number of assumptions, including the steady state approximation for ES which states that $d[\mathrm{ES}]/dt = 0$. The steady state velocity is given by:

$$v_{\mathrm{MM}} = \frac{d[\mathrm{S}]}{dt} = \frac{V_{max}[\mathrm{S}]}{K_{\mathrm{M}} + [\mathrm{S}]} \tag{10.5}$$

where the maximum velocity, V_{max}, and the Michaelis constant, K_{M}, are constants containing the rate constants of the mechanism. The maximum velocity is proportional to the enzyme concentration. The Michaelis–Menten equation is nonlinear, showing hyperbolic behavior of the velocity with respect to the substrate concentration. If a Michaelis enzyme is put into an open system where a source produced substrate and a sink eliminated product, neither oscillatory nor chaotic behavior is observed. While this system is nonlinear and open, it does not have feedback or enough variables to support complex behavior.

If a product activation step is included along with a Michaelis sink in Eq. 10.4, there is now enough nonlinearity in the system to support simple oscillations (Goldbeter and Dupont, 1990). This mechanism is given by:

$$\begin{aligned}
&\mathrm{A} \rightarrow \mathrm{S} \\
&\mathrm{E}_1 + \mathrm{S} \leftrightarrow \mathrm{E}_1\mathrm{S} \rightarrow \mathrm{E}_1 + \mathrm{P} \\
&\mathrm{E}_1 + \mathrm{P} \leftrightarrow \mathrm{E}_1\mathrm{P} \\
&\mathrm{E}_1\mathrm{P} + \mathrm{S} \leftrightarrow \mathrm{E}_1\ \mathrm{PS} \rightarrow \mathrm{E}_1\ \mathrm{P} + \mathrm{P} \\
&\mathrm{E}_2 + \mathrm{P} \leftrightarrow \mathrm{E}_2\ \mathrm{P} \rightarrow \mathrm{E}_2 + \mathrm{Q} \\
&\mathrm{Q} \rightarrow \mathrm{R}
\end{aligned} \tag{10.6}$$

The first and last steps are the source and sink, respectively. Two enzyme species, E_1 and E_2, are involved, and two products, P and Q, are formed. The first enzyme obeys a normal Michaelis–Menten mechanism (see second step), but can bind product, P (third step), at a site other than the catalytic site. The occupation of the regulatory site enhances the enzymatic activity via its effect on the rate constants in the fourth step. The third step provides a positive feedback. The second enzyme serves as the Michaelis sink and is an additional source of nonlinearity. The mechanism of Eq. 10.6 allows for oscillations, but does not support chaos. For chaos to occur, a stronger nonlinearity must be present. This can be accomplished with allosteric enzymes.

10.2.2 Mechanisms involving Allosteric Enzymes

Allosteric enzymes are multisubunit enzymes that show cooperativity as a result of interaction between enzymes within the complex. This cooperativity results in a sigmoidal response of the initial velocity to substrate concentration. Figure 10.7 compares the behavior of an allosteric enzyme with the hyperbolic behavior of a

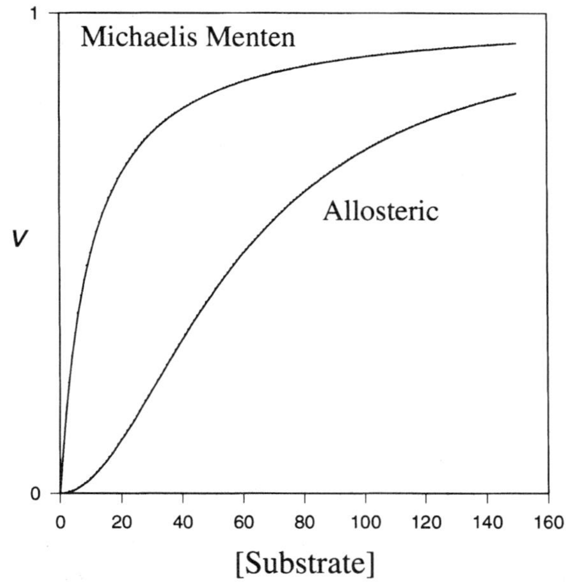

FIGURE 10.7 Comparison of the behavior of a Michaelis enzyme and an allosteric enzyme. Plot of the steady state rate of catalysis versus substrate concentration. The Michaelis enzyme shows simple, hyperbolic behavior. The allosteric enzyme shows sigmoidal behavior. Allosteric enzymes introduce a greater nonlinearity into a reaction sequence.

Michaelis–Menten enzyme. The cooperativity of this response is reflected by the steepness of the curves at the inflection point. This cooperativity is affected by a number of factors, including the number of units in the complex and the binding of small molecule effectors at other sites. The term allosteric is derived from the Greek and means "other site." Allosteric enzymes serve as metabolic "on–off" switches. Enzymatic activity is controlled by the substrate concentration, being very low at low substrate concentration and switching to high activity over a narrow range of increasing substrate. The "other sites" are not catalytic, rather they have a regulatory role. They work by providing an independent way of shifting the equilibria between active and inactive forms.

There have been a number of models used to describe this cooperativity and two popular ones are referred to as the (Monod–Wyman–Changeaux) (MWC) and (Koshland–Nemethy–Filmer) (KNF) (for a review, see Neet, 1983). These models have very different mechanistic details and, for any given enzyme, one model may be favored over the other. The main point for our purposes is the nonlinearity inherent in the enzyme's response. Both models lead to a mathematical form that adequately describes the nonlinearites in the experimental data. The issue of the most appropriate description may be sidestepped in favor of a purely empirical description of the rate process. This is adequate for modeling the overall sequence of reactions.

The MWC model provides a particularly simple functional form for the steady state rate and is the one that is often considered in computational models. In the MWC model the enzyme is a complex of N units that can exist in two conformations: active and inactive. For historic reasons, these were designated relaxed, R, and taut, T, respectively. The equilibrium for the two conformations favors the taut or inactive state. Therefore, at low substrate concentrations most of the enzyme is inactive and the overall activity is low. Substrate can bind at active sites in both the R and T forms. However, it prefers the R form and is more effectively converted to product when bound to this form. These processes are shown in the following equations:

$$
\begin{aligned}
&\mathrm{R}_N \leftrightarrow \mathrm{T}_N \\
&\mathrm{R}_N\mathrm{S}_i + \mathrm{S} \leftrightarrow \mathrm{R}_N\mathrm{S}_{i+1} \rightarrow \mathrm{R}_N\mathrm{S}_i + \mathrm{P} \\
&\mathrm{T}_N\mathrm{S}_i + \mathrm{S} \leftrightarrow \mathrm{R}_N\mathrm{S}_{i+1} \rightarrow \mathrm{T}_N\mathrm{S}_i + \mathrm{P} \\
&\mathrm{R}_N\mathrm{P}_i + \mathrm{S} \leftrightarrow \mathrm{R}_\mathrm{N}\mathrm{P}_i\mathrm{S} \rightarrow \mathrm{R}_N\mathrm{P}_i + \mathrm{P}
\end{aligned}
\tag{10.7}
$$

where the ability of the T state to produce substrate, the third line in Eq. 10.7, is slower than the R state, second line in Eq. 10.7. The subscript i represents the number of S molecules bound to the complex. This number can range from 0 to N–1. Thus, there are N equations represented by the last three reactions in Eq. 10.7. Because the substrate prefers the R state, the presence of increasing amounts of substrates pulls the R–T equilibrium (first line in Eq. 10.7) to the left, creating more open R sites. This generates more active enzyme and the overall enzymatic activity rises sharply, causing the sigmoidal behavior shown in Figure 10.7. In addition to substrate effects, allosteric sites (as opposed to catalytic ones) can shift the R–T equilibrium as well. In Eq. 10.7, one sees product activation as a result of preferential binding of product to an allosteric site in the R form. This creates a positive feedback situation for the enzyme.

The MWC model makes a number of simplifying assumptions regarding the equivalence of binding sites in the R and T forms. This allows for a particularly compact rate expression. In Eq. 10.8 the steady state rate law is presented for conditions where S does not bind at all to the T state (the third line does not exist in Eq. 10.7). This gives:

$$
v_{allo}(\mathrm{E,S,P}) = V_{max}\,\frac{([\mathrm{S}]/K_\mathrm{S})(1 + [\mathrm{S}]/K_\mathrm{S})^{N-1}(1 + [\mathrm{P}]/K_\mathrm{P})}{L + (1 + [\mathrm{S}]/K_\mathrm{S})^N(1 + [\mathrm{P}]/K_\mathrm{P})^N}
\tag{10.8}
$$

where K_S and K_P are the dissociation constants for binding of substrate and product to the R form, respectively, and L is the equilibrium constant for the R form going to the T form. Equation 10.8 provides a much stronger source of nonlinearity than the Michaelis–Menten equation (Eq. 10.5) and consequently allows for a wider range of oscillatory and chaotic behavior.

Using this form of the MWC model for allosteric enzymes, a number of different open systems have been investigated (see Goldbeter, 1995). If we merely put an allosteric enzyme in a flow system where substrate is continually added and

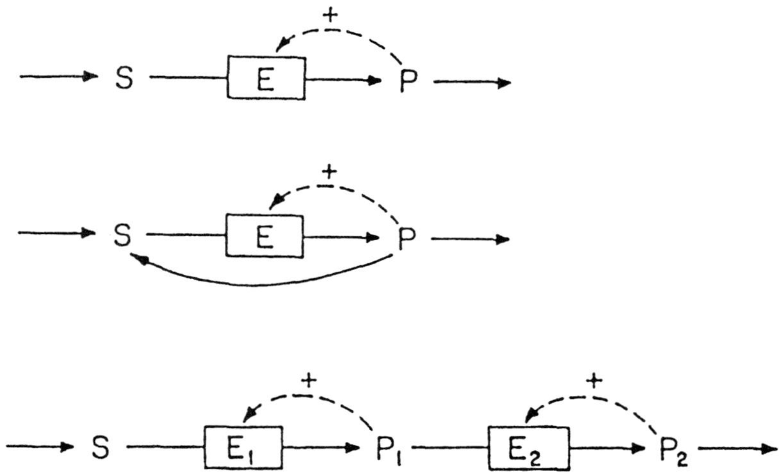

FIGURE 10.8 Models of enzymatic regulation that show complex behavior. E, Enzyme; S, substate; P, product. Solid arrows represent the reaction path and broken arrows indicate positive feedback on the enzyme. (Top) Product-activated allosteric enzyme supports simple oscillations. (Middle) Product-activated allosteric enzyme in which product can also be converted to substrate will permit birhythmic behavior. (Bottom) A succession of product-activated allosteric enzymes can show chaotic behavior. (Adapted from Goldbeter, 1992.)

product is continually removed, then the feedback and nonlinearity are sufficient to create oscillatory conditions. However, again there is not a sufficient number of variables to generate chaos. If two allosteric enzymes are considered in a sequential mechanism, a wide range of behavior is possible and chaos can be observed. These two mechanisms are represented schematically in Figure 10.8. The rate expressions for the two-enzyme mechanism are:

$$\frac{d[\mathrm{S}_1]}{dt} = v - v_{allo}(\mathrm{E}_1, \mathrm{S}_1, \mathrm{P}_1)$$

$$\frac{d[\mathrm{P}_1]}{dt} = v_{allo}(\mathrm{E}_1, \mathrm{S}_1, \mathrm{P}_1) - v_{allo}(\mathrm{E}_2, \mathrm{S}_1, \mathrm{P}_2) \tag{10.9}$$

$$\frac{d[\mathrm{P}_2]}{dt} = v_{allo}(\mathrm{E}_2, \mathrm{P}_1, \mathrm{P}_2) - k_\mathrm{S}[\mathrm{P}_2]$$

where Eq. 10.8 is used for the steady state rate for each allosteric enzyme, v is the rate of production of S_1 or the source rate, and k_s is the rate constant associated with the sink for the P_2 term. Each enzyme shows positive feedback from its product. In the investigation of the rate laws given Eq. 10.9, an N value of 2 was assumed (Decroly and Goldbeter, 1982).

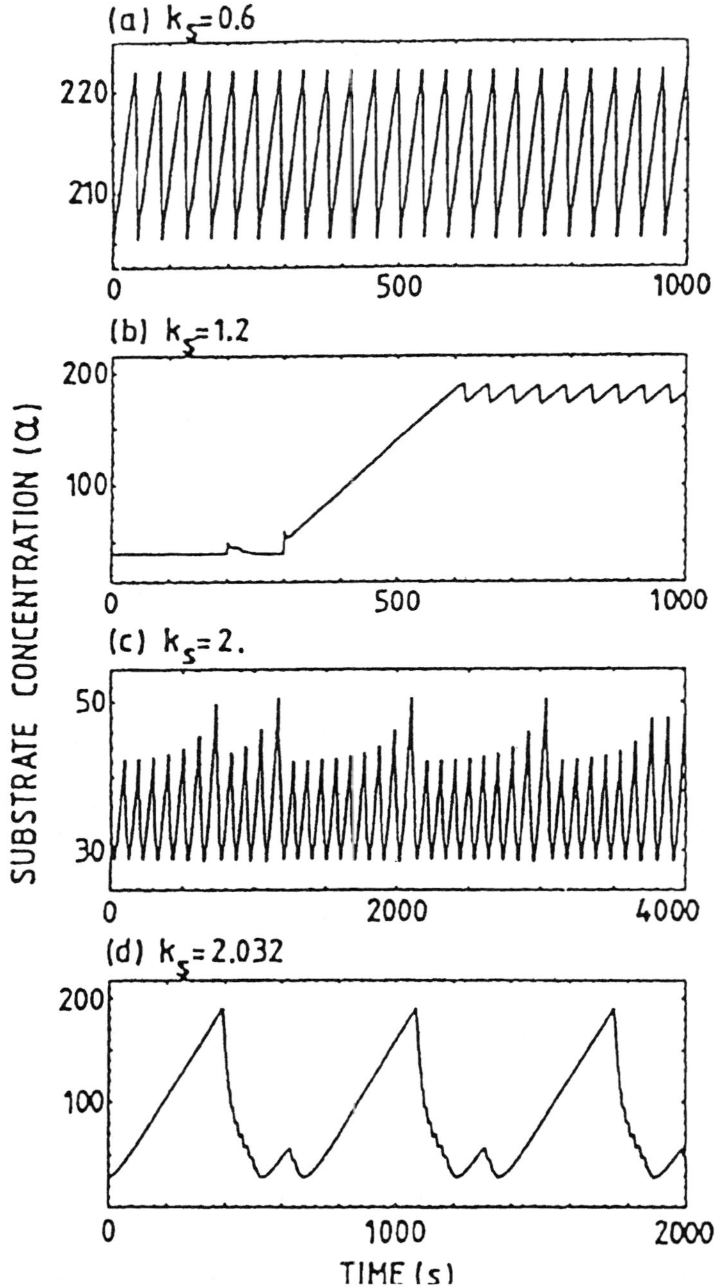

FIGURE 10.9 Using the mechanism shown at the bottom of Figure 10.8, a number of complex time courses are observed. Plots of substrate concentration versus time for different values of the rate of removal of product: (a) simple periodicity; (b) "hard" excitation; (c) chaotic; (d) complicated periodicity. (From Decroley and Goldbeter, 1982. Reprinted by permission of A. Goldbeter.)

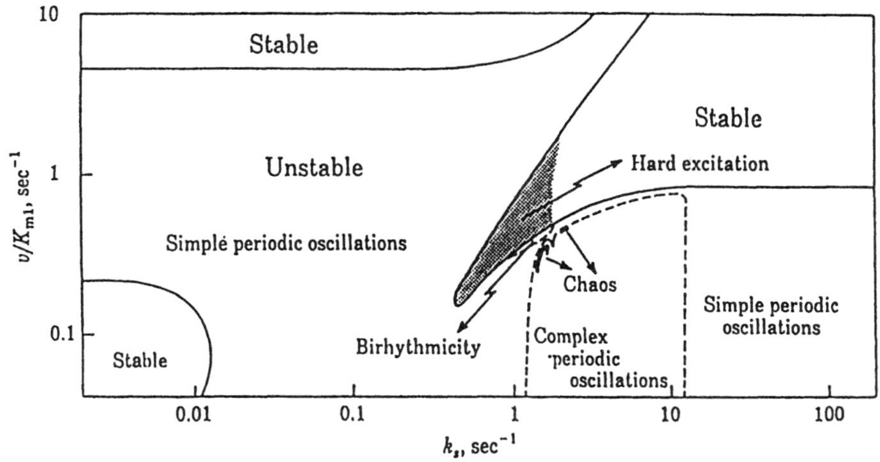

FIGURE 10.10 Plot showing regions of different behavior in parameter space. The mechanism shown at the bottom of Figure 10.8 is considered. *v* Flux rate of substrate; *k*, rate of removal of product. Domains were determined computationally. Chaos exists in only limited regions. (From Decroley and Goldbeter, 1982. Reprinted by permission of A. Goldbeter.)

The two enzyme mechanisms show a range of oscillatory and chaotic behavior. Figure 10.9 shows the time course of [S_1] as the sink rate constant is varied. Figure 10.9 gives examples of the different types of temporal behavior exhibited by the system. As can be seen, there is extreme sensitivity to changes in the sink rate constant, a hallmark of a chaotic system. This complex interplay of dynamic modes is well illustrated by the behavioral domains shown in Figure 10.10. In this figure, the behavior domains are shown in regions of a parameter space that consists of the source

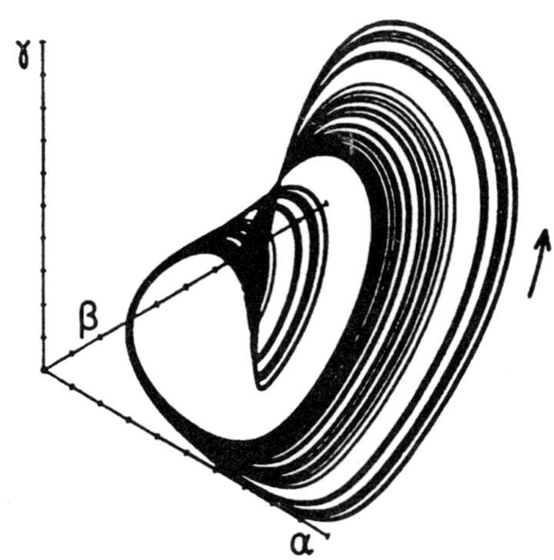

FIGURE 10.11 Strange attractor from the chaotic regime shown in Figure 10.10 (c). The phase space is in concentration coordinates of the system. (From Decroley and Goldbeter, 1982. Reprinted by permission of A. Goldbeter.)

velocity and the sink rate constant. The chaotic domain actually represents a small region of this parameter space. This suggests that a very specific set of conditions would be required to achieve chaos. Figure 10.11 shows the strange attractor of the chaotic regime. This trajectory of the time course in phase space (or concentration space) has a characteristic appearance for a chaotic system. Although not apparent from Figure 10.11, the attractor is a finely layered structure with scale-invariant regions.

10.3 Experimental Systems

From the previous theoretical discussion, it is seen that the conditions necessary for oscillations and chaos are not unusual in biochemical systems. Most metabolic pathways have control mechanisms that result in feedback and in kinetic nonlinearities. The feedback can be a result of both short- and long-range effects of a metabolic product. Such mechanisms provide a means for both local control and global integration of metabolic pathways. Nonlinearities commonly arise from allosteric enzymes, a popular motif in metabolic pathways. However, other sources of nonlinear kinetics exist, and these include zymogen activation, covalent modification, and channel gating. These latter effects have not been explored extensively in theoretical models.

Given these considerations, it would appear that metabolic systems provide ideal settings for the observation of chaotic dynamics. Despite this, there are a limited number of examples of enzymes that show oscillations. The two most studied cases, systems that involve the enzymes phosphofructokinase (PFK) and horseradish peroxidase, are considered here. PFK is responsible for oscillations in glycolysis, a central pathway in intermediary metabolism. The glycolytic pathway exhibits oscillations, but will not show chaos unless there is a periodic driving force. Horseradish peroxidase catalyzes the reduction of molecular oxygen to water. Peroxidase can show both oscillations and chaos under a wider range of conditions. Oscillations and complex kinetic behavior have also been studied in a number of cellular systems. These systems often involve signal transduction systems that are mediated by calcium fluxes (see Meyer and Stryer, 1991). While these will doubtless provide additional examples of chaotic behavior, their molecular origins are still controversial. In the present treatment, we focus on the better defined systems.

10.3.1 Glycolytic Oscillations

Glycolysis is a central process in the energy metabolism of the cell. It oxidizes glucose into two pyruvates. The process captures some of the energy of the "burning" of glucose in the form of energy-rich molecules such as adenosine triphosphate (ATP). ATP is the "energy currency" mediating the flow and utilization of energy between anabolic and catabolic pathways. As such, it is an ideal molecule for signaling the activation or deactivation of a given pathway. Thus ATP is used in the metabolism to exert a global control of energy production. The glycolytic pathway has a number of control points, the most important of which is at the enzyme PFK. PFK is an especially appropriate control site because it is at the committed step in the path. It catalyzes the essentially irreversible phosphorylation by ATP of fructose 6-phosphate (F6P) to fructrose 1,6-bisphosphate (F16P). Additionally, the reaction is a slow step in the path, and consequently is a more important point of regulation.

The glycolytic pathway provided the first observations of oscillations in biochemical systems (Ghosh and Chance, 1964). Many of these early studies involved the observation of sustained oscillations in the concentration of nicotinamide adenine dinucleotide (NADH), a product in glycolytic steps that follow the PFK step. This had the advantage of providing a useful fluorescence indicator. Figure 10.2 shows an example of such oscillations. Glycolytic oscillations have been observed in a wide variety of physiological settings (for a comprehensive bibliography, see Scott (1991)). Some progress has been made with reconstituted systems prepared from purified enzymes (Hess and Boiteux, 1971; Eschrich et al., 1983).

Glycolytic oscillations were first modeled with a mechanism involving product activation of PFK by F16P (Higgins, 1964). This product was subsequently removed with a Michaelis enzyme. Thus, this mechanism has two sources of nonlinearity: the activation of an allosteric enzyme, and the Michaelis removal of product. The mechanism supports limit cycle oscillations, but not chaos. A more sophisticated model that includes ADP activation was developed by Sel'kov (1968). This process involves the binding of adenosine diphosphate (ADP) at allosteric sites in the enzyme. This model is given by:

$$
\begin{aligned}
&\mathrm{X} \rightarrow \mathrm{ATP} \\
&\mathrm{ATP} + \mathrm{E(ADP)_2} \leftrightarrow \mathrm{(ATP)E(ADP)_2} \\
&\mathrm{(ATP)E(ADP)_2} \rightarrow \mathrm{ADP} + \mathrm{E(ADP)_2} \\
&\mathrm{E} + 2\,\mathrm{ADP} \leftrightarrow \mathrm{E(ADP)_2} \\
&\mathrm{ADP} \rightarrow \mathrm{Y}
\end{aligned} \tag{10.10}
$$

PFK is an allosteric enzyme consisting of two identical subunits. This is the motivation for having two binding sites for ADP. The production of ATP from X is the source term, and ADP is removed to Y as the sink. From the steady state enzyme kinetics of this mechanism, the following differential equations are obtained where the substrate x, and product, y, are presented in a dimensionless form:

$$\frac{dx}{dt} = \mu - f(x,y) \tag{10.11a}$$

$$\frac{dy}{dt} = f(x,y) - y \tag{10.11b}$$

with

$$f(x,y) = \frac{xy^2}{1 + y^2\,(1 + \alpha x)} \tag{10.11c}$$

where μ is the rate of ATP production from the source, and α is a constant containing rate constants for the mechanism. To create a dimensionless form, the

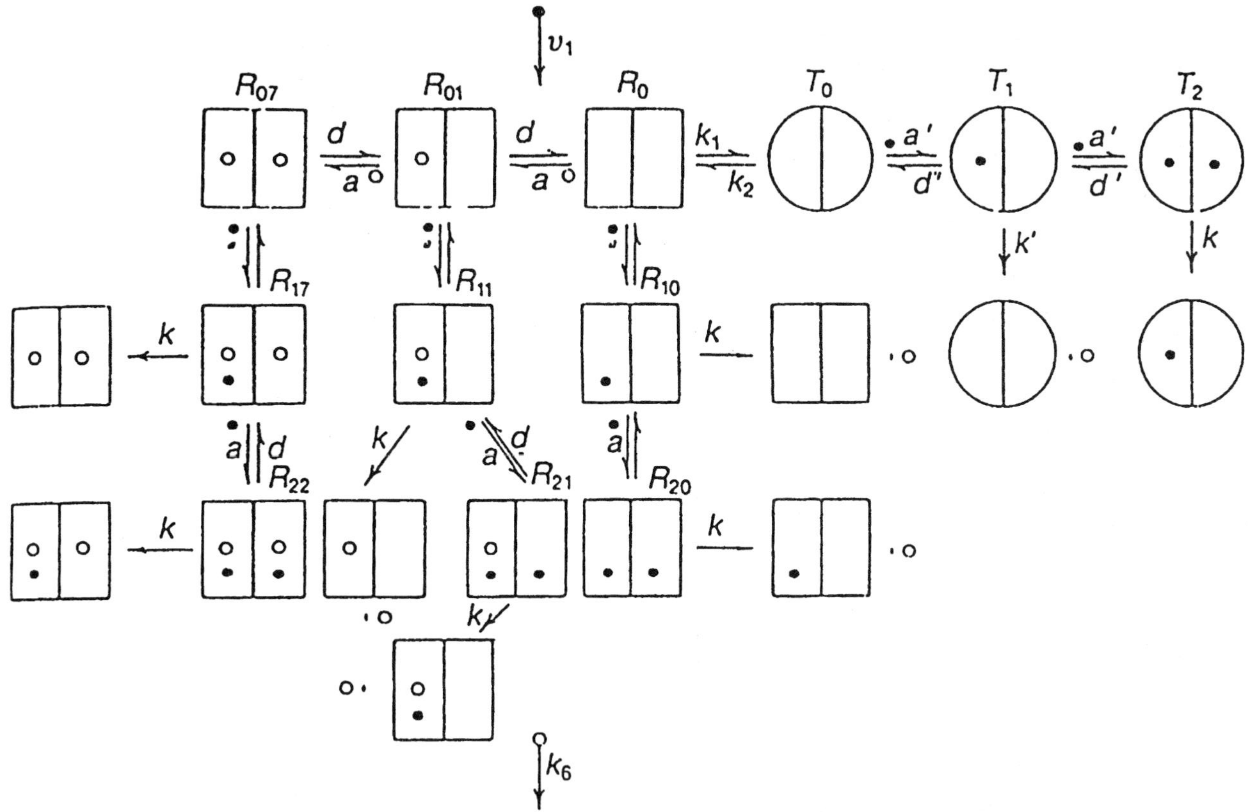

FIGURE 10.12 Equilibria and states for an allosteric model of phosphofructokinase. The model of Goldbeter and Levefer (1972) is used. Large circles represent active state conformations of the enzyme with low affinity for the substates. Squares represent inactive states with high substrate affinity. Small dark circles are substrate molecules, and small open circles are products. (From Goldbeter and Lefever, 1972. Reprinted by permission of the Biophysical Society.)

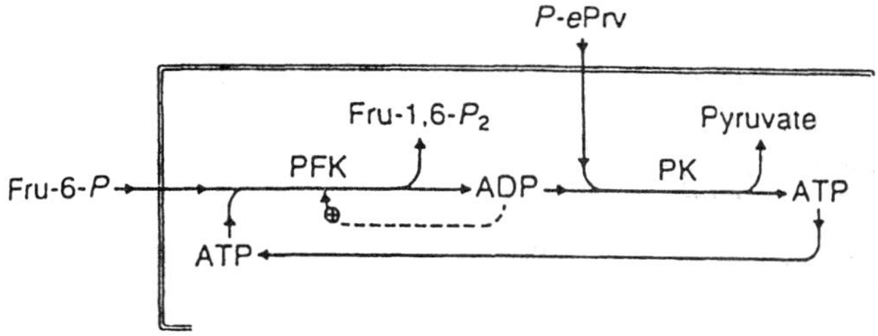

FIGURE 10.13 General metabolic reaction scheme of the Markus – Hess model for glycolysis. ADP feedback to the enzyme phosphofructokinase (PFK) results in oscillatory behavior of the Markus–Hess model for glycolysis. ADP feedback to the enzyme phosphofructokinase (PFK) results in oscillatory behavior. (Adapted from Markus and Hess, 1984.)

reactant concentrations are divided by an appropriately chosen quantity of rate constants and steady state concentrations. Such transformations do not change the behavior of the system, but rather create equations that are less cumbersome and more easily manipulated. For small values of μ, Eq. 10.11c can be approximated by the simple cubic form $f(x,y) = xy^2$ and fits into general models of cubic autocatalysis (see Scott, 1991). This mechanism shows limit cycle behavior. Unfortunately, it has some unrealistic physical features, such as unbounded growth in x for some conditions, that make it unacceptable.

A more general form of the allosteric model has been used by Goldbeter and Lefever (1972). This model is illustrated in Figure 10.12. Again, the crucial variables are the ATP and ADP concentrations. As can be seen in the figure, a much more complicated mechanism is considered, involving equilibrium processes for both the active and inactive states. The key equilibrium is the interconversion of these two forms of the enzyme when the enzyme has no bound activator (ADP). Subsequent binding of any species to either of these forms shifts the equilibrium further in its respective direction. The product (ADP activator) is assumed to bind only to the active form, but the substrate (ATP) can bind to either form. However, the catalytic rates for the inactive form are lower than for the active form.

This more complicated mechanism can again be reduced to a simple form using a steady state analysis for the enzymatic species. The functional form of Eq. 10.11 is retained, but now $f(x,y)$ shows behavior more akin to Eq. 10.8. Although this is a complicated nonlinear form, it still consists of only two coupled differential equations. Consequently, it will not support behavior more complicated than sustained oscillations. To achieve chaos, the system must be periodically forced through a variation of the parameter μ. This general theoretical result is observed experimentally as well. Many qualitative features of this model are in agreement with experiment, and regions in parameter space have been found that show birhythmicity if the product (ADP) is recycled into the substrate (ATP) by a regenerating system (Moran and Goldbeter, 1984). While this may seem like a convoluted set of conditions, it indeed does occur in the glycolytic pathway.

Markus and Hess (1984) have produced a different model (see Figure 10.13). This involves one feedback process, again through ADP, and two enzymatic species, PFK and the enzyme pyruvate kinase (PK). A key intermediate in the pathway is the energy-rich molecule, phosphoenolpyruvate (PEP). Pyruvate kinase is the final enzyme in the glycolytic pathway, and is also an allosteric enzyme and a control point. This system couples two different allosteric enzymes. Despite this, one again only observes simple oscillations unless the system is periodically forced. The equations for this system, including periodic driving forces, are:

$$\frac{d[\mathrm{ADP}]}{dt} = -\frac{d[\mathrm{ATP}]}{dt} = v_{allo}(\mathrm{PFK}) - v_{\mathrm{MM}}(\mathrm{PK})$$

$$\frac{d[\mathrm{F6P}]}{dt} = v_{\mathrm{IN}} + A \sin(\omega t) - v_{allo}(\mathrm{PFK}) \tag{10.12}$$

$$\frac{d[\mathrm{PEP}]}{dt} = v_{\mathrm{IN}} + A \sin(\omega t) - v_{mm}(\mathrm{PK})$$

where generalized forms of Eqs 10.5 and 10.8 are used for PK and PFK, respectively, and the flux term, v_{IN}, has a periodic forcing term added to it, $A \sin(\omega t)$. Again, this model can be reduced to two coupled nonlinear equations with dimensionless parameters. This system of equations shows a rich phenomenology with a variety of periodic and chaotic regimes. A qualitative correspondence exists between experimental and theoretical data for various regions of chaos at different forcing frequencies. Without the forcing function this glycolytic mechanism will not support chaos.

10.3.2 Horseradish Peroxidase

Unlike the glycolytic enzymes, horseradish peroxidase shows chaos even without periodic perturbations. The peroxidase reaction is also different from glycolysis in that allosterism does not play a crucial role. Rather, it is the complex interplay of redox states that create a system capable of supporting chaos. This enzyme has a long history, and chaos was actually observed in it before the Belousav–Zhabotinsky reaction. Peroxidase catalysis is the oxidation of electron donors by molecular oxygen. When this reaction takes place in a flow system with reduced NADH, the concentrations of reactants (oxygen and NADH) as well as some of the enzyme intermediates can be seen to oscillate with periods ranging from minutes to about an hour, depending on the experimental conditions. Aperiodic oscillations resembling random fluctuations are also observed for some conditions, and these irregular oscillations were shown to be chaotic in a study by Olsen and Degn (1977).

The peroxidase is the first step in a sequence of reactions in plants that eventually culminates in the production of lignin. The enzyme is also involved in the dark reactions of photosynthesis. Currently, the physiological significance of oscillations and chaos in the peroxidase flow reaction remains unclear. However, the existence of such chaotic behavior does suggest a number of interesting physiological roles and possibilities. An obvious question is how pervasive is chaos in physiological settings. If chaos can occur in a single enzyme reaction, then is it inevitable in complex, interconnected metabolic systems? Does enzymatic chaos provide a manageable control system? Is it beneficial to have such chaotic processes? Such questions have only recently been formulated, let alone answered. The final section of this chapter speculates on the significance of chaos in biological systems.

The overall reaction catalyzed by peroxidase is:

$$2\mathrm{NADH} + 2\mathrm{H}^+ + \mathrm{O}_2 \rightarrow 2\mathrm{NAD}^+ + 2\mathrm{H}_2\mathrm{O} \tag{10.13}$$

The simplicity of the above reaction belies the complexity of the mechanism. The enzyme itself will exist in five different forms: ferrous peroxidase, ferric peroxidase, and three other forms known as compounds I, II, and III. The native form of the enzyme is the ferric state. Compounds I, II, and III have oxidizing equivalents of +4, +5, and +6, respectively. They are critical intermediates in the mechanism leading to oscillations. The kinetics of reactions of III, in particular, are of interest in understanding the oscillations.

Another critical species in the oscillatory reaction mechanism (both regular and chaotic) is the free radical, NAD•. This species is thought to be produced autocatalytically by a sequence of reactions involving NADH, molecular oxygen, hydrogen peroxide and the superoxide radical ($O_2^{-\bullet}$). Several of the enzyme species are also implicated in this process. Autocatalysis in a reaction mechanism which also involves a termination pathway for the autocatalytic species is known to be a sufficient explanation for the existence of oscillatory behavior. Whether or not this structure exists in the peroxidase reaction remains to be seen.

Yamazaki et al. (1965) discovered that the peroxidase-catalyzed oxidation of NADH occurs via damped oscillations when oxygen is continuously bubbled into the reaction mixture. A very similar system where the oxygen is introduced through a gas–liquid interface was shown to exhibit bistability, i.e., the existence of two simultaneously stable steady states for the same oxygen concentration in the gas phase. Brief perturbations in the oxygen concentration in the gas phase could induce reversible switches from one steady state to the other. The bistability phenomenon is thought to be due to inhibition of the enzyme by O_2, and evidence was presented that the active enzyme intermediate is identical to compound III. Degn (1968) later showed that damped oscillations could also be obtained using the substrates dihydroxyfumaric acid and indoleacetic acid instead of NADH. The oscillations were accompanied by a chemiluminescence that was ascribed to free-radical intermediates. This was taken as evidence for autocatalysis in the reaction mechanism.

Sustained oscillations in the peroxidase reaction were first obtained (Nakamura et al., 1969) using NADPH as the substrate. NADPH was regenerated from the oxidized form ($NADP^+$) by glucose 6-phosphate and glucose-6-phosphate dehydrogenase. Sustained oscillations were found only when the modifiers 2,4-dichlorophenol (DCP) and methylene blue were present. Olsen and Degan later reported

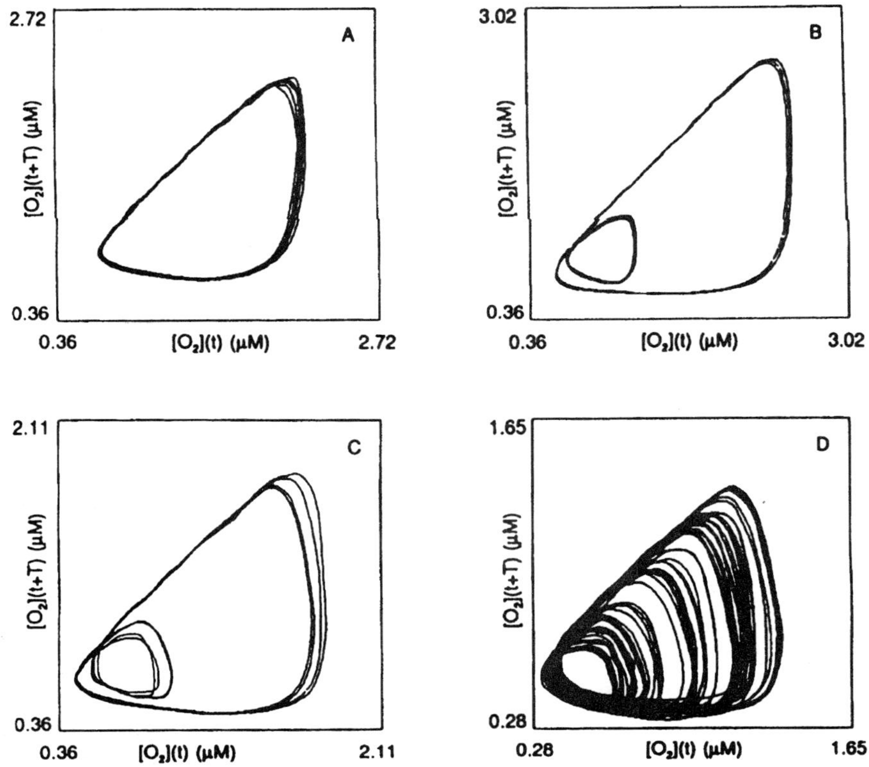

FIGURE 10.14 Delay plots determined from experimental data on the time series of oxygen concentration for the peroxidase reaction. Plots are for different oxygen flow rates, plot (A) being for the highest and plot (D) for the lowest flow. Plots show period one (A), period two (B), period four (C) and chaotic (D) behavior. (From Geest et al., 1992. Reprinted by permission of American Chemical Society.)

sustained oscillations with a constant infusion of NADH, thus demonstrating that glucose-6-phosphate dehydrogenase was unnecessary to sustain oscillations. However, these authors found that the presence of DCP (and, perhaps, methylene blue) was critical. Olsen and Degn (1978) provided further evidence that oscillatory behavior is more likely a result of autocatalysis than substrate inhibition by oxygen.

Olsen and Degn (1977) observed that the waveform of the peroxidase oscillations depends strongly on the concentration of the enzyme. Simple periodic oscillations with a period of about 5 minutes were obtained at enzyme concentrations of about 1 μM. Below 0.5 μM, bursting oscillations with periods of up to 60 minutes were seen. The oscillations were aperiodic and irregular at enzyme concentrations slightly above 0.5 μM. These results are illustrated in Figure 10.5. Analysis of a delay plot showed that a period-three cycle existed for certain initial conditions. A theorem of Li and Yorke (1975) states that the existence of a period-three oscillation implies chaos, so it can be argued that the irregular oscillations were in fact chaotic.

TABLE 10.1 Fractal dimension, d_f, of the chaotic attractor in simulated and experimental data from the peroxidase reaction (data from Larter et al., 1993)

Species measured	No. of Oscillations	d_f
Peroxidase III	46	2.45
O_2	48	2.64
Computer simulations	> 2000	2.09–2.26

For a considerable time, the observation of period-three oscillations provided only indirect evidence that the peroxidase reaction could support chaos. This experimental evidence was supported by numerical simulations that yielded delay plots very similar to the experimental ones. Recent work by Geest et al. (1992) has demonstrated that chaos in the peroxidase reaction arises by the well-known period-doubling route as well. The attractor in the delay plots for this range is shown in Figure 10.14. A similar period-doubling route to chaos has been predicted from simulations with a detailed model of the reactions for variations in the enzyme concentration. This result has yet to be verified experimentally.

The fractal information dimension of the attractor was computed in a later study (Geest et al., 1992) and compared with those obtained from computer models of the reaction. These results are shown in Table 10.1. The fractal dimension measures the information needed to specify the position of a point on an attractor to within a given accuracy, and hence in some sense expresses the complexity of the motion. Simple types of motion have low-integer dimensions. For example, the dimension of a steady state is zero, whereas a limit cycle oscillation has a dimension of 1. Chaotic motions usually have finite but noninteger dimensions. The dimension of 2.45–2.7 obtained for the experimental data is indicative of chaotic motion and is also strikingly in accordance with theoretical values previously predicted from a simple model of the peroxidase reaction. These experimental results thus confirm that the irregular oscillations observed by Olsen and Degn (1978) are indeed chaotic. Recent results have revealed the existence of quasi-periodic oscillations in the peroxidase reaction using the same experimental configuration as Olsen and Degn. Whether or not this quasi-periodicity is associated with a different route to chaos is presently unknown. Both period doubling and quasi-periodicity are well-known routes to chaos, and both have been found to be associated with chaos in models of the peroxidase reactions.

10.4 Physiological Advantages of Chaos

Since Claude Bernard introduced the concept in 1878, homeostasis has become ingrained in the physiological literature. The basic idea is that an organism must maintain some sort of balance or steady state via a series of control and feedback mechanisms. This feedback confers a stability to the organism. An example might be the sugar concentration in the blood. The organism can suffer dietary swings, yet this quantity is reasonably insensitive to such changes. Homeostasis provides a framework with which to view a range of physiological data (Cannon, 1929). Recent, controversial work on chaos has challenged the underlying tenets of "homeostasis" in favor of the new concept of "homeodynamics" (see Bassingthwaighte et al., 1994). Homeodynamics views organisms as dynamical systems that must respond with great flexibility to challenges from the environment. Complex and even unstable systems can respond more quickly and more effectively to such changes (see Goldberger et al., 1990). While homeostatic systems are rigidly constrained to steady state and concomitant dynamical paths, the homeodynamic view offers a system that is free to move about phase space on a strange attractor. The strange attractor provides the stability and at the same time allows the system to respond dynamically. These ideas are not without controversy.

As we have seen in this chapter, the conditions for chaos in biochemical reactions are not stringent. Certainly such conditions could exist in vivo. Thus it is obvious to ask what, if any, physiological advantage a chaotic system can offer. There have been a number of suggestions on the benefits of chaos. As mentioned above, chaotic systems provide the flexibility to adjust to a changing environment in a dynamic yet stable fashion. The variables of a chaotic system explore a much wider range of values than do those of a system in rhythmic oscillation. Yet, at the same time, they are constrained by the structure of the attractor. This provides a certain flexibility within structure. Recently there has been considerable interest in and work done on the control of chaotic systems. Such systems can be controlled more finely and more quickly than linear systems. In linear systems, the response of the output depends linearly on the input. Small changes in a parameter of a linear system produce only small changes in the output. Chaotic systems, on the other hand, can have bifurcations, i.e., qualitative changes in behavior, as a parameter is varied. Thus small changes in a parameter of a chaotic system can produce very large changes in the output. Hence a chaotic system is under much finer control. Also, the changes of state in a linear system happen slowly. The sensitivity to initial conditions of a chaotic system means that very small changes in the value of the variables soon result in very different states of the system. Thus, a chaotic physiological system can switch very rapidly from one physiological state to another. This feature can have both good and bad aspects to it. Certainly, fast response is important in some physiological settings. On the other hand, extreme sensitivity could also be undesirable. Thus, there must be some tuning of the system for it to be effective. The advantages of sensitivity and speed in the control of chaotic systems have only recently been appreciated, yet a number of medical and engineering applications are within reach. It remains to be seen whether the homeodynamic view will overtake the traditional idea of homeostasis.

10.5 Summary

Chaotic systems are deterministic systems that show seemingly random behavior. They are marked by extreme sensitivity to initial conditions. For chaotic systems, neighboring trajectories in phase space will diverge from each other exponentially.

Often, chaotic trajectories will move toward a fractal region of phase space known as a "strange attractor." Fractals provide an important descriptor for such dynamical systems. Chaos has been observed in a number of different chemical and biochemical systems. For chaos to occur, the system must be open and have specific feedback and nonlinear reaction steps. These features appear abundantly in biochemical reaction sequences, especially in metabolism. A number of theoretical models have shown chaos in simple biochemical reaction sequences involving allosteric enzymes. Oscillations have been observed in the glycolytic pathway, and this reaction sequence can be driven into chaos with a periodic perturbation. Chaos also appears in the reaction mechanism of a single enzyme, horseradish peroxidase. There may be a physiological role for chaotic systems because they are sensitive and respond quickly. However, this concept of homeodynamics remains controversial.

References

Baker, G. L., and Gollub, J. P. 1990. *Chaotic Dynamics: An Introduction* Cambridge University Press, Cambridge.

Bassingthwaighte, J. B., Liebovitch, L. S., and West, B. J. 1994. *Fractal Physiology.* Oxford University Press, New York.

Bernard, C. 1878. *Lectures on the Phenomena of Life Common to Animals and Plants.* J. Libraire, Paris.

Cannon, W. B. 1929. Organization for Physiological Homeostasis. *Physiol. Rev.* **9**, 399–431.

Decroly, O., and Goldbeter, A. 1982. Birhythmicity, Chaos, and Other Patterns of Temporal Self-Organization in a Multiply Regulated Biochemical System. *Proc. Natl. Acad. Sci. USA* **79**, 6917–6921.

Degn, H. 1968. Bistability Caused by Substrate Inhibition of Peroxidase in an Open Reaction System. *Nature* **217**, 1047–1050.

Eschrich, K., Schellenberger, W., and Hofmann, E. 1983. Sustained Oscillations in a Reconstituted Enzyme System Containing Phosphofructokinase and Fructose 1,6-Biphosphate. *Arch. Biochem. Biophys.* **222**, 657–660.

Field, R. J., and Györgyi, L. 1993. *Chaos in Chemistry and Biochemistry.* World Scientific, Singapore.

Geest, T., Steinmetz, C. G., Larter, R., and Olsen, L. F. 1992. Period-Doubling Bifurcations and Chaos in an Enzyme Reaction. *J. Phys. Chem.* **96**, 5678–5680.

Ghosh, A., and Chance, B. 1964. Oscillations of Glycolytic Intermediates in Yeast Cells. *Biochem. Biophys. Res. Commun.* **16**, 174–181.

Goldberger, A. L., Rigney, D. R., and West, B. J. 1990. Chaos and Fractals in Human Physiology. *Sci. Am.* **262**, 42–49.

Goldbeter, A. 1995. From Periodic Behavior to Chaos in Biochemical Systems. In *Chaos in Chemistry and Biochemistry,* ed. R. J. Field and L. Gyögyi. World Scientific, Singapore.

Goldbeter, A., and Dupont, G. 1990. Allosteric Regulation, Cooperativity, and Biochemical Oscillations. *Biophys. Chem.* **37**, 341–353.

Goldbeter, A., and Lefever, R. 1972. Dissipative Structure for an Allosteric Model. *Biophys. J.* **12**, 1302–1315

Gray, P., and Scott, S. K. 1990. *Chemical Oscillations and Instabilities.* Oxford University Press, Oxford.

Hess, B., and Boiteux, A. 1971. Oscillatory Phenomena in Biochemistry. *Ann. Rev. Biochem.* **40**, 237–258.

Higgins, J. 1964. A Chemical Mechanism for Oscillation of Glycolytic Intermediates in Yeast Cells. *Proc. Natl. Acad. Sci. USA* **51**, 989–994.

Larter, R., Olsen, L. F., Steinmetz, C. G., and Geest, T. 1993. Chaos in Biochemical Systems: The Peroxidase Reaction as a Case Study. In *Chaos in Chemistry and Biochemistry*, ed. R. J. Field and L. Györgyi. World Scientific, Singapore.

Li, T.-Y., and Yorke, J. 1975. Period Three Implies Chaos. *Am. Math. Mon.* **82**, 986–992.

Lichtenberg, A. J., and Lieberman, M. A. 1983. *Regular and Stochastic Motion* Springer-Verlag, New York.

Markus, M., and Hess, B. 1984. Transition Between Oscillatory Models in a Glycolytic Model System. *Proc. Natl. Acad. Sci. USA* **81**, 4394–4398.

May, R. M. 1976. Simple Mathematical Models with Very Complicated Dynamics. *Nature* **261**, 459–467.

Meyer, T., and Stryer, L. 1991. Calcium Spiking. *Ann. Rev. Biophys. Biophys. Chem.* **20**, 153–174.

Moran, F., and Goldbeter, A. 1984. Onset of Birhythmicity in a Regulated Biochemical System. *Biophys. Chem.* **20**, 149–156.

Nakamura, S., Yokota, K., and Yamazaki, I. 1969. Sustained Oscillations in a Lactoperoxidase, NADPH and O_2 System. *Nature* **222**, 794.

Neet, K. 1983. In *Contemporary Enzyme Kinetics*, ed. D. L. Purich. Academic Press, New York, pp. 267–320. Cooperativity in Enzyme Function: Equilibrium and Kinetic Aspects.

Olsen, L. F., and Degn, H. 1977. Chaos in an Enzymatic Reaction. *Nature* **267**, 177–178.

Olsen, L. F., and Degn, H. 1978. Oscillatory Kinetics of the Peroxidase–Oxidase Reaction in an Open System. Experimental and Theoretical Studies. *Biochim. Biophys. Acta* **523**, 321–334.

Rasband, S. N. 1990. *Chaotic Dynamics of Nonlinear Systems* Wiley, New York.

Rössler, O. E. 1976. An Equation for Continuous Chaos. *Phys. Lett.* **57A**, 397–398.

Scott, S. K. 1991. *Chemical Chaos*. Oxford University Press, Oxford.

Sel'kov, E. E. 1968. Self-Oscillations in Glycolysis. 1. A Simple Kinetic Model. *Eur. J. Biochem.* **4**, 79–86.

Willamowski, K. D. and Rössler, O. E. 1980. Irregular Oscillations in a Realistic Quadratic Mass Action System. *Z. Naturforsch.*, *Terla* **35**, 317–318.

Yamazaki, I., Yokota, K., and Nakajima, R. 1965. Oscillatory Oxidations of Reduced Pyridine Nucleotide. *Biochim. Biophys. Acta* **105**, 301–312.

Index